Emerging Technologies Shaping New Communication and Electronic Infrastructures

Eugene Ogbodo
University of Hertfordshire, UK

IGI Global
Scientific Publishing
Publishing Tomorrow's Research Today

Vice President of Editorial	Melissa Wagner
Director of Acquisitions	Mikaela Felty
Director of Book Development	Jocelynn Hessler
Production Manager	Mike Brehm
Cover Design	Jose Rosado

Published in the United States of America by
IGI Global Scientific Publishing
701 East Chocolate Avenue
Hershey, PA, 17033, USA
Tel: 717-533-8845 | Fax: 717-533-7115
Website: https://www.igi-global.com E-mail: cust@igi-global.com

Library of Congress Cataloging-in-Publication Data

Names: Ogbodo, Eugene (Eugene Amobichukwu), 1980- editor
http://id.loc.gov/rwo/agents/n2025008765
Title: Emerging technologies shaping new communication and electronic infrastructures / edited by: Eugene Ogbodo.
Description: Hershey PA : IGI Global Scientific Publishing, [2026] | Includes bibliographical references. | Summary: "The objective of this book is to provide a comprehensive examination of the technologies driving the evolution of global communication networks and electronic infrastructures"-- Provided by publisher.
Identifiers: LCCN 2024060457 (print) | LCCN 2024060458 (ebook) | ISBN 9798337308524 hardcover | ISBN 9798337308531 paperback | ISBN 9798337308548 ebook
Subjects: LCSH: Wireless communication systems--Technological innovations--Case studies | Artificial intelligence--Industrial applications--Case studies | LCGFT: Case studies.
https://id.loc.gov/authorities/genreForms/gf2017026140
Classification: LCC TK5103.2 .E5154 2025 (print) | LCC TK5103.2 (ebook) | DDC 621.382--dc23/eng/20250313
LC record available at https://lccn.loc.gov/2024060457
LC ebook record available at https://lccn.loc.gov/2024060458

British Cataloguing in Publication Data
A Cataloguing in Publication record for this book is available from the British Library.

To
Patience Florence Ogbodo,
Alvin Obiora Ogbodo
&
Lexie Obiageli Ogbodo

Editorial Advisory Board

Table of Contents

Detailed Table of Contents

and 14.81 GHz. At the observed frequency points, the reported peak gain is 3.15, 4.93, 3.56, 5.13, and 8.11 dB, respectively. Section 8 compiles the entire findings.

In this study, a microstrip patch antenna is designed for 5G applications and is presented, with a resonance frequency of 3.33 GHz. The antenna, created with CST Microwave Studio and built on a FR4-lossy substrate with a dielectric constant of 4.3, is suitable for satellite and WiMAX communications. The square patch features three well-positioned slots, which enhance the antenna's electromagnetic performance. The slots measure 12mm x 30mm, 12mm x 8mm, and 12mm x 12mm, and are crucial for optimizing impedance matching, bandwidth, and radiation characteristics. The fully grounded patch ensures a broader bandwidth, and the antenna achieves a high gain of 7.63 dBi, a return loss of -28.457 dB, and a VSWR of 1.467. The simulation results confirm the antenna's potential for 5G applications, enabling fast data transfer and reliable communication. The proposed antenna is a viable option for 5G-enabled devices due to its compact size, high gain, and low return loss, addressing the need for efficient and space-saving solutions in modern wireless communication systems.

The sixth generation (6G) wireless communication networks have the potential to revolutionize customer services and applications through the Internet of Things (IoT), opening the way for fully intelligent and autonomous systems. This chapter explores the new prospects that 6G technology has brought to IoT networks and applications across important sectors like unmanned aerial vehicles, satellite IoT, healthcare IoT, automotive IoT and autonomous driving, and industrial IoT. A novel circularly polarized 6G antenna is proposed for Internet of Things applications that operates in the frequency range of 6.4 GHz to 8.4 GHz with a bandwidth of 2 GHz. To accomplish miniaturization, metamaterial approaches guarantee compactness without sacrificing performance. The antenna's parameters are fine-tuned using the

Chameleon Swarm Algorithm, a bio-inspired optimization method that achieves the required bandwidth of 2 GHz, return loss S11 <-25 dB, high gain, and an excellent radiation pattern. Simulation results validate the antenna's potential for next-generation 6G IoT applications.

This chapter discuss the regression-based machine learning approach to achieve the resonance by computing the line length, impedance matching width and substrate thickness of the Implantable Microstrip Antennas in the frequency band of 2.4–2.5 GHz. In the design, a meandered antenna with microstrip feed is loaded. The resonant frequencies of 1,000 IMAs were simulated using HFSS by varying line length, width, substrate thickness. These variations adhered to the specifications required for implantable medical device (IMD) for biotelemetry. Various regression-based ML algorithms, like Artificial Neural Networks were studies. The study found that the Gaussian Process Regression (GPR) model has the most accurate predictions parameters at resonance. The proposed GPR model was validated by simulating and characterizing an implantable microstrip antenna. The antenna has achieved a size of the antenna achieved is 3.5x3.5 mm2. The designed antenna's performance, including its bandwidth characteristics, closely matched both the simulation results and the GPR predictions.

This research explores the applicability of employing the concepts of multi-band slicing proposed in the context of 5G networks to Wi-Fi networks, thus, opening the way to adopting enhanced resource management strategies that will be essential for 6G networks. Multi-band slicing enables the dynamic supply of resources at multiple communication bands, which is a crucial feature when it comes to meeting the complex and often very high 6G requirements such as ultra-reliable low-latency communication (URLLC), enhanced mobile broadband (eMBB), and massive machine-type communication (mMTC). These techniques accommodate the increase in next-generation wireless communication while efficiently managing bandwidth and offering adaptable control over traffic flow. This research is a comprehensive and primary suggestion for applying slicing mechanisms in 6G architectures, as it provides guidance for constructing efficient and scalable networks for the next generation of communications based on the needs and requirements posed by new technologies and services.

The integration of multiple wireless technologies within the same system on a chip has significantly enhanced connectivity and device interoperability. However, the close proximity of antennas and the non-linear behavior of RF front-end components introduce signal coupling, leading to the generation of intermodulation and harmonic distortions. These distortion not only degrade system performance but also cause spectral leakage. As modern wireless systems increasingly rely on multi-radio coexistence, understanding the underlying mechanisms of these distortions becomes crucial. This chapter provides a comprehensive analysis of intermodulation distortions, examining their spectral characteristics, contributing factors, and impact on system efficiency. Additionally, it explores key design parameters that influence distortion severity and discusses mitigation strategies aimed at minimizing interference. By addressing these challenges, the chapter offers valuable insights into optimizing wireless communication systems for improved reliability and regulatory compliance.

The next generation of nanosheet-based semiconductor devices marks a pivotal evolution in modern electronics, as industry and research converge to address the scaling limitations of traditional FinFET architectures. Nanosheets, with their enhanced electrostatic control and high on-off current ratios, are emerging as the cornerstone of advanced semiconductor technologies. This chapter explores the advancements in nanosheet transistor design, materials innovation, with an emphasis on their scalability beyond 3nm technology nodes. Furthermore, the discussion extends to device reliability, energy efficiency, and potential applications in artificial intelligence (AI), high-performance computing, and edge devices. By delving into the synergistic collaboration between process technology and device architecture, this chapter underscores the transformative role of nanosheet technologies in shaping the future of semiconductor devices and redefining the boundaries of Moore's Law.

Wireless sensor networks (WSNs) face challenges due to limited human intervention post-deployment and finite battery life at each node, where the failure of one node's battery can jeopardize the entire system. To address this, we proposed a novel routing technique called Specific Arbitrary Hierarchical Clustering Entrenched Advanced Swastika Routing (SAHCEASR). This technique combines clustering with an advanced version of Swastika Routing, employing specific arbitrary clustering for optimal cluster selection. Implemented in MATLAB using a power model, SAHCEASR significantly reduces power consumption at both system-wide and individual node levels. Our adaptive routing approach demonstrated nearly a 50% reduction in battery power consumption, thereby enhancing the network lifetime of WSNs.

Wide area networks, which are the basis of access to information and communication today, form the backbone of the internet. The ever-increasing data traffic, different application types, and service quality requirements of different systems used in local area networks cause problems such as congestion and loss of performance in networks due to the limited bandwidth received from the internet service provider. This study focuses on the effects of QoS functions and critical network management

Preface

INTRODUCTION

Modern society is increasingly dependent on advanced communication and electronic infrastructures that enable seamless connectivity, high speed data exchange, and intelligent system integration. The evolution of wireless communication technologies, from early cellular networks to 5G and the emerging vision of 6G, is transforming how devices, systems, and individuals interact. At the same time, breakthroughs in antenna design, semiconductor technologies, machine learning, and network management are redefining the capabilities and performance of next generation communication systems.

The rapid expansion of connected devices, autonomous systems, smart cities, and advanced healthcare technologies requires communication infrastructures that are not only faster but also more reliable, energy efficient, and adaptable. Achieving these goals demands interdisciplinary innovations spanning radio frequency engineering, antenna design, semiconductor device architecture, wireless networking, and artificial intelligence driven optimization.

This volume, *Emerging Technologies Shaping New Communication and Electronic Infrastructures*, brings together contributions from researchers and industry professionals who are actively advancing the frontiers of communication technologies. The chapters explore key developments across the entire communication ecosystem, from physical layer innovations such as antennas and radar systems to higher level network management strategies and energy efficient communication architectures.

Particular emphasis is placed on the enabling technologies that support emerging applications including autonomous vehicles, Internet of Things (IoT) systems, smart healthcare devices, wireless sensor networks, and next generation wireless communication standards. These technologies require highly optimized hardware and software infrastructures capable of supporting massive connectivity, ultra low latency communication, and intelligent resource management.

The book also highlights the growing role of machine learning and advanced computational techniques in optimizing communication systems. From antenna parameter estimation to resource allocation in wireless networks, intelligent algorithms are becoming essential tools for improving system performance and scalability.

By presenting both theoretical insights and practical engineering solutions, this volume aims to provide researchers, engineers, and students with a comprehensive overview of the emerging technologies shaping the future of communication and electronic infrastructures. Below is a chapter by chapter summary that you may use to supplement your reading.

CHAPTER OVERVIEWS

Chapter 1: Automotive RADARs: From Antennas to Applications

The first chapter provides an in depth overview of automotive radar systems, which have become a fundamental component of modern vehicles and Advanced Driver Assistance Systems (ADAS). The authors examine the complete radar signal chain, beginning with the fundamental principles of radar operation and progressing through antenna design, radio frequency (RF) components, and signal processing stages. The chapter explains how raw in phase and quadrature (IQ) data is processed to derive critical parameters such as range, velocity, and angle of detected objects. Key concepts such as the radar cube and angle resolution are explored, particularly within the context of multiple input multiple output (MIMO) radar systems. Additionally, the chapter discusses radar calibration, communication protocols, and regulatory standards that influence the deployment and performance of automotive radar technologies.

Chapter 2: Characterization and Implementation of Penta Band Modified Circular Patch Antenna Employing Machine Learning for Parameter Estimation

This chapter investigates the design and characterization of a penta-band modified circular patch antenna enhanced with a symmetrical X shaped slot structure. The antenna design, measuring $30 \times 30 \times 1.6$ mm^3, is developed to support multiple frequency bands within wireless communication systems. The chapter provides a comprehensive overview of antenna design principles, followed by a detailed explanation of the antenna structure and design methodology. Machine learning techniques, specifically Artificial Neural Networks (ANN), are employed to estimate and optimize antenna parameters. Simulation results demonstrate that the antenna achieves resonant frequencies at 2.85 GHz, 5.30 GHz, 10.76 GHz, 13.14 GHz, and 14.81 GHz with corresponding peak gains across these bands. The integration of machine learning techniques highlights the potential of intelligent optimization methods in modern antenna engineering.

Chapter 3: Design and Performance Evaluation of Microstrip Patch Antenna for 5G Applications

Chapter 3 presents the design and simulation of a microstrip patch antenna optimized for 5G wireless communication systems. Designed using CST Microwave Studio and fabricated on an FR4 substrate, the antenna operates at a resonance frequency of 3.33 GHz. The proposed design incorporates three strategically positioned slots within a square patch structure, enhancing bandwidth, impedance matching, and radiation performance. Simulation results indicate a high gain of 7.63 dBi, return loss of −28.457 dB, and a voltage standing wave ratio (VSWR) of 1.467, demonstrating strong electromagnetic performance. The chapter highlights how compact antenna designs can support the growing demand for efficient, space saving components in modern wireless devices.

Chapter 4: Forging the Path to 6G: Next Gen Antennas for Seamless IoT Communication

This chapter explores the role of advanced antenna technologies in enabling the future vision of 6G wireless networks and large scale Internet of Things (IoT) ecosystems. The authors propose a circularly polarized antenna operating between 6.4 GHz and 8.4 GHz, designed specifically for next generation IoT applications. Metamaterial based techniques are used to achieve miniaturization while maintaining high performance. Additionally, a bio inspired optimization method known as the Chameleon Swarm Algorithm is employed to fine tune antenna parameters, achieving a bandwidth of 2 GHz and return loss values below −25 dB. The chapter highlights the importance of antenna innovation in supporting emerging applications such as autonomous vehicles, satellite IoT systems, and industrial automation.

Chapter 5: ML Optimized Biotelemetry Antennas Redefining Next Gen Healthcare Connectivity

The fifth chapter examines the application of machine learning techniques in the design of implantable microstrip antennas for medical biotelemetry systems. These antennas operate within the 2.4–2.5 GHz band and are intended for implantable medical devices used in healthcare monitoring and diagnostics. The study simulates 1,000 antenna configurations using HFSS, varying parameters such as line length, impedance matching width, and substrate thickness. Several regression based machine learning models are evaluated to predict optimal antenna parameters, with Gaussian Process Regression (GPR) demonstrating the highest predictive accuracy. The resulting implantable antenna achieves a compact size of 3.5×3.5 mm^2 while maintaining performance characteristics consistent with simulation predictions. This chapter demonstrates how machine learning can accelerate the design process for biomedical communication devices.

Chapter 6: Adapting 5G Network Slicing Approaches for Wi Fi: Strategic Insights for Advancing Multi Band Resource Management in 6G Networks

Chapter 6 addresses the growing complexity of wireless networks and the need for efficient resource management strategies. The authors investigate the application of 5G network slicing concepts to Wi Fi environments, enabling dynamic allocation of resources across multiple frequency bands. Such techniques are critical for supporting the diverse service requirements expected in future 6G networks, including ultra reliable low latency communication (URLLC), enhanced mobile broadband (eMBB), and massive machine type communication (mMTC). By extending slicing principles to Wi Fi systems, the study proposes a framework for improving bandwidth management and traffic control in next generation communication infrastructures.

Chapter 7: Analysis of Intermodulation Distortion in Integrated RF Coexistence

As modern electronic devices increasingly incorporate multiple wireless communication technologies within a single platform, signal interference becomes a critical challenge. This chapter examines the phenomenon of intermodulation distortion caused by the nonlinear behavior of RF front end components and the proximity of multiple antennas within integrated systems. The authors analyze the spectral characteristics of these distortions and their effects on system efficiency and signal quality. Additionally, the chapter explores design parameters that influence distortion severity and discusses mitigation strategies to reduce interference and improve wireless system reliability.

Chapter 8: Advancements in 5G and 6G Technologies: The Next Generation of Nanosheet Devices

This chapter focuses on nanosheet transistor technology as a promising solution to the scaling limitations of traditional FinFET architectures. Nanosheet devices offer improved electrostatic control and higher on–off current ratios, making them well suited for advanced semiconductor nodes beyond the 3 nm scale. The chapter explores innovations in nanosheet device design, materials engineering, and fabrication techniques that enhance device performance, reliability, and energy efficiency. The authors also discuss the potential applications of nanosheet technologies in artificial intelligence systems, high performance computing, and edge devices, highlighting their importance in the evolution of next generation electronic infrastructure.

Chapter 9: Optimizing Power Efficiency in Wireless Sensor Networks Through Specific Arbitrary Hierarchical Clustering and Enhanced Swastika Routing

Wireless sensor networks (WSNs) often operate in environments where battery replacement is difficult or impossible, making energy efficiency a critical design consideration. This chapter introduces a novel routing method known as Specific Arbitrary Hierarchical Clustering Entrenched Advanced Swastika Routing (SAH-CEASR). The proposed method combines hierarchical clustering techniques with enhanced routing strategies to reduce energy consumption within sensor networks. Simulation results conducted in MATLAB demonstrate that the approach can reduce overall power consumption by nearly 50%, significantly extending network lifetime. The chapter highlights the importance of efficient routing algorithms for sustainable large scale sensor network deployments.

Chapter 10: Traffic Shaping and Flow Management in Wide Area Networks

The final chapter examines traffic management strategies within wide area networks (WANs), which form the backbone of global internet connectivity. As network traffic continues to grow due to diverse applications and services, efficient bandwidth utilization becomes increasingly important. This chapter explores quality of service (QoS) mechanisms such as traffic shaping, traffic control, bandwidth allocation, and data prioritization techniques. Through practical analysis, the authors demonstrate how these network management tools can mitigate congestion, improve performance, and maintain stability in networks with limited bandwidth capacity.

CONCLUDING REMARKS

The chapters presented in this volume collectively illustrate the technological advancements that are shaping the next generation of communication and electronic infrastructures. From innovative antenna designs and radar technologies to machine learning driven optimization, semiconductor breakthroughs, and advanced networking strategies, the research contributions highlight the multidisciplinary efforts required to support the rapidly evolving digital landscape.

As communication systems continue to expand in scale and complexity, the integration of intelligent design methodologies, efficient hardware architectures, and adaptive network management will become increasingly critical. The insights presented in this book aim to support researchers, engineers, and practitioners working to develop robust, scalable, and efficient communication infrastructures capable of supporting the demands of the 5G, 6G, and IoT driven future.

The editor hopes that the ideas and findings presented in this volume will inspire continued research and collaboration, ultimately contributing to the advancement of communication technologies that will power the next generation of digital innovation.

Acknowledgement

This book, *Emerging Technologies Shaping New Communication and Electronic Infrastructures,* is the result of the collective effort, expertise, and encouragement of many individuals and institutions.

First and foremost, I express my deepest gratitude to God Almighty for the strength, clarity, and perseverance granted throughout this project.

I would like to sincerely thank the members of the Editorial Advisory Board—Dr. Azunka N. Ukala, Dr. Kisalaya Chakrabarti, *Dr. Augustine Chidiebere Onuora, Dr. Taiwo Adisa Oyeniran, Dr. Smrity Dwivedi, etc*—for their intellectual guidance, critical insights, and unwavering support in shaping the vision and direction of this book.

Special appreciation goes to all the chapter contributors whose time, research, and passion brought depth and richness to the topics explored in this volume. Your individual contributions have created a cohesive and meaningful resource for students, professionals, and researchers alike.

To my colleagues, mentors, and students who provided feedback, asked thought-provoking questions, and helped refine my understanding—thank you.

To my family and close friends, whose patience, encouragement, and belief in me never wavered, I owe you more than words can express.

Finally, I acknowledge the efforts of the editorial and publishing teams who made the production of this book a smooth and rewarding experience.

May this work serve as a valuable contribution to the growing field of wireless communication and inspire further innovation and collaboration.

Eugene Ogbodo
University of Hertfordshire, UK

Chapter 1
Automotive RADARs: From Antennas to Applications

Sunil Javali
http://orcid.org/0009-0007-5835-4724
IEEE, USA

Ashish Sheikh
http://orcid.org/0009-0000-3393-1328
IEEE, USA

ABSTRACT

Automotive radar has become a key technology in modern vehicles, driving advancements in Advanced Driver Assistance Systems(ADAS).This chapter takes a comprehensive look at how the radar systems work, starting from the fundamentals and moving through the full signal chain—from antennas to real-world applications. The discussion begins with the basics of radar technology and the different types of automotive radars in use today. It then dives into the critical components including RF and the signal processing chain, explaining how raw IQ data is converted into meaningful range, velocity, and angle measurements using simple examples. The concept of the radar cube is introduced, showing how it organizes radar data and why angle resolution is crucial, especially in MIMO radar systems. Beyond the technical details, the chapter also covers radar calibration, communication protocols, and the global regulatory standards shaping the deployment of automotive radars.

INTRODUCTION

The advent of automobiles has given rise to the new definition for the automotive and transportation industry. As humanity walks towards the new generation of advanced technology leading to the autonomous cars, the automakers are relying

DOI: 10.4018/979-8-3373-0852-4.ch001

more on sophisticated technologies governing the safety of passengers by integrating the radio technologies such as radars, lidars, cameras both interior and exteriors, IR camera, crash sensors, V2X communication, Ultrasound sensors, GNSS, Mobile communication to enhance the safety of drivers and occupants. Automotive radars use radio waves to detect objects including pedestrians, cars, traffic posts, traffic signals etc., The modern automotive radars can communicate with each other, with the surroundings and also sense about the things happening around them over radio frequencies. The Figure 1 depicts the placement of all sensing technologies in an automotive vehicle (Aptiv Corporation, 2022). Apart from the components shown in Figure 1, next-generation automotive vehicles — i.e., connected autonomous vehicles (CAVs) — are equipped with RADAR, LiDAR, vision cameras, infrared sensors, GNSS, cellular modules, Vehicle-to-Vehicle (V2V) radios, and Wi-Fi/BLE radios. Sensor fusion of the outputs from all these technologies is used to autonomously navigate the vehicle while enhancing passenger safety and road security.

The Radar is abbreviated for Radio Detection and Ranging. The Radar works by transmitting a known RF signal and captures the reflected signals where it is processed to determine the range, distance and velocity of an object. The automotive radars are also known as millimeter wave sensors as the wavelength of electromagnetic signal is in the order of millimeter. The automotive radars operate at three different frequency ranges i.e., 24GHz, 51GHz to 60GHz and 76GHz to 81GHz. The 24GHz frequency band will be discontinued due to the higher interference with radio astronomy and earth exploration applications, whereas, 60GHz and 76GHz band is predominantly used and accepted by various governing bodies across the world to develop most advanced automotive radars requiring better range and resolution (Renesas Electronics, 2023).

Figure 1. The placement of all sensing technologies in an automotive vehicle

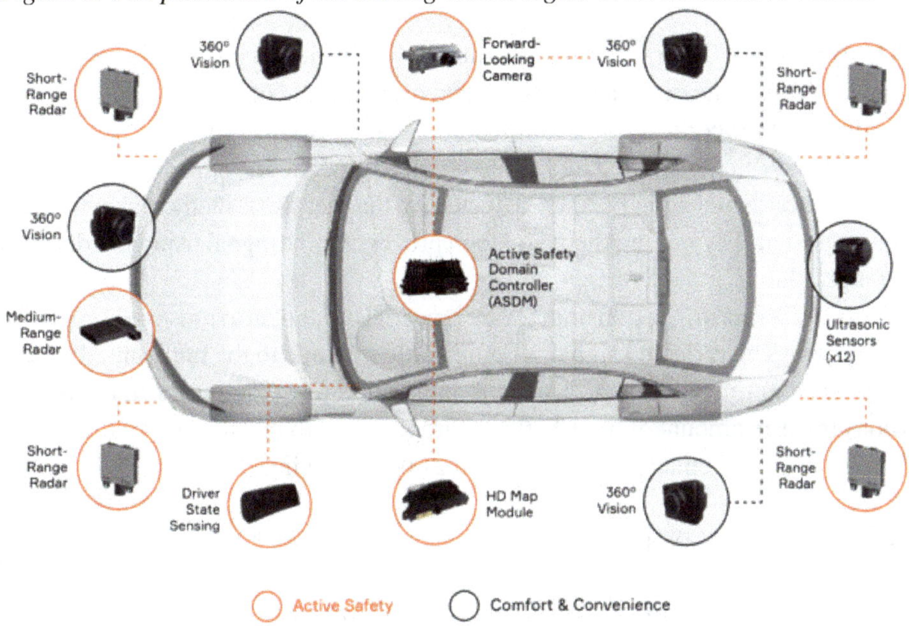

AUTOMOTIVE RADARS - TYPES

The automotive radars are classified into four major categories based on the application such as Long-Range Radar (LRR), Medium Range Radar (MRR), Short Range Radar (SRR) and Ultra-Short-Range Radar (USRR). In order to facilitate the distance requirements, the modern automotive radar MMIC uses two types of millimeter wave technology such as Frequency Modulated Continuous Wave (FMCW) and Pulse Radar. The Figure 2, depicts the simplified block diagram of a radar operation (Renesas Electronics, 2023).

Figure 2. Radar – high level operation

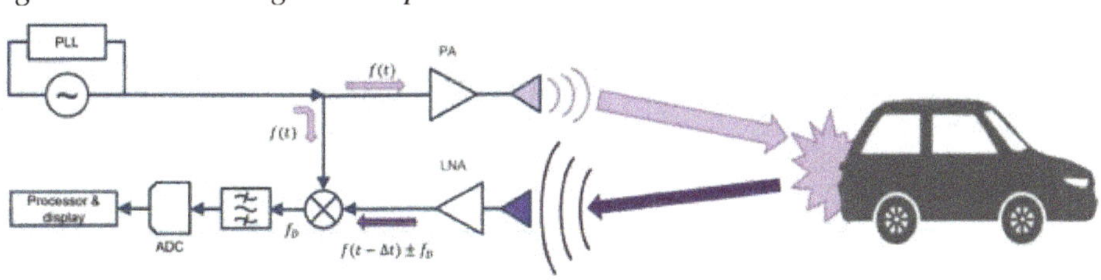

FMCW Radar

The typical FMCW radar uses continuous wave electromagnetic signals at a certain known frequency which is modulated over a period of time to determine the target or object properties. In the automotive radar world, the object is also known as a target and the transmitted and received electromagnetic signals are known as Chirps. The reflected signal is then detected by the receivers, then compared with the original signal by mixing them and then processing the signal using the defined set of algorithms.

In an FMCW radar, the transmitters are powered off and on to preserve the power of the system. The reflected signal is comparable in shape to the transmitted signal but displaced in time as the two-way communication from the sensor to the object has consumed an amount of time **Δ**t that is proportional to the distance of the object R. This can be mathematically represented as the following

$\Delta t = \frac{2R}{c}$; c is the speed of EM wave i.e., 2.997 * 10^8 m/s

The Figure 3, shows the transmitted and received signal which is displaced in time. The pink line indicates the transmitted signal and the purple line indicates the received signal. By superimposing the reflected wave with the original signal at any given instant of time, the signal contains a frequency displacement Δf. This displacement allows determining the range R for each period of the "chirp". If the signal is monitored over several chirps, an additional frequency displacement will be detected for a target moving towards or away from the radar, due to the doppler effect. This allows for determining the velocity of the target.

Figure 3. Frequency vs time FMCW signal

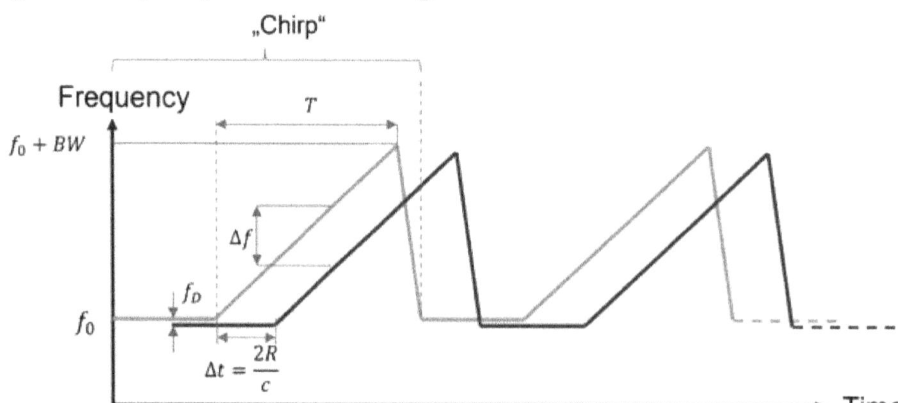

Pulse Radar

Among various automotive radar systems, the pulsed radar is also most commonly used for long distance applications such as Blind spot detection, Collision avoidance, Lane Change assistance, Rear cross traffic alert. The basic principle of a pulsed radar is that it transmits and receives short RF pulses on every rising clock edge to detect the objects or targets. The pulse width can be varied based on the type of object in the field of vision. More often, the pulses of high power are used to detect targets farther in the environment (Wolff, n.d.).Unlike FMCW radar, the transmitters are switched off after the transmission is completed. The drawback of the pulsed radar is its poor range resolution. The transmitted signal can be mathematically represented as follows

$$x(t) = A(t) * \sin[2\pi f(t)*t + \Theta(t)]$$

Here, x(t) is the transmitted pulse at an instant of time t, A is the amplitude of the signal, f is the frequency of the transmitted pulse and Θ is the phase shift of the signal. The figure 4 shows a simple radar pulse (FURUNO, n.d.).Some simple terminologies to remember when understanding the pulse radar are as follows

1. The Pulse Repetition frequency (PRF) is the duration of time (τ) at which the pulses are repeated.
2. Pulse width is the length of time of a single pulse.
3. Pulse Repetition Interval (PRI) is the duration between the start of two consecutive pulses. PRI is often referred to as pulse repetition period of time.
4. Duty Cycle is the duration of time, when the radar pulses are actively transmitted.
5. Dwell time is the duration of the time the radar pulse is spent on a target.

There are some useful relations between the PRF, PRI and Duty Cycle.

1. Pulse Repetition Frequency (PRF) = 1/PRI i.e., as PRF increases, PRI decreases.
2. Duty cycle = Pulse Length/ Pulse repetition Interval
3. Average Power = Peak Power * Duty Cycle

Figure 4. Radar Pulse

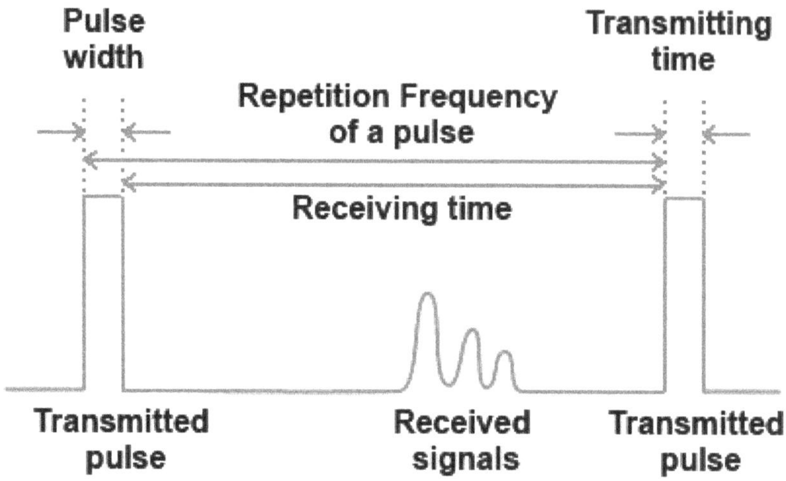

RADAR RANGE EQUATION

The Radar range equation describes the relationship between the received power -vs – transmitted power and can be mathematically represented as

$$\mathbf{P_r} = \frac{\mathbf{P_t G_t G_r \lambda^2 \sigma}}{(4\pi)^3 \mathbf{R}^4}$$

$\boldsymbol{P_r}$ is power of reflected signal, $\boldsymbol{P_t}$ is power of the transmitted signal, $\boldsymbol{G_t}$ is the gain of the transmitter antenna, $\boldsymbol{G_r}$ is the gain of the receiver antenna, λ is the wavelength of the transmitted signal, σ is the radar target cross section, R is the radar range. In any radar system, the detection range can be increased by simply increasing the transmit output power and also increasing the transmit output power; the target with the smallest Radar Cross Section (RCS) can be detectable. These parameters are helpful in performing the link budget analysis and also can be used by engineers to design a better transmitter and receiver antennas. The term $\lambda^4/(4\pi R)^4$ represents the two-way free-space loss, while $(4\pi\bullet\sigma)/\lambda^2$ accounts for the reflection on the target.

$$R_{max} = \sqrt[4]{\frac{P_t G_t G_r \lambda^2 \sigma}{P_{rmin}(4\pi)^3}}$$

$\mathbf{R_{max}}$ is the maximum achievable range for a given receiver sensitivity $\mathbf{P_{rmin}}$

Using this equation, the detection range can be modified based on the sensor which can detect objects in Long Range, Medium range, short range and Ultra Short range. For example, 1dB more of output power in a Long-Range Radar (LRR) will give us 2.8m to detect a bike, 6m for a motorbike and 11m for a typical car. Another important aspect of designing a better receiver is understanding the Noise figure of the receiver. The rule of thumb for noise figures is lower the noise figure the better the receiver performance. The noise figure is the ratio between the reflected signal power and the noise power.

$$SNR = \frac{P_r}{N}$$

Here, the received power is given by the equation mentioned above. Thus, the detection range can be increased by reducing the noise figure of the radar receiver. In a Long-Range Radar (LRR), reducing the noise figure by 1dB will increase the range by 4m (bike), 10m (motorbike) and 13m (car).

THE AUTOMOTIVE RADAR – BLOCK DIAGRAM

After exploring the different types of automotive grade radars, this section will be focusing on the internal architecture at a very high level. The radar building block contains the Microwave Monolithic Integrated Circuit (MMIC) and Microprocessor. The MMIC consists of a transmitter chain, receiver chain, a local oscillator, high speed communication interfaces to handle complex data and Microprocessor block which consists of computing cluster, signal processing toolbox and antennas at the end of transmitter and receiver which helps in the transmission and reception of the radar waves. We will address each of these blocks individually.

Figure 5. Automotive Radar – High level block diagram

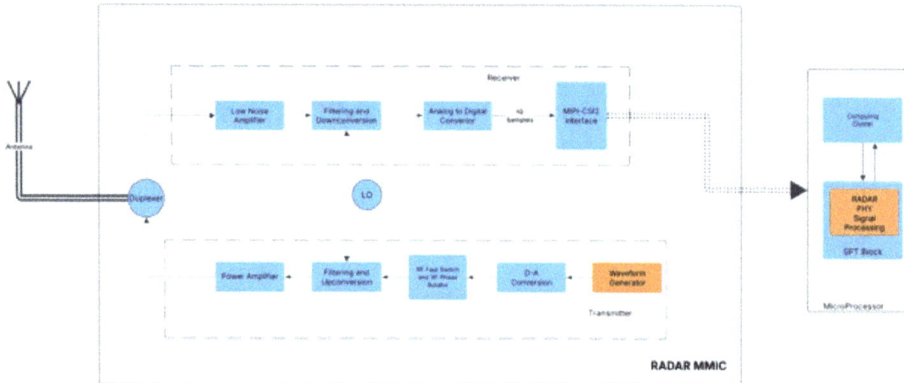

The transmitter consists of a waveform generator also known as chirp generator, which generates a low frequency sample in the order of 25GHz to 27GHz with the power level as per the country-based regulations, and the digitized samples are converted to analog domain using digital to analog converter. The analog signal is fed into the phase rotator which rotates the signal by 0 or 180 degrees and the output is multiplied by the input signal from a local oscillator by a factor of 3 to generate a microwave frequency component in the range of 76GHz to 81GHz. The phase rotator and RF fast switch blocks are used for MIMO operations, where the multiple transmit, chains are used to increase the angular object detection of the radar sensor (NXP Semiconductors, 2025).

As discussed in the prior sections, the signal to noise ratio i.e., SNR is directly proportional to the transmitted power and inversely proportional to the system temperature and system losses. Therefore, digital designers must carefully optimize these three parameters to ensure a high SNR and overall system efficiency while adhering to regulatory standards. Additionally, during the amplification process, the key criteria is to maintain the low noise and minimum distortion of the input signal before the signal is fed into the power amplifier (O'Donnell, 2001).

The received RF signal first passes through a Low-Noise Amplifier (LNA), which minimizes noise and enhances weak signals. After amplification, the signal goes through a filtering stage, typically consisting of a high-pass filter (HPF) and a low-pass filter (LPF), to remove unwanted frequency components. The filtered signal is then fed into a downconverter (mixer stage), which converts the high-frequency RF signal (76GHz to 81GHz) into a lower frequency signal (0 to 40MHz) and known as the Intermediate Frequency (IF) signal. Down conversion is necessary because ADC performs better at lower frequencies, offering a higher dynamic range and improved resolution. Finally, the IF signal is sampled by the ADC, and the resulting digital samples are processed in the Radar PHY signal processing block for further analysis (Skolnik, 2008).

The local oscillator is used to feed low frequency signals into the upconverter block, usually the mixer, in the RF transmitter front End and down converter block in the receiver front end. The local oscillator signal can come from the crystal oscillator which is usually 40MHz in Automotive radars applications. The ADC data is passed through the MIPI CSI-2 interface to the microprocessor. MIPI CSI-2 subsystem is explained in greater detail in the further sections. Finally, the microprocessor handles the computing of ADC samples to form the radar cube and determine the target characteristics.

RADAR ANTENNAS

The radiating element present in the transmitter and receiver chain is more often referred to as Radar antenna. The basic principle of the antenna is to facilitate the transmission and reception of the electromagnetic waves through the free space by reducing the loss of RF energy. At millimeter wave frequency, any kind of transitions or interfaces can lead to degraded performance of Radar antennas. Reducing the transmission loss and maintaining RF signal integrity is the key challenge in designing a robust antenna (O'Donnell, 2001). The important terminologies associated with the antennas are as follows

1. Antenna Gain is defined as a radiation intensity over that of an isotropic source.

Antenna Gain = $4\pi A/\lambda^2$,

where A is the antenna aperture and λ is the wavelength of center frequency. Thus, antenna gain is directly proportional to antenna aperture. Gain increases as the aperture increases (O'Donnell, 2001).

2. Antenna pattern is defined as the graphical representation of gain versus angle.
3. Antenna Polarization is defined by the behavior of the electric field vector as it propagates in time. Three kinds of polarization are vertical, horizontal and circular polarization.
4. Antenna Field regions: There are two kinds of regions based on the propagation of electromagnetic waves and are mainly far field and near field (O'Donnell, 2001).
    a. Near Field or Reactive region: It is defined as the region where RF energy is stored in the vicinity of the antenna. Important qualities include input

impedance and mutual coupling. The near field is also known as the Fresnel zone.

$$R < 0.62\sqrt{D^3/\lambda}$$

Where R is the Near field region, D is the diameter of antenna, λ is the wavelength of center frequency.

b. Far- field Region: It is defined as a region far from the antenna where all the RF energy or power is radiated out. Where the radiated wave is the plane wave. Far Field antenna parameters include radiation pattern, gain, directivity, polarization, and radar cross section (RCS). The far-field region is also known as the Fraunhofer region.

$$R > 2D^2/\lambda$$

Where R is the Near field region, D is the diameter of antenna, λ is the wavelength of center frequency.

5. Antenna Input impedance: It is defined as the ratio of voltage to current at antenna feed port. The antenna must be designed to maximize power transfer from the transmission line. Input impedance usually defines antenna bandwidth. If antenna input impedance is zero, the incident power is delivered to the antenna, if input impedance is unity, all incident power is reflected.
6. Antenna Spacing: It is defined as the spacing between the radiating elements in array antennas (Texas Instruments, 2023).
   a. RX-Rx spacing: The distance between receive elements
   b. TX-TX spacing: The distance between transmit elements
   c. TX-RX spacing: The distance between transmit and receive elements.

The maximum field of view that two antennas spaced L apart is represented by

$$\theta_{max} = sin^{-1}\left(\frac{\lambda}{2L}\right)$$

Thus, the maximum FOV is achieved when L = $\lambda/2$

At millimeter wave frequencies, due to the shorter wavelengths, the chirp to board transitions is complex to achieve without losing RF energy. The most common design is using the Ball Grid Array (BGA) to carry the RF from the chip to a microstrip line. Additionally, Strip lines, Coplanar waveguides (CPW) or substrate integrated waveguides (SIW) can also be utilized based on the applications (Renesas Electronics, 2023).

Radar antennas – Types

The radar antennas types can be classified based on the coverage, directivity, signal-to-noise ratio and frequency range. There are various antennas namely grid parabolic antenna, horn antenna, phased array antenna, array antenna, conformal antenna used in applications such as satellite communication, radio astronomy, airplanes, 5G technology, ships, medical imaging, automotive applications. Most frequently used automotive FMCW radar antennas are as patch array and slotted waveguide antennas which are explained in the following sub-sections.

Figure 6. Simulation model of patch array antenna

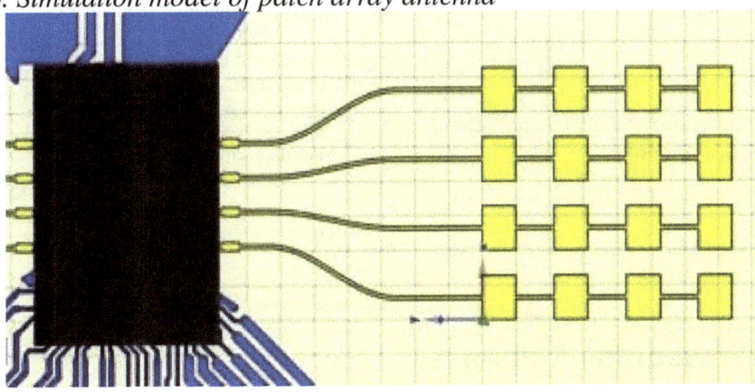

Patch array antennas

The microstrip patch antennas are the most commonly used radiating element in FMCW radars due to their compact size, low profile and directional radiation pattern. The advantages of patch array antennas include ease of integration, minimal components used to manufacture, and thus low-cost design. The metallic structures present in the patch antennas are usually PCB over a ground plane. The popular radiating elements are series-fed linear patch arrays, comb-line antennas and corporate-fed patch arrays.

In linear patch array antennas, the frequency of operation can be controlled by changing the length of the individual patch elements, whereas the amplitude contribution to the radiation pattern can be determined by varying the patch width. The length of the line and spacing between the array elements will contribute to the phase distribution. The lower side lobes in the radiation pattern can help in reducing the interference, which can be achieved by varying the patch width and spacing. The figure 6 highlights the simulation model of the 16-element patch array antenna (Renesas Electronics, 2023).There are various design aspects and many factor that impacts the performance of patch array antennas operating in higher frequency bands and these are listed in Table (1).

Table 1. Design characteristics Patch Array Antennas

Design characteristics	**Details**
High-Frequency Challenges	Losses in feeding lines to the antennas Loss of efficiency and gain Dielectric loss of the PCB High resistance of metallic lines *Note: Keeping the lines between patches short helps reduce losses.*
Factors Impacting Bandwidth	Bandwidth of individual patch elements Feeding lines Phase of the signal Length of line between patch elements Distance between patch elements
Factors Impacting Performance	Manufacturing tolerances Etching tolerances Defective surface finish Material parameter variations

Slotted Waveguide Antennas

The modern FMCW radars usually integrate slotted waveguide antennas due to their good performance, high reliability and ease of integration. Slotted waveguide antennas can operate at microwave frequencies with low losses, as the propagation medium in the waveguide is air, which eliminates all dielectric losses. The waveguides are usually made of copper or aluminum conductive material with slots etched using high precision machinery. In modern days, the 3D printing technologies and metalized plastic injection molding are contributing to the advances in manufacturing technologies. The figure 7 depicts the simulation model of the 16-element slotted waveguide array (Renesas Electronics, 2023) .

Figure 7. Simulation model of 16-element slotted waveguide array

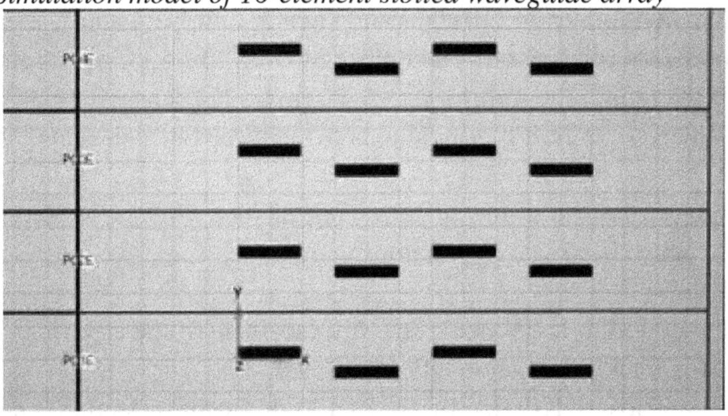

The slotted waveguide antennas are etched with more than one rectangular waveguide with narrow slots cut into the wider walls to create slot-like structure. These slots will act as a radiating element, which is fed with the transmitted signal through the waveguide. The coupling of the waveguide modes and frequency of operation is determined by size, shape and position of individual slots. The sidelobe level and radiation characteristics of the array is dependent on the distance between the slots. Major advantages of slotted waveguide over patch array antennas are listed in Table (2).

Table 2. Design characteristics: Slotted Waveguide Antennas

Design characteristics	Details
Advantages over patch array Antennas	Higher power handling capacity Lower transition losses Increase in overall gain and efficiency Facilitates wider bandwidth
Design Challenges	Requires more spacing between the slot elements Poor design can reduce angular resolution Highly sensitive to small geometrical changes Fabrication tolerances Requires high precision positioning over boards

RADOMES

Radomes are the structures used to protect the antennas from the severe environmental effects. The typical radome must be completely lossless or immune to the RF energy or the radiation from and to the antenna. This allows the radome to withstand the environmental effects such as wind, rain, hail, snow, ice, sand, salt spray, lightning, thermal, erosion and other unknown factors influencing the performance of the radar. The millimeter wave sensors are sensitive to all the factors in the surroundings. Thus, it is critical to note the mechanical design of the radome (Skolnik, 2008).

The designer must be aware of following factors which effects the antennas performance which is caused by radomes:

1. Beam deflection: It is defined as the displacement of the electrical axis which impacts the tracking functionality of the radar.
2. Transmission loss: It is defined as the RF loss due to the reflection and absorption
3. Reflected power: The antennas mismatch and higher sidelobes in both small and large radomes.

The radomes can be designed in several shapes such as planar, spherical, geodesic etc., the share of the radome has higher impact on the radiation pattern or field of view and the detection range. The materials used in the radome design are as follows: fiberglass, PTFE-coated fabric, and polycarbonate based on the radar applications. The distance the millimeter wave travels through the radome is angle dependent, thus higher angle of arrival could cause the larger error (Skolnik, 2008).

RADAR SIGNAL PROCESSING PIPELINE

To understand the signal processing in the receiver pipeline, the following discussion considers a commonly used automotive FMCW Radar. The fundamental principles of FMCW Radar are well explained in the prior sections along with the antennas and RF front end architecture. This section mainly focuses on the Analog to Digital Converter (ADC) and subsequent blocks that help to convert the In-phase and Quadrature-Phase (IQ) time domain samples to meaningful information that the host layer can utilize to track and classify various objects. There are various implementations across the automotive industry, but the core concepts of converting IQ samples to range, velocity and angle FFT remains the same. Apart from signal processing operations, there also exist architectures that implement the De-chirping operation in software (in comparison with hardware), where the beat frequency (*fo*) is estimated by mixing the digitized received signal with the digitized reference chirp signal that is stored in the memory.

Once the received signal is converted into IQ samples by the ADC, the signal processing pipeline broadly consists of the following stages. These stages may be processed immediately within the MMIC (Monolithic Microwave Integrated Circuit) or transferred to a microprocessor via a high-speed interface, where further processing takes place.

1. Range FFT (Fast Fourier Transform) – Extracts the range of detected objects.
2. Doppler FFT (Velocity Estimation) – Determines the velocity of detected objects.
3. CFAR (Constant False Alarm Rate) Detection – Identifies real targets while suppressing noise.
4. Angle of Arrival Estimation (AoA) – Determines the direction of detected objects.

The Final Outputs to the Application Layer depending on the radar type can include one or more of the following:

1. Range-Doppler Map – A 2D representation of detected objects in range and velocity space.
2. Radar Cube Formation – A 3D matrix that stores range, Doppler, and angle data for further processing. The radar cube typically consists of complex IQ samples across all dimensions, depending on the stage at which it is extracted.
3. Point Cloud Generation – A set of (X, Y, Z, velocity) points representing objects in Cartesian space.

The host layer further processes these outputs for higher-level functions such as:

1. Tracking – Continuously monitoring detected objects over time.
2. Classification – Identifying object types (e.g., vehicles, pedestrians).

Figure (8) illustrates the complete radar signal processing pipeline, starting from the point where IQ samples are received from the ADC for each channel. This specific implementation is widely used in automotive radars in current-generation vehicles. The ADC data is transferred to the signal processing section via high-speed interfaces.

Figure 8. RADAR PHY Processing

Range FFT (Range Estimation)

As briefly discussed in the prior sections, the beat frequency (f_0) is directly proportional to the range of the target due to the linear increasing frequency of the transmitted signal. To understand the Range calculation, the analog transmitted and received signal can be mathematically represented as the following

$$S_{tx}(t) = A_{tx} cos\left(2\pi\left(f_c + \frac{BW_{chirp}}{T_{chirp}} t\right) t\right)$$

$$S_{rx}(t) = A_{rx} cos\left(2\pi\left(f_c + \frac{BW_{chirp}}{T_{chirp}}(t - t_{delay})\right)\left(t - t_d\right)\right)$$

Here,
f_c is carrier frequency (e.g., 77 GHz for FMCW radar)
BW_{chirp} is chirp bandwidth (e.g., 200 MHz)
T_{chirp} is chirp duration (e.g., 50 μs)
t_{delay} is the receive signal delay due to the round-trip travel i.e., $t_{delay} = \frac{2 * Range}{c}$)

As the received signal is a delayed version of the transmitted signal, there exists a frequency difference which we define as a beat frequency. At any moment, the received chirp signals lags behind the transmitted chirp by constant time delay (t_{delay}) resulting in a fixed-frequency offset.

When the signal is received and processed through the RF hardware chain, it is mixed with the transmitted signal to produce a baseband signal which removes the carrier frequency (f_c) and the signal only contains the beat frequency (f_{beat}). The baseband signal is then digitized using an Analog to digital convertor (ADC) leaving digitized IQ complex samples. The IQ samples are converted to frequency representation by applying a fast-Fourier transform (FFT) and the output peak represents the beat frequency (f_{beat}).

In software, the bins with maximum peaks are determined which gives us the f_{beat} and initial Range can be calculated as the following

$$Range = \frac{c * f_{beat}}{2 * BW_{chirp}}$$

The Range FFT output is a 1D array, where each bin corresponds to a distance bin. If multiple peaks exist in the FFT output, it indicates the presence of multiple targets at different distances. The FFT size (N-point FFT) is determined by the number of ADC samples collected per chirp. Mathematically, the higher the N-point FFT, the better is the range resolution, but this also increases the implementation cost.

Doppler FFT (Velocity Estimation)

The Range-FFT derives only the distance to the target, but doesn't provide any information on the speed/velocity of the target. An accurate estimation of velocity is required to support target detection and classification tasks performed by the host layers. The velocity estimate is dependent on the doppler frequency shifts ($f_{doppler}$) which occurs due to relative motion between a transmitter and receiver and this is quite common to every wireless communication system and the impact is higher in high-speed automobiles.

The Doppler shift occurs when the RADAR (or the target) is moving and the perceived frequency of the electromagnetic wave changes due to the wavefront's compression/expansion. As the target approaches, the waves get compressed which increases the perceived frequency (Wolff, n.d.).

Consider an example of an automotive RADAR operating at 77GHz and a target (a car) is moving towards the RADAR with 10m/s. The doppler shift can be calculated as the following

$$\lambda\ (wavelength) = \frac{c}{f_c} = \frac{3 * 10^8\ m/s}{77 * 10^9\ Hz} = 3.9\ mm$$

$$f_d\ (doppler\ shift) = \frac{2 * v}{\lambda} = \frac{2 * 10\ m/s}{3.9 * 10^{-3}} = 5.12\ KHz$$

The calculations indicate that the received signal is shifted by 5.12 kHz higher than that of the transmitted signal. With RADAR point of view, the velocity of the target (or any source) is not available and needs to be estimated and this can be achieved by running a Fast-Fourier transform on the sequence of chirps within a single RADAR return.

After the Range-FFT is computed for each chirp, the process is repeated to get a stack of Range-FFTs that are computed at every received chirp thus creating a Range-Doppler matrix which is 2D in nature. Each column in the 2D matrix corresponds to the single range bin across chirps and FFT is applied across each column of the range-doppler matrix and the peak magnitude corresponds to the doppler frequency. This estimated doppler frequency is used to compute the velocity of the particular target. The output of the Doppler FFT process can be summarized as follows:

1. Output is a set of peaks per column.
2. Each peak corresponds to the target moving at a different velocity.
3. Typical cases would be the multiple peaks indicating multiple moving targets at similar range with different velocities.

The key difference with respect to Range-Doppler Matrix/Map before and after applying the FFT needs to be addressed as the units are different. Consider an example of Range-Doppler Matrix before velocity is estimated, in other words, a 2D Range bin data collected across chirps.The Range-FFT Map from a RADAR transmitting 4 chirps as shown in Table (3)

Table 3. Range-FFT Map

Chirp Index → Range Bin ↓	**Chirp 1**	**Chirp 2**	**Chirp 3**	**Chirp 4**
Range Bin 1	1	10	12	15
Range Bin 2	2	5	6	7
Range Bin 3	3	1	2	3

After the Doppler FFT is applied across each row which converts the time-domain variation of each range bins into frequency domain to extract the doppler shift (f_d), the Table (4) is the absolute values of each Doppler FFT bin and the values are rounded to two decimals.

Table 4. Doppler- FFT Map

Doppler Bin → **Range Bin ↓**	**Bin 0**	**Bin 1**	**Bin 2**	**Bin 3**
Range Bin 1	38	12.08	12	12.08
Range Bin 2	20	4.47	4	4.47
Range Bin 3	9	2.23	1	2.2

The Range-Doppler Map is further converted to Range-Velocity Map by using chirp-repetition frequency and FFT size (in this case, 4) and doppler frequency is calculated by taking the peaks which are nothing but 38 for bin0, 12.08 for bin1, 12 for bin2 and 12.08 for bin3 and this indicates that "range bin1" has strongest peaks and the dominant frequency component is at velocity of 0 m/s. Considering the chirp repetition frequency of 100 kHz,

$$f_d = \frac{K_{peak} * f_{chirp\_rep}}{N\_fft} = \frac{0 * 100KHz}{4} = 0 \text{ m/s}$$

By running similar calculations, range-velocity is shown in Table (5) by converting the doppler frequency to corresponding velocity bins.

Table 5. Doppler-FFT Map with velocity bins

Velocity Bin → **Range Bin ↓**	**Bin 0** **(0 m/s)**	**Bin 1** **(48.75 m/s)**	**Bin 2** **(-97.50 m/s)**	**Bin 3** **(-48.75 m/s)**
Range Bin 1	38	12.08	12	12.08
Range Bin 2	20	4.47	4	4.47
Range Bin 3	9	2.23	1	2.2

The above numbers are highly sparse due to the fact of using random Range numbers to create a Range-FTT and 4-point FFT was used to discuss/compute the data. From the above range-velocity map, we can derive the following conclusions

1. There is a stationary target near the RADAR as map indicates a strong peak at 0 m/s
2. A fast-approaching target with 48.75 m/s velocity towards the RADAR
3. A target moving away from the RADAR at a velocity of -97.5 m/s and -48.75m/s
4. A weak stationary target at 0m/s near the RADAR.

CFAR (Constant False Alarm Rate) Detection

The Range-Doppler (or Range-Velocity) Map contains the information about actual target and noise as these are computed using RADAR return and additional FFT processing leads to various high-frequency fluctuations in the map which will be in-efficient for target classification. CFAR attempts to filter such noise by utilizing various methods and is applied directly on the Range-Doppler Map.

CFAR uses a sliding window across the map and uses an adaptive threshold to detect the targets among the clutter. The threshold needs to be adaptive as a too high threshold misses the weak targets and too low threshold increases the false alarms. The sliding window considers the cells (the data in the map) as cell-under-test, guard cells and reference cells. The Guard cells protect the influence of strong targets on the cell-under-test and reference cells are used to calculate the average noise power.

The adaptive threshold is calculated by using detection threshold and average noise power of the reference cells

$$N_{avg\_power} = \frac{1}{N}\sum_{i=1}^{N} P_i$$

$$T_{CFAR} = \propto * N_{avg\_power}$$

Here, P_i represents the power in reference cell I and N represents the number of reference cells.The Target is deemed to be a valid if the Power of cell-under-test is higher than the adaptive threshold T_{CFAR} and there are various methods to compute the adaptive threshold (O'Donnell, 2001).

1. Cell Averaging CFAR (CA-CFAR)
   a. Averages all reference cells to estimate noise
   b. Typically used for uniform noise environments

$$N_{avg\_power} = \frac{1}{N}\sum_{i=1}^{N} Pi$$

$$T_{CFAR} = \propto * N_{avg\_power}$$

2. Greatest of CFAR (GO-CFAR)
   a. Divide the reference cells into groups and use the group with the highest noise estimate.
   b. Typically used where one group has a stronger target.

$$T_{CFAR} = \propto * max\left(N1_{avg}, N2_{avg}\right)$$

3. Smallest of CFAR
    a. Divide the reference cells into groups and use the group with the lowest noise estimate.
    b. Typically used where noise is uneven

$$T_{CFAR} = \propto * min\left(N1_{avg}, N2_{avg}\right)$$

MIMO in RADAR Signal Processing

Before exploring MIMO RADAR, it is important to understand radar resolution, as it plays a fundamental role in the design and development of radar systems. Radar resolution is generally classified into three categories: range, velocity, and angle resolution. Range and velocity resolution determines how accurately a radar system can estimate a target's distance and speed, but does not give any information on the direction of the target. However, Angle resolution is essential for determining the exact direction of the target (Texas Instruments, 2023).

Range Resolution

Range resolution is defined as an ability of the radar system to distinguish two or more targets at different ranges. As discussed in the earlier sections, when a radar receives a signal, it passes through the RF front end before being digitized by an ADC. The digitized baseband signal is then used to estimate the target's range by running a FFT on the IQ samples (Chaudhari, n.d.).

Consider an FMCW radar operating at a 77 GHz carrier frequency with a chirp sweeping from 76.7 GHz to 78.3 GHz that results in an RF bandwidth of 600 MHz and a chirp duration of 500 μs. If a target is located 50 meters away, the corresponding beat frequency can be calculated by

$$f_{beat} = \frac{2 * BW_{sweep} * target\_range}{c * sweep\_time} = 400 \text{ KHz}$$

To correctly sample this 400 kHz beat frequency, the ADC must follow the Nyquist criteria which states that the sampling rate must be at least twice the highest frequency component

$$f_{adc\_samping\_rate} \geq 2 * 400\ KHz\ i.e, \geq\ 800\ KHz$$

Thus, an ADC with a minimum sampling rate of 800 kHz is required to properly detect the target at 50 meters. However, if a higher sampling rate of 40 MHz is used, the radar can effectively detect targets at a much wider range of distances with improved accuracy. Using similar calculations, Table (6) outlines the detectable target distances with an ADC sampling rate of 40 MHz

Table 6. Impact of ADC Resolution on Range Calculations

Target Distance (meters)	Beat Frequency	ADC Sampling Rate (MHz)	Can Target be detected?
50	400 kHz	40	True
100	800 kHz	40	True
500	4 MHz	40	True
1000	8 MHz	40	True
2000	16 MHz	40	True
4000	32 MHz	40	False

The current generation of radar systems utilizes sufficiently high sampling rates to make sure the performance is not compromised when detecting targets at various distances, even with different combinations of sweep time and RF sweep bandwidth. However, selecting a very high sampling rate increases the computational load on the DSP processor, which in turn raises the overall system cost.

When a high sampling rate cannot be implemented due to hardware constraints, various interpolation techniques are used in software to compensate for the missing data due to under sampling. This approach introduces additional noise into the system impacting accuracy. Therefore, system designers carefully evaluate these factors and maintain a tradeoff between efficiency and cost while designing a radar system.

Velocity Resolution

Velocity Resolution is an ability of the RADAR system to distinguish between multiple targets moving at different velocities. As explained in the prior sections, doppler frequency is estimated by running a FFT on the stacked Range Maps over multiple chirps which is then used to convert to velocity bins. The baseband waveform that is used has a sequence of N-chirps called a RADAR frame and time between two consecutive chirps is pulse repetition rate.

When the target is in motion, there will be a phase difference over multiple chirps and the rate of phase change determines the doppler shift. The two angular frequencies will be distinct if the difference is greater than $\frac{2\pi}{N}$ and velocity resolution is inversely proportional to the chirp time (T_{chirp}). With N chirps in a frame, velocity resolution can be mathematically represented as

$$V_{resolution} = \frac{\lambda}{2 * N * T_{chirp}}$$

Angle Resolution

As briefly discussed in the section 1.3.1, radio transceivers are often classified based on the number of transmit and receive antennas. A radar system with just one receiver antenna can't determine the direction of a target because it picks up signals from all directions without any spatial reference. To figure out the direction of the target, the system needs at least two receive antennas. Angle resolution refers to how well a radar can distinguish two targets that are close together at the same range, and it depends on both the wavelength of the signal and the size of the antenna array—the larger the array, the finer the resolution (Rao, 2018)

Consider a uniform antenna array with N number of antennas spaced at a certain distance $d_{antenna}$, the antenna aperture D is the distance between the first and last antenna and is given by,

$$D = (N) * d_{antenna}$$

The spacing between the antenna i.e., $d_{antenna}$ is typically used as $\lambda/2$ and with these, angle resolution of a system with equispaced antennas can be mathematically given as

$$\theta_{res} \approx \frac{2}{N} \, radians$$

The Table (7) shows the angle resolution with respect to the number of received antennas.

Table 7. Angle Resolution -vs- Number of Rx Antennas

No. of Rx Antennas	Approximate Angle Resolution (degrees)
1	No aperture
2	60
4	30
8	15
16	7.5

With increase in the number of antennas, the angle resolution decreases but comes with a hardware implementation cost. Each additional antenna needs a separate RF front-end processing chain and this leads to the implementation of MIMO based FMCW RADAR.

Consider a radar with N_{rx} = 4 and N_{tx} =1 and the received signal at each antenna has subsequent phase shift(α) as it travels an additional distance .The phase difference between each antenna will be a linearly increasing wrt the first antenna i.e. 0, α, 2α, 3α, 4α etc and the phase shift can be estimated by sampling the signal across N_{rx} antennas and running an FFT on the sequence and this is called angle-FFT which forms as one of the dimensions in the RADAR data cube. The peak output of the FFT determines the presence of the target at that angle and with increase in antennas, the radars can determine the direction of similar range targets.

Figure 9. RADAR Cube formation

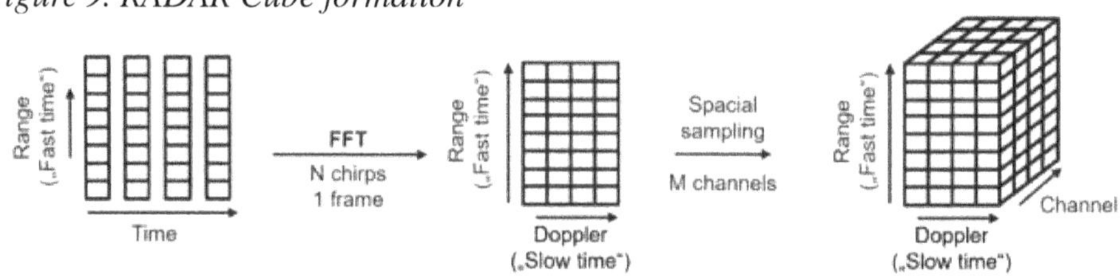

MIPI-CS2 Subsystem

The digitized data samples from the receiver are serialized through the high-speed communication interfaces known as Camera Serial Interface - 2 (CSI-2). CSI-2 is the standard defined and governed by Mobile Industry Processor interface (MIPI) alliance group and usually known as MIPI CSI-2 data interface (MIPI Alliance, 2024). The CSI-2 data interface is the automotive standard camera interface for handling serialized data via high-speed communication bus between the Radar MMIC and microprocessor. The ADC samples from the receivers are multiplexed and passed through the CSI-2 transmitter block through the PCB. In order to reduce the loss of data packets, the CSI-2 transmissions occur point-to-point from Radar MMIC to microprocessor without any extended trace length through buried via.

CSI-2 transmitter multiplexes the ADC data samples and CSI-2 receiver demultiplexes the ADC samples after reception. A simplified block diagram of CSI-2 communication is shown in the Figure (10)

Figure 10. MIPI -CSI2 Subsystem

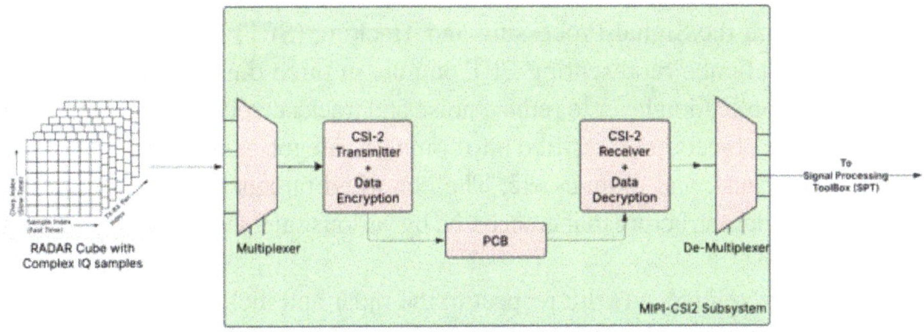

One of the major advantages of using the MIPI CSI-2 interface is that it is low power and high transition speed to match the high sampling rate of ADC blocks in the each receive chain.

Radar Applications

This section deals with the real-time application of automotive RADARs as they are crucial in modern driver-assistance systems, enabling a range of safety and convenience features. Out of many standardized automotive safety warning applications, the following are the commonly used applications (Miucic, 2018).

1. Adaptive Cruise Control (ACC): Maintains safe distances between vehicles.
2. Blind Spot Detection (BSD): Alerts drivers about vehicles in their blind spots.
3. Forward Collision Warning (FCW): Warns of imminent frontal collisions.
4. Lane Change Assistance (LCA): Helps drivers make safe lane changes.
5. Rear Cross-Traffic Alert (RCTA): Detects approaching vehicles when reversing.
6. Automatic Emergency Braking (AEB): Applies brakes in emergencies.
7. Parking Assistance: Assists in tight parking spaces.
8. Left Turn Assist (LTA): Prevents collisions during left turns at intersections.

Among these applications, the chapter will focus on the Left Turn Assist (LTA) functionality, a crucial feature for intersection safety. In a typical vehicle scenario, when multiple targets are detected/tracked, there needs to be a system in place where the targets needs to be classified with respect to the vehicle.

Radar Target Classification

The data from the Signal Processing and Tracking (SPT) block—specifically, the RADAR data cube representing FFT outputs in three dimensions—is further utilized at the application layer to reduce noise and track surrounding objects. Once processed, these objects are classified into common categories such as automobiles, cyclists, pedestrians, and animals etc. The classified targets are then stored in a comprehensive data structure that is accessed by various safety warning applications in the user space.

Each target is referenced with respect to the radar and includes key parameters such as speed, direction, range, and the type of return signal. Given the wide range of applications that rely on this data, the host processor classifies targets using a target classification matrix, which organizes objects/targets based on their range and direction. This classification reduces the computational burden by limiting the number of applications that need to process and verify each radar-detected target (Miucic, 2018).

Figure (11) illustrates a typical highway scenario where the host vehicle (depicted in blue) is equipped with an automotive radar sensor, while all other vehicles are radar-detected targets. Consider the Blind Spot Warning (BSW) system as an example where it primarily monitors remote vehicles located in the behind-left and behind-right zones, which are critical for identifying objects in the driver's blind spot. Other remote vehicles outside these zones are irrelevant to the BSW application.

Target classification is based on a combination of estimated vehicle speed, range, and direction relative to the radar sensor. A standard lane width of three meters is used to classify automobile-type radar returns, dividing the road into twelve distinct zones. Each zone is associated with one or more warning applications. For instance, vehicles detected ahead of the host vehicle primarily trigger applications such as front-end collision detection and intersection collision warning systems (Lui, 2015). This classification approach significantly reduces processing time and ensures efficient execution, as it is performed at every sampling period of the host processor.

Figure 11. Generic representation of surrounding targets

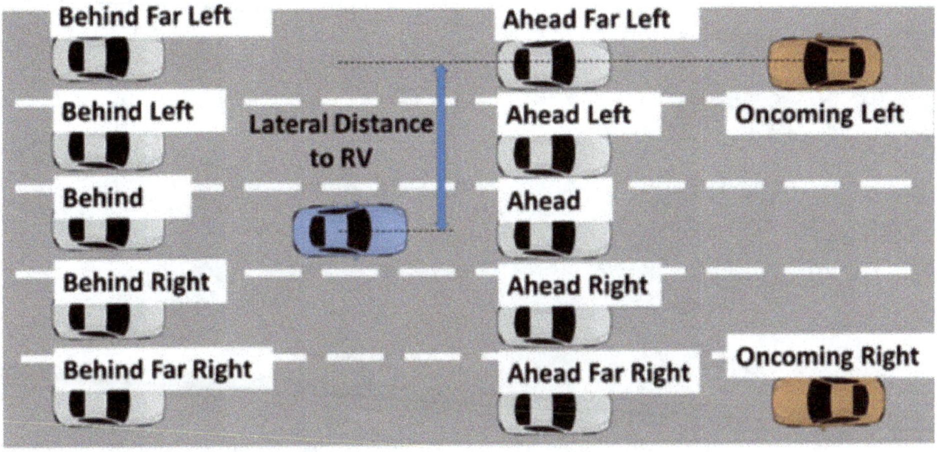

Left-Turn Assist (LTA) technology

Left turns at intersections can be tricky and dangerous. Drivers face challenges like limited visibility, unpredictable traffic, and potential errors. According to the NHTSA (ref), most automotive/pedestrian accidents reported at intersections are caused by vehicles taking an unsafe left turn. This droves the need to develop an algorithm that warns the driver of potential collisions with oncoming vehicles while making a left turn.

Figure 12. Left Turn Assist Description

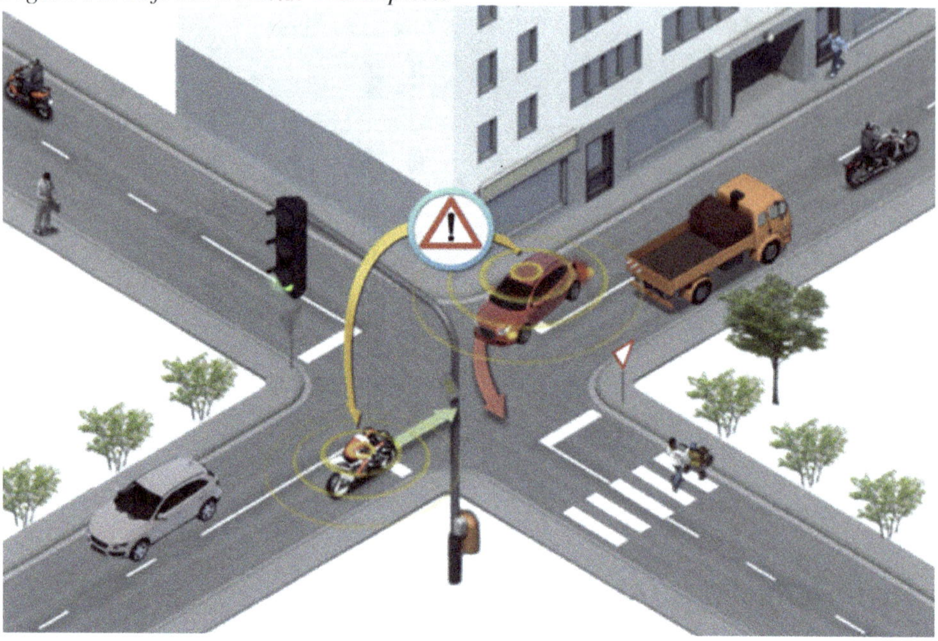

When designing a safety warning system for vehicles, it's important to integrate multiple wireless technologies, such as LiDAR, V2X communication, GNSS, 5G, Wi-Fi, and Millimeter Wave. Each of these plays a role in improving decision-making and enhancing safety. The Figure (12) shows a red car trying to make a safe left turn at an intersection. To do this, it relies on several sensors that provide real-time information about the surroundings that are shown in Table (8).

Table 8. Sensors used for LTA application

High-Definition Map	**Lane Level Map Points of the intersection**
GNSS Receiver	Absolute GNSS Latitude/Longitude Coordinates with lane level accuracy
FMCW Radar	Detects nearby vehicles, classifies them, and estimates their range, speed, and direction.

When the driver (or the autonomous system) activates the left-turn signal on the host vehicle, the application layer gets the data from all the sensors mentioned in table (8). The Radar cube data is used to classify different targets and track them and the approaching vehicles are used for further calculations. From this, it calculates how long each approaching vehicle (T_{remote}) to reach the intersection.

The host vehicle uses the high-definition map to understand the intersection's layout and calculates the best possible turn path using a spline function, ensuring the turn stays within safe lane boundaries. Based on typical intersection speeds in the U.S. (around 10 to 15 mph), it estimates how long it will take to complete the turn (T_{host}). Once these are calculated,

1. If T_{host} is shorter than T_{remote} plus a safety buffer, the turn is considered safe.
2. Otherwise, a safety warning is triggered, alerting the driver (or autonomous system) to wait.

If we consider the Host Vehicle (HV) and Remote Vehicle (RV) i.e., the target identified by FMCW radar as mentioned in the Figure (8), The minimum time taken by remote vehicle (RV) can be calculated by using the kinematics equation given by

$$Range_{RV1} = v_{initial\_RV1} * t_{minimum} + \frac{1}{2} * accel_{target} * t^2_{minimum}$$

$$t_{minimum} = \frac{-v_{initial\_RV1} + \sqrt{v_{initial\_RV1}{}^2 + 2 * accel_{target} * Range_{RV1}}}{accel_{target}}$$

Here, the acceleration of the target needs to be estimated based on the velocity returns from multiple measurements. As depicted in the figure (8), If we consider the remote vehicle approaching intersection has a velocity of 10m/s at 2 m/s^2 acceleration at a range distance of 100 meters, the $t_{minimum}$ is approximately 6.2 seconds

To determine the time taken by the host vehicle, we need to calculate the vehicle path using 4th order spline function using the starting position in the current left-lane to the ending position. As high-definition map database provides the GNSS positions of the lane-edges, the algorithm can run a 4th order polynomial to find the interpolated points that can be used as path for the lateral controls of the vehicle.

The typical turn distance in the united states road intersection is round 20 ~30 meters and with the recommended constant speed of 12 miles per hour (i.e., 5.36 m/s),

$$T_{host} = \frac{Distance_{max\_turn}}{Speed} = \frac{30\ meter}{5.36\ m/s} = 5.6 \text{ seconds}$$

$$T_{host} = 5.6 \text{ seconds} + T_{threshold}\left(3\ sec\right) = 8.6 \text{ seconds}$$

As the T_{host} is higher compared with $T_{minimum}$, the safety warning application is triggered to alert the driver. The above example doesn't consider all road conditions and other vehicles. The system typically needs to calculate the $T_{minimum}$ for each approaching target as per the RADAR data and have to take the minimum value and then comparison is done. The actual implementations also consider the road curvature in determining the range of the target.

Radar ECU Calibration and Communication Protocol

The Electronic Control Units (ECU's), a common terminology used in synonymous with automotive radars in embedded systems and often use a communication protocol known as XCP which provides read and write access to internal memories. XCP is a standard protocol governed by the Association for Standardization of Automation and Measuring systems (ASAM) working group. This group constantly reviews and updates the protocol standard. The term XCP is abbreviated for "Universal Measurement and Calibration Protocol" and the word 'X' indicates the support to all the transport layers such as CAN, CAN-FD, Ethernet (TCP/IP, UDP/IP), Flex Ray, SXL (SPI, SCI), USB.

The major use cases of XCP are during the Radar ECU development, testing and functional development. The basic features of the XCP protocol are as follows:

1. Synchronous data acquisition.
2. Synchronous data stimulation.
3. Memory calibration.
4. Data calibration.
5. Flash programming the Radar ECU's.

The XCP communication protocol relies on a single master multi slave architecture. The master is the computer application which serves as a measurement and calibration tool and slaves are usually the embedded systems such as Radar MMIC. Thus, a master can communicate to multiple slaves at a given instant of time whereas slaves can communicate to only a single master (ISO, 2025).

The XCP communication model

The master establishes contact with the slave using the Command Transfer Objects (CTO) and the exchange of synchronous measurement and stimulation data is accomplished using the Data Transfer Objects (DTO). The XCP protocol layer or message consists of three categories such as header, the packet and tail (Vector Informatik, 2022).

Figure 13. XCP Message Format

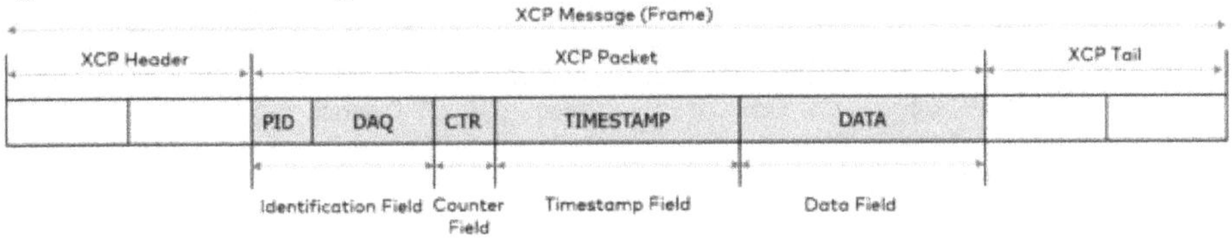

1. The identification field helps in recognizing which information was shared by a master or slave device.
2. The counter field is appended at the end of the identification field, which defines the property of the event for which the DAQ list is configured.
3. Timestamp field provides timestamp of the packet in byte, word and Dword data types.
4. The data field provides the actual ECU data, which can be measured values from slaves.

Consider the following problem scenario where the Radar MMIC is reporting that the transmit power amplifier level is too low in one of the MMIC transmit chains to the ECU that the. In order to query and troubleshoot the MMIC, the XCP communication protocol can be efficiently used to identify the problem and determine which transmitter chain is reporting a failure.

The following procedure can be followed to determine the problem in Radar MMIC:

1. Using the connect command 0xFF, the user can establish communication with the ECU.
2. Using the short upload command 0xF4, the user can query the status of the Radar transmitter to verify if the transmitter is enabled or disabled.
3. The short upload command will send a response with the data indicating that the transmitters are enabled.
4. Using the Short upload command 0xF4, the user can verify the threshold of the transmit PA level.

Thus, XCP commands can be a great tool to debug the Radar MMIC and ECU failures in real time. The various commands types which support XCP communication is shown in the table (9)

Table 9. Radar XCP communication protocol commands

Standard	**Calibration**	**Page switching**	**Cyclic data exchange**	**Flash Programming**	**Time synchronization**
Connect - 0xFF Disconnect - 0xFE Get Status - 0xFD Get_ID - 0xFA Set_Request - 0xF9 Get_Seed - 0xF8 Unlock - 0xF7 Set_MTA - 0xF6 Upload - 0xF5 Short_Upload - 0xF4	Download - 0xF0 Download Next - 0xEF Download Max - 0xEE Short Download - 0xED Modify Bits - 0xEC	Get Processor Info - 0xE9 Get Page Info - 0xE7 Get Segment Mode - 0xE5	Free DAQ (Data acquisition) - 0xD6 Allocate DAQ - 0xD5 Allocation ODT - 0xD4 Allocation ODT Entry - 0xD3	Program start - 0xD2 Program Clear - 0xD1 Program reset - 0xCF Program format - 0xCB Program prepare - 0xCC	Time correlation properties - 0xC6

AUTOMOTIVE SAFETY PROCESS

In modern days, the automotive industry is marching towards autonomous transportation by integrating various sensing technologies, which uses radio waves to establish robust communication with the surroundings. The complexity in the technologies are drastically increasing to address the more crowded environments and reduce the fatalities caused by road accidents throughout the year. That said, the need for safety architecture in every component of an automotive vehicle has become crucial. Thus, the international organization for standardization (ISO), which is the worldwide federation of national standard bodies, defines a set of guidelines and processes to be implemented by all automakers. The ISO 26262 series of standards is the excerpt of International Electrotechnical Commission (IEC) series of standards to guide the sector specific needs of electrical and electronics systems within a vehicle. The immense use of hardware and software components in controlling the vehicle poses a crucial risk to the safety of the passenger. The risk can be caused by the systematic and random hardware failures. Thus, the ISO 26262 defines which Automotive Safety Integrity Level (ASIL) are applicable to reduce unreasonable risk. Most of the automotive electronic components including radars are categorized as ASIL-B safety level.

The safety level is determined by considering the direct impact of the operation of electronic systems to the safety of the passenger. The levels are classified from A to D based on the criticality of the risk, with A being the lowest degree of hazard and D being the highest degree of hazard. The Table 10 lists different types of ASIL categories indicting the risk and associated components.

Table 10. ASIL Categories and Components

ASIL Category	Component
ASIL-D (Highest risk to passenger during failure)	Vehicle Engine or Battery Management Module in Electric Vehicles Steering Wheel Braking System Gear Shift Levers or Control Airbag
ASIL-C (Reduced risk but still critical)	Active Suspension Engine Management
ASIL-B (Reduced risk but critical to vehicle operation)	Radars or Millimeter Wave Sensors Cameras Ultrasonic Sensors High-Definition Map Module Driver State Sensing Telematic Module Wiper Blades System
ASIL-A (Least impact on passenger safety)	Rear Lights In-Cabin Lighting Systems Front and Rear-View Mirrors

In prior sections, we have reviewed the radar building block, which consists of the transmitter, receiver blocks, data interfaces and microprocessor. The following components in the radar building block is categorized as ASIL-B compliant.

1. Supply lines
2. Signal from the clock source
3. Transmitter antennas
4. Receiver antennas
5. Signal source from the Local oscillator
6. Communication data interfaces like Serial Peripheral Interface (SPI) and Camera Serial Interface - 2.

Additionally, the Radar MMIC consists of various monitoring circuits within the MMIC to determine the failure and report the status to the ECU. Later, the ECU takes necessary defined actions to mitigate the failure and warn the passenger about the criticality of the product. The safety monitoring circuits present in the Radar MMIC are as follows

1. Frequency monitor in the waveform generator block.

2. ADC clipping monitors.
3. IF voltage level monitors.
4. Phase monitors in the phase rotators
5. Power level monitors in the amplifier block
6. Clock signal monitor in the Local oscillator

For example, During the normal sensor operation, the radar MMIC sends a report to the microprocessor stating that the MMIC is in the failure state. The safety monitoring circuits are further analyzed to determine the kind of MMIC failure reported to the ECU. Upon reviewing the status, power level within the amplifier circuit is below the known threshold set during the design of the MMIC. Thus, indicating a critical hardware failure affecting the safe operation of the radar module. This scenario indicates the need for safety monitoring circuits to prevent the Radar module from being damaged and impact the safety of the passenger.

GOVERNMENTAL REGULATIONS FOR AUTOMOTIVE RADARS

The predominant use of radio waves has paved the way to the development of various consumer electronics and automotive products including radars or mm-wave sensors. As automotive radars are vastly developed and enhanced their performance over the years to support advanced safety applications. As Automotive radars use radio waves to detect the objects, countries around the world have established various organizations to regulate the use of radio frequencies based on frequency range, maximum bandwidth, maximum transmit power, maximum burst duty cycle and active chirping time limit.

The various governing bodies includes Federal Communications Commission (FCC) in United States, State Radio Regulatory Commission (SRRC) in China, European Telecommunications Standards Institute (ETSI) in European Union, Radio Waves Act, Ministry of Science and Information & communication technology (ICT) in Korea, Association of Radio Industries and Businesses (ARIB) in Japan. The Table (11) shows the power limits, maximum burst duty cycle in every frequency range as defined by the US FCC is shown. Similar regulation tables for different countries can be found at (KLRI, 2022), (SRRC, 2008), (AIRB, 2025) and (European Regulation, 2021).

Table 11. US FCC Regulations

FCC Clause	Frequency Range (GHz)	Use Cases	Power Limit	Max Burst Duty Cycle
15.255 (c) (2)(v)	61-61.5	All except drone applications	40dBm avg EIRP 43dBm peak	None
	57-71		- 10dBm conducted 10dBm peak EIRP	
15.255 (c) (2)(i)	57.0 - 59.4		20 dBm peak EIRP for indoor 30 dBm peak EIRP for outdoor, including all vehicular, applications	
15.255 (c) (2)(ii)	57.0 - 61.56		3 dBm peak EIRP	
	57.0 - 61.56		20 dBm peak EIRP	50% (16.5/33.0ms)
15.255 (c)(2) (iii)(A)	57.0-64.0		14 dBm peak EIRP	23% (7.5/33.0ms)
15.255 (c)(2) (iii)(B)	57.0-64.0	Fixed outdoor or vehicular uses external to cabin	20 dBm peak EIRP	50% (16.5/33.0ms)
15.255 (b)(3)	60-64	Drones/UAVs (Altitude limited to 121.92m)	20 dBm peak EIRP	50% (16.5/33.0ms)

REFERENCES

Aptiv Corporation (Ed.). (2022) Placement of sensing technologies, Satellite architecture. Available at: https://www.aptiv.com/docs/default-source/infographics/aptivsatellitearch_infographic.pdf?sfvrsn=4804163d_2

Chaudhari, Q. (n.d.). FMCW Radar Part 1 – Ranging. Wireless Pi. https://wirelesspi.com/fmcw-radar-part-1-ranging/

Christian Wolff Radar basics, Radar tutorial. Available at: https://www.radartutorial.eu/02.basics/Pulse%20Radar.en.html

Commission, F. C. (2023) Federal Communications Commission FCC 23-35, New opportunities for unlicensed field disturbance sensors. Available at: https://docs.fcc.gov/public/attachments/FCC-23-35A1.pdf

Commission, F. C. (2023a) Federal Communications Commission, FCC Empowers Short-Range Radars in the 60 GHz Band. Available at: https://www.federalregister.gov/documents/2023/07/24/2023-15367/fcc-empowers-short-range-radars-in-the-60-ghz-band

Commission, F. C. (2025) The Federal Communications Commission, 15.255 Operation within the band 57-71 GHz. Available at: https://www.ecfr.gov/current/title-47/chapter-I/subchapter-A/part-15/subpart-C/subject-group-ECFR2f2e5828339709e/section-15.255

European regulation (2021) Delegated regulation - 2022/30 - en - EUR-lex, COMMISSION DELEGATED REGULATION (EU) 2022/30. Available at: https://eur-lex.europa.eu/eli/reg_del/2022/30/oj

FURUNO Radar basics, FURUNO. Available at: https://www.furuno.com/en/technology/radar/basic/

ISO 26262 series of standards. Available at: https://www.iso.org/obp/ui/en/#iso:std:iso:26262:-1:ed-2:v1:en

Korea (2022) Radio Waves Act. Available at: https://elaw.klri.re.kr/eng_service/lawView.do?hseq=7162&lang=ENG

Liu, B., & El Kamel, A. (2015). V2X-based decentralized cooperative adaptive cruise control in the vicinity of intersections. *IEEE Transactions on Intelligent Transportation Systems*, *17*(3), 644–658. DOI: 10.1109/TITS.2015.2486140

MIPI Alliance. (2024) Camera serial interface 2 (MIPI CSI-2) v4.1, MIPI Alliance. Available at: https://www.mipi.org/specifications/csi-2

Miucic, R., Sheikh, A., Medenica, Z., & Kunde, R. (2018, August). V2X applications using collaborative perception. In 2018 IEEE 88th Vehicular Technology Conference (VTC-Fall) (pp. 1-6). IEEE. doi:DOI: 10.1109/VTCFall.2018.8690818

O'Donnell, R. M. (2001) Radar: Introduction to radar systems - online course, Radar: Introduction to Radar Systems - Online Course | MIT Lincoln Laboratory. Available at: https://www.ll.mit.edu/outreach/radar-introduction-radar-systems-online-course

Overview of ARIB Standards (STD-T48) Association of Radio Industries and Businesses. (n.d.). https://www.arib.or.jp/english/std_tr/telecommunications/desc/std-t48.html

Qian, K., He, Z., & Zhang, X. (2020). 3D point cloud generation with millimeter-wave radar. *Proceedings of the ACM on Interactive, Mobile, Wearable and Ubiquitous Technologies*, *4*(4), 1–23. DOI: 10.1145/3432221

Rao, S. & Texas Instruments Incorporated. (2018b). MIMO Radar. https://www.ti.com/lit/an/swra554a/swra554a.pdf?ts=1739548406903

Ratnam, D. V., Pasha, A., Swathi, P., & Rao, M. V. G. "Acquisition of GPS L1 signals using Cooley-tukey FFT algorithm," *2013 IEEE International Conference on Signal Processing, Computing and Control (ISPCC)*, Solan, India, 2013, pp. 1-4, DOI: 10.1109/ISPCC.2013.6663397

Renesas Electronics. (2023) Automotive radar, Millimeter wave automotive radars. Available at: https://www.renesas.com/en/blogs/how-connect-antennas

Ruppel, C. D., Javali, S. L., & Ltd, A. T. (2019, May 14). *US10938487B2 - Monitoring a vehicle-to-everything communication system - Available on Google Patents*. https://patents.google.com/patent/US10938487B2/en

Semiconductors, N. X. P.Automotive and Industrial Radar Systems, NXP Semiconductors. Available at: https://www.nxp.com/applications/RADAR-SYSTEMS

Shi, W., Tong, P., & Bi, X. (2025). Moving-Least-Squares-Enhanced 3D Object Detection for 4D Millimeter-Wave Radar. *Remote Sensing (Basel)*, *17*(8), 1465. DOI: 10.3390/rs17081465

Skolnik, M. I. (2008). Radar handbook. *IEEE Aerospace and Electronic Systems Magazine*, *23*(5), 41–41.

SRRC. (2008) General Micro-power (short range) radio transmission equipment, Regulation -SRRC type approval-radio certificate of P.R China. Available at: http://www.srrccn.org/srrc-approval-new2.htm

Tektronix (2018) Fundamentals of radar measurement and signal analysis -- part 1, Fundamentals of radar measurement and signal analysis. Available at: https://www.tek.com/en/blog/fundamentals-radar-measurement-and-signal-analysis-part-1

Texas Instruments (Ed.). (2023) Texas Instruments Radar academy, Ti developer zone. Available at: https://dev.ti.com/tirex/explore/node?node=A__AEIJm0rwIeU.2P1OBWwlaA__RADAR-ACADEMY__GwxShWe__LATEST&placeholder=true

Valiente, R., Raftari, A., Mahjoub, H. N., Razzaghpour, M., Mahmud, S. K., & Fallah, Y. P. (2023). Context-aware target classification with hybrid Gaussian process prediction for cooperative vehicle safety systems. *IET Intelligent Transport Systems*, *17*(7), 1344–1362. DOI: 10.1049/itr2.12327

Vector Informatik Gmb, H. (2022) XCP – the standard protocol for ECU development. Available at: https://cdn.vector.com/cms/content/application-areas/ecu-calibration/xcp/XCP_ReferenceBook_V3.0_EN.pdf

APPENDIX: ABBREVIATIONS

2D – Two dimensional
3D – Three dimensional
ACC - Adaptive Cruise Control
ADC – Analog to Digital Converter
AEB - Automatic Emergency Braking
AOA – Angle of Arrival
ARIB - Association of Radio Industries and Businesses
ASAM - Association for Standardization of Automation and Measuring systems
ASIL - Automotive Safety Integrity Level
BGA - Ball Grid Array
BLE – Bluetooth Low Energy
BSD - Blind Spot Detection
BSD – Blind Spot Detection
BSW – Blind Spot Warning
C2X – Car to Everything
CA – Collision Avoidance
CAN – Controller Area Network
CAN FD – Controller Area Network with Fast Data Rate
CAV – Connected Autonomous Vehicles
CFAR – Constant False Alarm Rate
CPW – Coplanar Waveguide
CSI – Camera Serial Interface
CTO - Command Transfer Objects
DSP – Digital Signal Processing
DTO – Data Transfer Objects
ECU – Electronic Control Unit
EIRP – Effective Isotropic Radiated Power
ETSI - European Telecommunications Standards Institute
FCC - Federal Communications Commission
FCW - Forward Collision Warning
FFT – Fast Fourier Transform
FMCW – Frequency Modulated Continuous Wave
FOV – Field Of Vision
GHz – Giga Hertz
GNSS – Global Navigation Satellite systems
HPF – High Pass Filter
HV - Host Vehicle
ICT - Information & Communication Technology
IEC - International Electrotechnical Commission
IF – Intermediate Frequency
IQ – In-Phase and Quadrature Phase
IR – InfraRed
ISO – International Organization for Standardization
LCA - Lane Change Assistance

LiDAR – Light Detection and Ranging
LNA – Low Noise Amplifier
LPF – Low Pass Filter
LRR – Long Range Radar
LTA - Left Turn Assist
MIMO – Multiple Input Multiple Output
MIPI – Mobile Industry Processor Interface
MMIC – Monolithic Microwave Integrated Circuit
MRR – Medium Range Radar
NHTSA – National Highway Traffic Safety Administration
PA - Parking Assistance
PA – Power Amplifier
PCB – Printed Circuit Board
PLL – Phase Locked Loop
PRF - Pulse Repetition Frequency
PRI – Pulse Repetition Interval
PTFE – Polytetrafluoroethylene
RADAR – Radio Detection and Ranging
RCS – Radar Cross Section
RCTA - Rear Cross-Traffic Alert
RF – Radio Frequency
RV – Remote Vehicle
RX – Receiver
SIW – Substrate Integrated Waveguides
SNR – Signal Noise Ratio
SPI - Serial Peripheral Interface
SPT – Signal Processing Toolbox
SRR – Short Range Radar
SRRC - State Radio Regulatory Commission
TCP/IP – Transmission Control Protocol/Internet Protocol
TX – Transmitter
USB – Universal Serial Bus
USRR – Ultra Short Range Radar
V2V – Vehicle to Vehicle
Wi-Fi – Wireless Fidelity
XCP – Universal Measurement and Calibration Protocol

Chapter 2
Characterization and Implementation of Penta-Band Modified Circular Patch Antenna Employing Machine Learning for Parameter Estimation

Manpreet Kaur
http://orcid.org/0000-0003-2307-0477
Guru Kashi College, Punjabi University, India

ABSTRACT

The presented book chapter focuses on the characterization of a modified ring shaped antenna loaded with symmetrical X shaped pattern. The book chapter includes eight Sections. Section 1 offers general overview on wireless communication, whereas Section 2 illustrates the role of antennas in wireless communication. Section 3 covers a comprehensive literature review related to numerous antenna designs. The entire construction process of the realized structure is illustrated in Section 4. The overall dimensions of the investigated geometry are 30 x 30 x 1.6 mm3. During antenna designing, outer ring represents the base design and then slots loaded X shaped pattern is incorporated onto the parent ring. A brief overview on the ANN approach is elucidated in Section 5. Section 6 presents the software generated outcomes in an effective manner. Results reveal that the resonant points fall at 2.85, 5.30, 10.76, 13.14, and 14.81 GHz. At the observed frequency points, the reported peak gain is 3.15, 4.93, 3.56, 5.13, and 8.11 dB, respectively. Section 8 compiles the entire findings.

DOI: 10.4018/979-8-3373-0852-4.ch002

1. INTRODUCTION TO WIRELESS COMMUNICATION

In the present days, significant revolution in the Wireless Communication (WC) sector has caused fundamental changes in networking and telecommunication sector (Dash et al., 2022). WC technologies have shaped our civilization significantly and enhanced economic growth globally (Kaur, 2024). Effective WC techniques help in the production of future generations of wireless systems. Currently, this mode of communication is experiencing an unprecedented boom. It eliminates the requirement of wires/cables for dedicated connection between the communicating end point devices (Sharma et al., 2018). It also gives users more flexibility and opportunities to access the required information, services and perform complex tasks. This sector has developed quickly and becoming one of the strong drivers of business and economy development. Even today, this technology is progressing continuously especially with digitization including 5G and Internet of Things. Moreover, this technology aims to fulfil the contemporary application needs, focussing on suitable antenna size reduction for compact, multiresonant devices. The future of these wireless technologies is bright but bit challenging. Every sector including health care, transportation, household, agriculture, banking, business and so on relies on this mode of communication (Singh, Kumar, Singh et al, 2024). The latest radio frequency (RF) circuit fabrication techniques, large-scale circuit integration and other size reduction methods have influenced this communication sector to a large extent. Innovative advancements in the modern WC have massive impact on human lives. This led to distinct antenna designers to work in this direction (Bhatia et al., 2018). This sector require portable, highly reliable, miniaturized and high gain antennas for efficient communication. Modern WC technology requires reduced antenna size and enhanced antenna characteristics. The growing number of wireless devices throws a challenging task for antenna designers to design multifrequency antennas in very less time to cater the needs of the present times. Nowadays, reduction in antenna dimensions and design complexity is the prime objective of antenna design community. Multiband antennas show multiple resonant bands, and cover numerous wireless applications. Therefore, it is urgent to produce miniaturized antenna structures with multiband functionality

2. ROLE OF ANTENNAS IN WIRELESS COMMUNICATION

Owing to diverse demands of WC sector, robust multifunctional antenna systems with decent compactness have become a necessity. Antenna are vital in wireless systems, as they provide enhanced functionality and better performance (Jamil et al., 2022). Therefore, it has been stated that the antenna technology has a huge contribution in the WC sector. Data broadcasting across a wireless network is typically done using a small hardware device, named as antenna. For instance, from a TV remote (a few metres) to satellite communication (thousand kilometres), transfer of data is through electromagnetic signals (Sharma & Bhatia, 2021). Microstrip Patch Antennas (MPAs) are the preferred choice because these antennas exhibit low profile characteristics including easy fabrication, small size, conformablity and low manufacturing cost (Elabd & Al-Gburi, 2024). However, such antennas with aforementioned features are required by several portable wireless devices. The designed antenna structures and the associated dimensions are very crucial for the achievement of operative resonant frequencies (Kaur, Krishan, Kaur et al, 2024). The demand of multiband/wideband covering antenna designs is creeping up day-by-day. The attributes such as compactness, high performance make them highly suitable for multiple portable and wearable devices. Designing a well-defined, compact, multifunctional antenna needs lot of tedious work. Traditional methods employ well-formulated exact mathematical expressions for accurate antenna designing with suitable compactness. Modern wireless networks rely on antenna design configuration. Therefore, an optimized and simple design is needed for such WC systems. MPAs can be easily modified and these modifications are done on the basis of system specifications and application type (Wang et al., 2023). With the rapid technological advancements, the need for multiband antennas in WC systems is increasingly day by day. An antenna with multiband characteristics with suitable bandwidth can perform the operations of numerous single band antennas (Arif et al., 2019). MPAs are widely employed in domestic, medical, industrial, aeronautical and space applications (Masroor et al., 2022). The designed shapes of MPAs and the associated dimensions are very crucial for the achievement of operative resonant frequencies. Selection of design parameters is a crucial task in the optimum design as this leads to distinct antenna designers to work in the particular direction.

3. LITERATURE REVIEW

The main purpose of an effective literature survey is to get wonderful ideas for new design creations. Numerous antenna designs exhibiting multi-functional features have been introduced in the recent past (Aboualalaa et al., 2017; Elabd & Al-Gburi, 2024; Kumar & Dileepan, 2017; Misala & Mosa, 2023; Panda & Ghosh, 2019; Wang et al., 2024). In the past decade, the compact antenna designing with significant size reduction has been an inspiring approach, and has the capability to generate multi-functional structures with wideband performance. The authors of (Kaur & Malik, 2021) pioneered the construction of an elliptical patch fractal antenna. Rogers RT Duroid based realized structure gave resonances at 2, 4, 6, and 8.2 GHz. The related

Impedance bandwidth (IBW) values were 410, 1070, and 4840 MHz. Based on the computations, it was undoubtedly confirmed that the geometry showed multiband functionality along with wideband features. A flexible wideband slot antenna was created in (Regulagadda & Kumari, 2023). The low-profile, portable structure had an area of 52 x 31 mm^2. Furthermore, it was elucidated that the resonant points were 2.45, 4.34, 5.96, and 7.94 GHz. The designers of the geometry stated that their projected design is a strong contender for biomedical and IoT applications. In (Verulkar et al., 2024), a high gain, compact dual band monopole antenna of electrical dimensions $0.746\lambda_0$ x $0.746\lambda_0$ x $0.128\lambda_0$ was demonstrated effectively. The researchers suggested that gain improvement was attained by the employment of frequency selective surface. Moreover, it was claimed that the prospective structure can be utilized for lower 5G and upper Wireless Local Area Network (WLAN) band. A multi-resonant monopole antenna incorporated with distinct type of slots such as U slots, k-shaped slots as well as split ring resonators and annular rings was mentioned in (Zade & Khade, 2024). The recommended structure offered triple resonances within the spanning range of 2.9 to 6 GHz. Maximum gain of 3.28 dBi was found at 5.6 GHz with a corresponding efficiency of 90%. A staircase pattern based hexagonal ring-shaped antenna was projected by the researchers in (Kaur et al., 2020). The wideband geometry produced maximum gain of 6.9 dB and the covered IBW was 7.74 GHz. All the simulated performance indicator values were in the acceptable range. II-K. Kim et al. (II-K. Kim et al., 2002) characterized and analyzed the behavior and performance features of a Koch Island MPA. Detailed experimentation illustrated that by following a fractal iteration factor of 2.5, patch size reduction of 45% was accomplished. Afterwards, it was proved that the new fractal antenna works well for mobile wireless systems. An X-shaped fractal radiator backed by recessed ground of area 88 x 20 mm^2 was designed and analyzed practically in (Gupta et al., 2017). Software generated results showed good accord with the practically tested outcomes. The generated resonant points were 3.6, 5.5, 5.95, and 6.5 GHz and the associated total efficiency was 45%, 69%, 68%, and 64%. A novel, quad-band hexagonal nested loop antenna supported by dumb bell shaped ground was critically verified in (Mark et al., 2018). The adopted size was 40 x 32 x 1.6 mm^3. Results demonstrated that five distinct resonances were produced and the provided radiation features were adorable. The authors in (Yu et al., 2017) projected a five-iterated fractal conductor similar to an ancient Chinese copper coin. At the considered bands, the examined gain values were in the 1.16-3.75 dBi range and the efficiency values were in the 40-72% range. In addition, it was proposed that the coin like structure covers numerous useful wireless bands. M. Gupta et al. (Gupta & Mathur, 2017) critically scrutinized a unique circular fractal patch antenna (shape like a wheel). The suggested simple design was assembled on a FR4 epoxy material and validated its functional behavior in the range of 2.9-9.5 GHz. For more clarity, parametric analysis was also included. Huge number of published papers discuss about the application of specific optimization approach in antenna engineering (Dhaliwal et al., 2020; Dhaliwal et al., 2019; Dhaliwal & Pattnaik, 2017; Kaur & Sivia, 2019). It is important to remember that optimization approach is merely a tool and strictly based on the designer's experience and intuition. In (Kaur & Sivia, 2019), the antenna designers deeply analyzed the operational features by employing two specific soft-computing strategies, namely Artificial Neural Network (ANN) and Firefly algorithm. J. S. Sivia et al. (Sivia et al., 2013) described the characteristics of an ANN based circular fractal antenna. Step-by-step generation process was terminated at 2nd iteration. For analysis and synthesis purposes, the adopted ANN technique was Feed Forward Back Propagation Network (FFBPN), and in the concuding remarks, it was revealed that the FFBPN predicted outcomes were well suitable for the design. In (Sran & Sivia, 2021), a multi-fractal curves based geometry was constructed with high precision. The utilized soft-computing methods showed appreciable results. A modified circular patterns embedded ring shaped fractal antenna was implemented and validated through experimental analysis in (Kaur, Krishan, Sivia et al, 2024). Measurements confirmed that at the appeared resonant points, performance metrics showed superior values. At the end, from the assessment of above literature, it is noticed that more research is needed to produce compact antenna geometries with high gain and wide/multiband characteristics.

The main contribution of this research work is to characterize and analyze an ANN based penta-band modified circular patch antenna by utilizing the selected physical dimensions. Due to the numerous advantages such as widespread availability, consistent performance and loss-less features, FR4 substrate is employed. The present research work has a novelty of compactness and simple geometry, operational over four important wireless bands with significant radiation features. The organization of the manuscript is summarized as: Section 1 is related to the importance of WC, whereas Section 2 effectively describes the role of an antenna in WC. Section 3 demonstrates the extensive literature survey required for the designing. Section 4 offers an in-depth explanation about the propounded antenna design process. Section 5 gives brief knowlegge about ANN method. Software engineered results are explained and discussed in Section 6. For a detailed explanation, Section 7 compares the outcomes of the created MCPA against the latest published state-of-the-art. Finally, all the findings are summarized in Section 8.

4. PROPOUNDED ANTENNA DESIGN

The goal behind the employment of the projected design approach is to attain multi-operational features with significant bandwidth values. The complete design process exhibits three design stages, named as Model I, Model II, and Model III, as depicted in Fig. 1.

Figure 1. Design stages of Propounded MCPA (a) Model I, (b) Model II, and (c) Model III

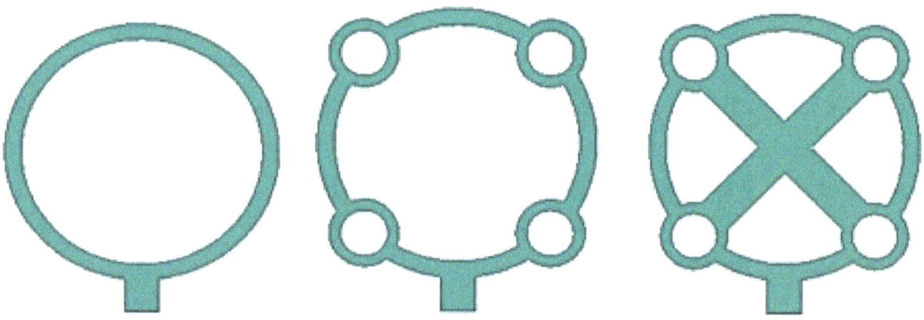

(a) (b) (c)

The structure is compact and ensures the desired design characteristics and functionality. The projected design has been premediated on an affordable FR4 substrate bearing an overall size of L_{SB} x W_{SB} x h mm^3. The prescribed design comprises of a modified circular-ring radiator and a defected ground. The gap between the upper and lower ends is 1.6 mm (Kaur, 2024). The design evolution steps are performed until the aforementioned features are achieved. Initially, a circular ring is created and is positioned at the centre of the substrate. Antenna at this stage is termed as Model I, as described in Fig 1(a). The value of R_1 is examined numerically with the Eqs. (1,2) (Sivia et al., 2013). In the subsequent stage, four circular patterns are embedded onto the parent ring and is named as Model II, as shown in Fig. 1(b). Model III (Fig. 1(c)) is created by adding X shaped pattern in the previously designed model, i.e., Model II. The design at this stage shows the desired performance characteristics. Therefore, this design is represented as the final model and is named as Modified Circular Patch Antenna (MCPA). Ground plane is tailored in such a way that it provide suitable impedance matching (Kaur, Krishan, Sivia et al, 2024). A square-shaped defect in incorporated in the ground at the upper portion. These modifications have been carefully crafted. The employed line feed of dimensions 'L_F' and 'W_F' mm^2 is placed at the estimated point (obtained using ANN approach). The small area structure is recommended for multi-functional wireless systems. Fig. 2 displays the systematic diagram of the Propounded MCPA. The dimensions associated with the Propounded MCPA are illustrated in Table 1.

Numerical way to compute the radius of the parent circle is mentioned below (Sivia et al., 2013):

$$a = \frac{F}{\left\{1 + \frac{2h}{\pi \varepsilon_r F}\left[ln\left(\frac{\pi F}{2h}\right) + 1.7726\right]\right\}^{1/2}} \quad (1)$$

Where

$$F = \frac{8.791 \text{ x } 10^{9}}{f_r \sqrt{\varepsilon_r}} \quad (2)$$

where f_r indicates the selected operational frequency, ε_r corresponds to the relative permittivity value and h specifies the chosen substrate height.

Figure 2. Propounded MCPA (a) Final patch geometry, and (b) Defected ground plane

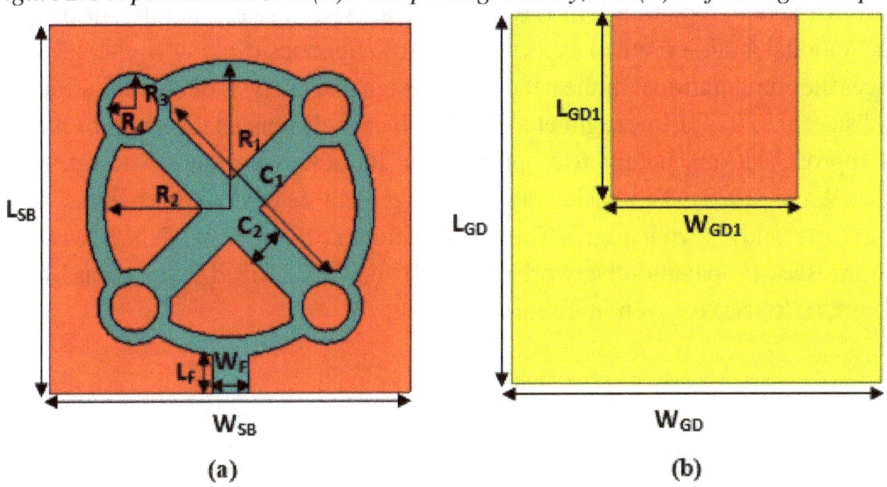

Table 1. Dimensions associated with the Propounded MCPA

L_{SB}	W_{SB}	C_1	R_1	C_2	R_2	R_3
30	30	9.5	12	2.46	10.5	3
R_4	L_F	W_F	L_{GD}	W_{GD}	L_{GD1}	W_{GD1}
2	3.5	3	30	30	15	15

All the belonging dimensions are in mm

5. ANN FRAMEWORK

Artificial Neural Networks (ANNs) are considered as the driving force for modeling, simulations and optimization. These tools improve the performance of an engineered system by mechanizing the acquisition of knowledge gained through experience (Dhaliwal et al., 2020; Sharma et al., 2020). Therefore, they are tremendously utilized in electromagnetics and microwave domain. ANN is represented as self-adaptive modelling tool that alters its geometrical structure according to the communicating external/internal information during the learning phase (Dhaliwal & Pattnaik, 2017). In the recent research, this methodology is adopted widely because the traditional ways of antenna designing are computationally expensive or efficient analytical formulaes are not available (Rodriguez-Soto et al., 2025; Singh, Rathor, & Kumar, 2024). Moreover, these models provide efficient designs and accurate optimal values within the training bounds (Dhaliwal et al., 2019). It ensures a speedy design process by eliminating the unnecessary time-consuming electromagnetic simulations. In the present era, ANN is very helpful in dealing computational and theoretical aspects of electromagnetics (Sran & Sivia, 2021). It reduces the computational timne needed for design and analysis of antenna structure (Bukhsh et al., 2023; Elmezughi et al., 2023). It also eliminates the need for hit and trial approach during parametric optimization for achieving the set design goals (Shi & Wang, 2022). The basic way of arrangement of layers is termed as ANN architecture, and shown in Fig. 3. The specifications of the chosen ANN types (Feed Forward Back Propagation Network (FFBPN) and Generalized Regression Neural Network (GRNN)) is given in Table 2.

Figure 3. ANN architecture

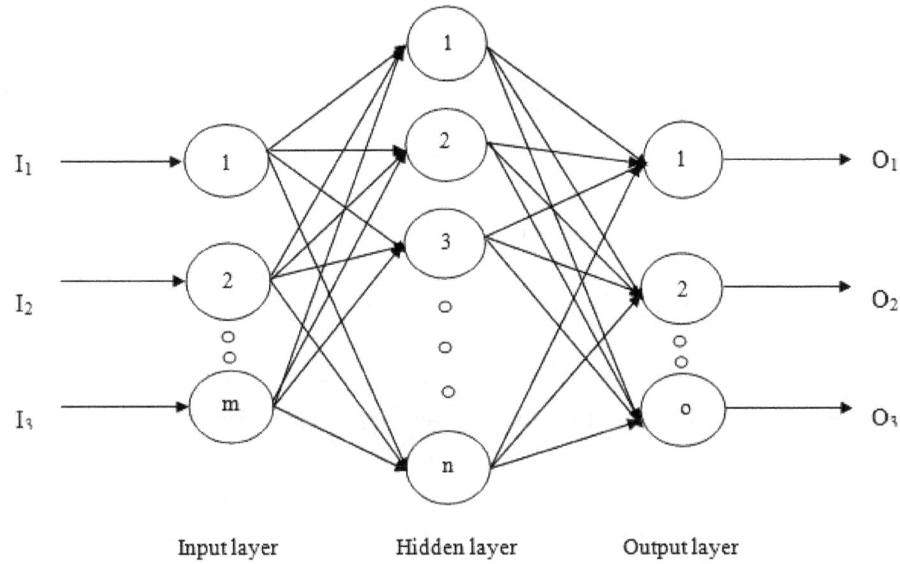

Table 2. Specifications of the chosen ANN types

S. No.	Network used	Number of hidden layers/ hidden layer neurons	Other details
1	FFBPN	2/10	Training function = Trainlm Data distribution = Random Adaptive learning function = Learngdm Transfer function = Tansig
2	GRNN	2/20	Spread constant = 1

6. RESULTS AND DISCUSSION

This Section demonstrates the evaluations carried out the projected structure in detail. The complete design and simulation criteria are performed and validated using Ansoft High-Frequency Structure Simulator (HFSS) v.15 software (Kaur & Sivia, 2019).

6.1. Comparative Analysis of Selected ANN Types

In the Sub Section, performance of the two selected ANN types (FFBPN and GRNN) is compared in terms of feed position estimation. The inputs are considered as: 'f_{r1}', 'f_{r2}', 'f_{r3}', 'f_{r4}', and 'f_{r5}'. The output assumed is the value of feed position. For this purpose, 40 samples are created by taking different feed position values. This pool of samples is generated using parametric analysis. Estimation of feed position using two different ANN types is revealed in tabulat form in Table 3. In this table, 18 samples are presented and the computations are done with respect to these 18 samples. The estimated values of feed position obtained by FFBPN and GRNN are compared with the desired values (Kaur & Sivia, 2019). The difference between these values corresponds to the computed error. After deep analysis, it is found that Mean Computed Error values obtained using FFBPN and GRNN are 0.0111 and 0.2691, respectively. The estimated values of feed position obtained using FFBPN and GRNN are graphically shown in Fig. 4. The examined values of MCE obtained using FFBPN and GRNN in terms of feed position are delineated in Table 4. In statistical form, MCE values are demonstrated in Fig. 5. From these values, it is concluded that FFBPN shows less MCE in comparison to the GRNN. Therefore, the value obtained using FFBPN is preferred in this work.

Table 3. Estimation of feed position using two different ANN types

S. No.	Resonant frequencies (Inputs)					Feed Position(mm)			Computed error	
	f_{r1}	f_{r2}	f_{r3}	f_{r4}	f_{r5}	Desired output	FFBPN	GRNN	FFBPN	GRNN
1	3.260	5.338	10.312	13.344	15.388	10.5	12.383	11.110	-1.883	-0.610
2	0	0	10.653	13.276	15.218	11.3	11.301	16.153	-0.001	-4.853
3	0	8.472	10.619	13.310	15.320	11.8	11.802	11.524	-0.002	0.275
4	0	5.338	10.721	13.344	15.218	12.3	13.470	12.715	-1.170	-0.415
5	2.953	5.270	10.721	13.310	15.252	12.5	12.593	11.786	-0.093	0.713
6	5.304	8.507	10.653	13.208	14.809	12.8	13.662	11.722	-0.862	1.077
7	5.304	8.472	10.823	13.174	14.775	13.1	13.629	12.160	-0.529	0.939
8	2.851	5.304	10.755	13.140	14.809	13.5	12.801	13.053	0.698	0.446
9	3.328	5.338	10.653	13.106	14.877	13.9	12.728	12.925	1.171	0.975
10	3.260	0	10.517	13.140	14.741	14.2	14.198	13.547	0.001	0.652
11	0	5.270	10.414	13.072	15.286	14.8	13.623	13.927	1.176	0.872
12	0	15.338	10.380	13.072	15.382	15.1	15.1	14.363	0.0001	0.736
13	5.304	9.392	10.346	13.242	15.422	15.6	14.704	13.078	0.895	2.522

continued on following page

Table 3. Continued

S. No.	Resonant frequencies (Inputs)					Feed Position(mm)			Computed error	
	f_{r1}	f_{r2}	f_{r3}	f_{r4}	f_{r5}	Desired output	FFBPN	GRNN	FFBPN	GRNN
14	6.124	9.326	11.224	13.128	15.341	16	15.112	16.320	0.887	-0.320
15	3.260	5.338	9.460	10.414	13.446	16.8	16.795	16.856	0.004	-0.055
16	9.358	10.312	11.539	13.242	15.354	17.6	18.293	18.648	-0.693	-1.048
17	8.164	10.548	11.264	12.958	14.853	18.5	18.574	16.103	-0.074	2.396
18	8.671	10.426	11.426	13.245	15.426	19.2	18.522	18.658	0.677	0.541
Mean Computed Error									0.0111	0.2691

Figure 4. Comparison of estimated values obtained using FFBPN and GRNN with desired value

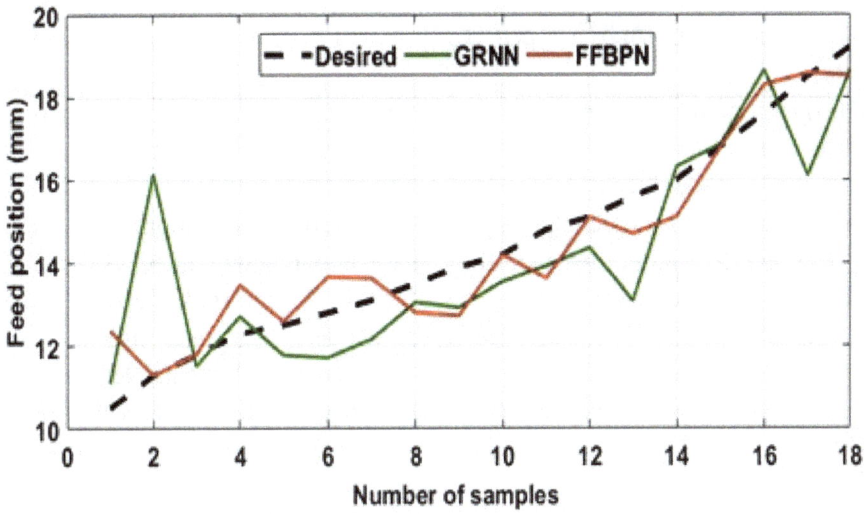

Table 4. Mean Computed Error obtained using FFBPN and GRNN in terms of feed position

Parameter	Mean Computed Error	
Feed position(p)	FFBPN	GRNN
	0.0111	0.2691

Figure 5. MCE value corresponding to FFBPN and GRNN

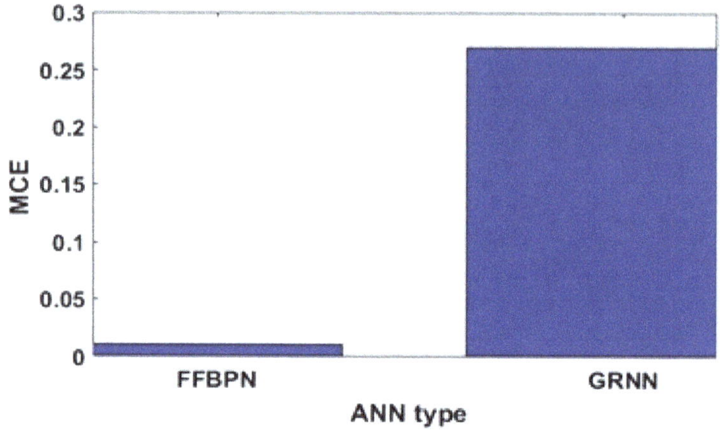

6.2. Propounded MCPA Results

The operational behavior of the Propounded MCPA, i.e., Model III in terms of all performance metrics is evaluated step-wise in the range of 2 to 19 GHz. S11 characteristics of the three different models (Model I, Model II, and Model III) described in Section 4 is depicted in Fig. 6. Upon careful examination of this Figure, it is noted that the Model I produces four resonant bands with good S11 behavior. The determined frequency points are 9.63, 11.98, 14.30, and 16.48 GHz with related S11 of -29.68, -20.27, -28.80, and -28.75 dB. Model II illustrates three frequency points (8.51, 11.13, and 12.22 GHz) with poor gain values. The final design (Model III) generates five resonances, i.e. 2.85, 5.30, 10.76, 13.14, and 14.81 GHz. The -10 dB IBW corresponding to these bands are are 450, 2270, 340, 400, and 520 MHz, which can serve S, C, X, and Ku bands. The related VSWR values are ≤ 2. Fig. 7 provides a visual depiction and description of the VSWR characteristics belonging to the above-said models. The propounded structure undergoes multiple geometrical adjustments to improve the number of operational bands (Kaur, Krishan, Kaur et al, 2024). Table 5 presents the intended performance parameters of Model I, Model II, and Model III (Propounded MCPA). By scrupulously visualizing the simulated outcomes, it can be advocated that the suggested geometry particularizes pentavalent frequency bands with adorable S11 values. The Propounded MCPA with different combinations, i.e., Model III and Model IV are presented in Fig. 8. The S11 values associated with these two models are portrayed in Fig. 9. For more clarity, the detailed performance comparison of Model III and Model IV are demonstrated in Table 6. Results confirm that the Model IV gives only three frequencies. The produced frequencies exhibit higher values such as 11.50, 15.53, and 18.39 GHz. The related S11 values are very weak and bandwidth values are very small. After a rigorous analysis, it can be contemplated that the Model III claims better performance behavior than the Model IV.

Figure 6. S11 characteristics associated with the Model I, Model II, and Model III (Propounded MCPA)

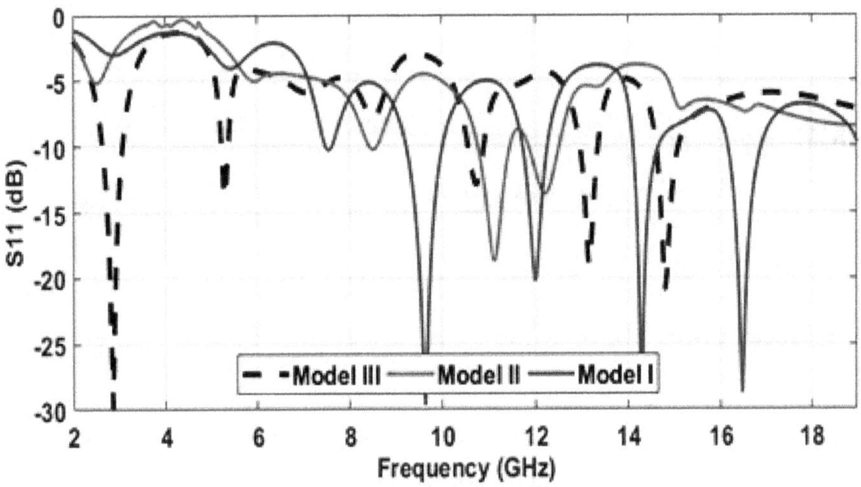

Figure 7. VSWR values associated with the Model I, Model II, and Model III (Propounded MCPA)

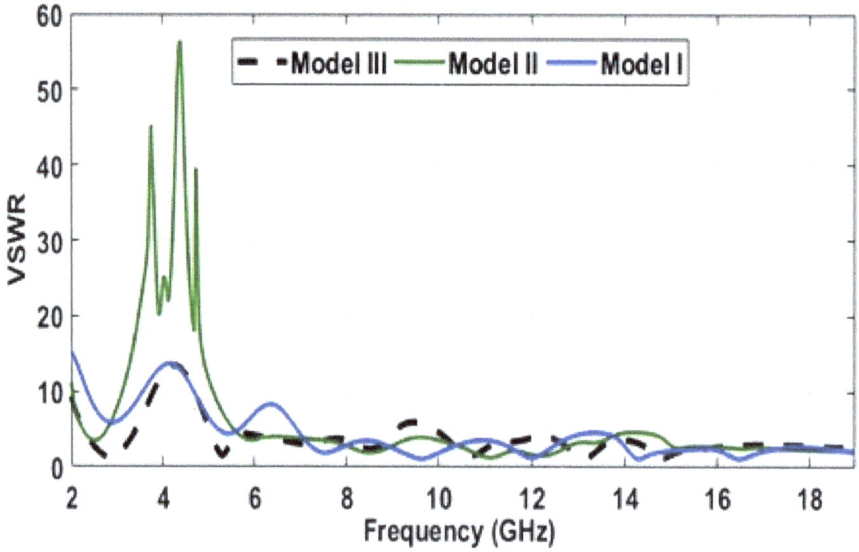

Table 5. Intended performance parameters of Model I, Model II, and Model III (Propounded MCPA)

Model no.	'F_L', GHz	'F_C', GHz	'F_H', GHz	-10 dB IBW, %	S11, dB	VSWR
Model I	9.22	9.63	10.10	9.1	-29.68	1.08
	11.70	11.98	12.25	4.5	-20.27	1.22
	14.09	14.30	14.67	4.0	-28.80	1.07
	16.13	16.48	16.92	4.7	-28.75	1.07
Model II	8.50	8.51	8.52	0.2	-10.25	1.93
	10.76	11.13	11.47	6.4	-18.72	1.28
	11.81	12.22	12.52	5.8	-13.60	1.73
Model III (Propounded MCPA)	2.62	2.85	3.07	15.8	-29.85	1.06
	3.10	5.30	5.37	53.6	-13.87	1.51
	10.55	10.76	10.89	3.2	-12.81	1.59
	12.94	13.14	13.34	3.0	-19.00	1.25
	14.60	14.81	15.12	3.5	-20.85	1.20

Figure 8. Propounded MCPA with different combinations (a) Model III, and (b) Model IV

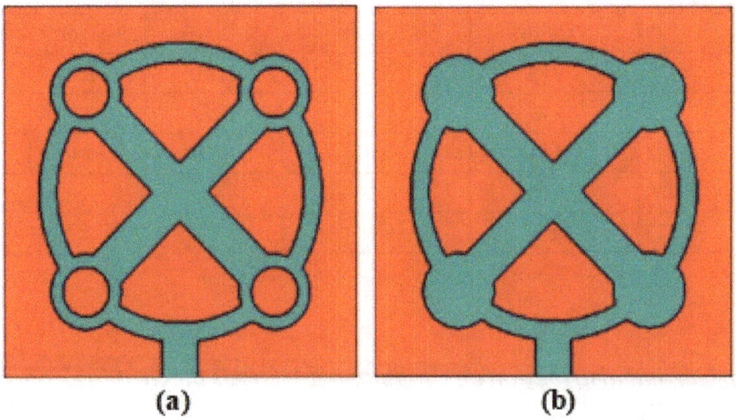

Figure 9. S11 characteristics associated with the Model IV and Model III (Propounded MCPA)

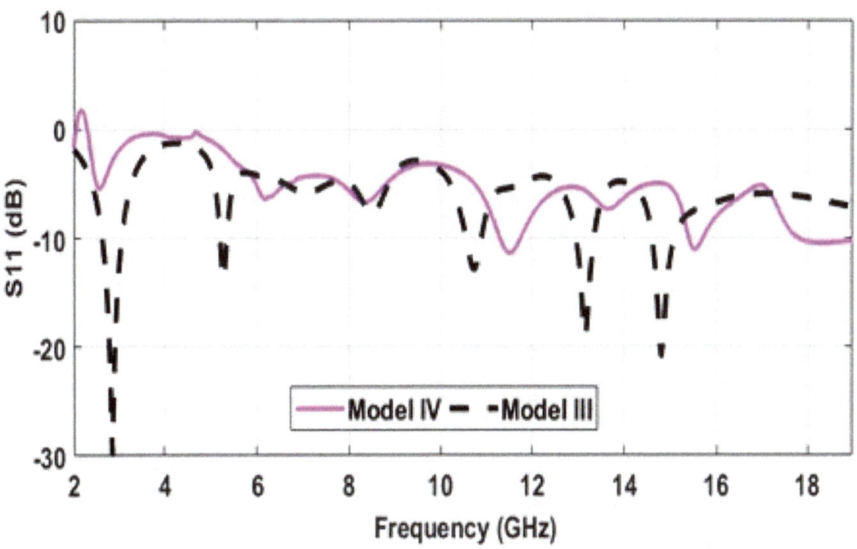

Table 6. Performance comparison of Model III (Propounded MCPA), and Model IV

Propounded Antenna incorporated with circular slots, Model III (Propounded MCPA)			**Propounded Antenna without circular slots, Model IV**		
'F_C', GHz	**S11, dB**	**Gain, dB**	**'F_C', GHz**	**S11, dB**	**Gain, dB**
2.85	-29.85	3.15	11.50	-11.39	1.46
5.30	-13.87	4.93	15.53	-11.11	2.28
10.76	-12.81	3.56	18.39	-10.46	1.71
13.14	-19.00	5.13			
14.81	-20.85	8.11			

Figure 10. Current distribution on Propounded MCPA's patch at (a) 2.85, (b) 10.76, and (c) 14.81 GHz

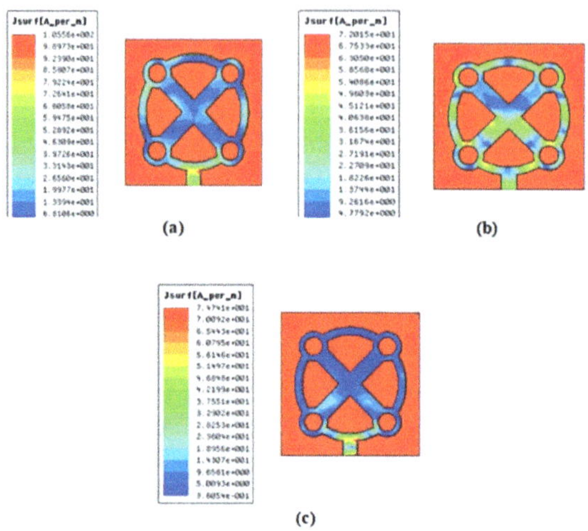

Figure 11. Vector current distribution on Propounded MCPA at (a) 2.85, (b) 10.76, and (c) 14.81 GHz

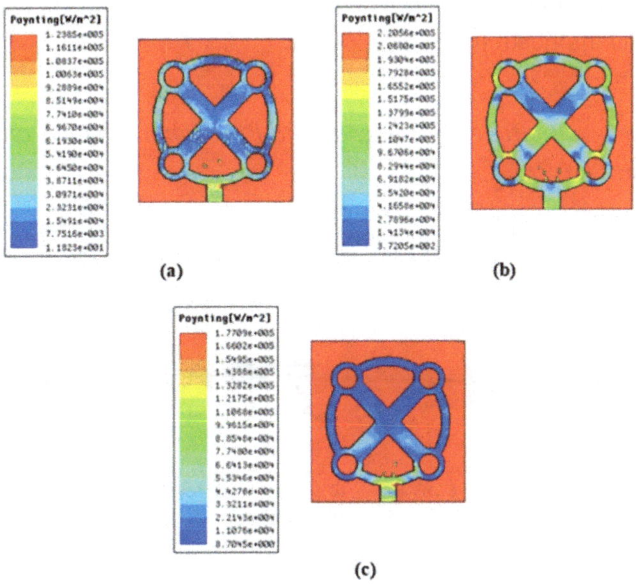

Analysis of current distribution is vital to actualize the electromagnetic properties. Fig. 10 reveals the current distribution behavior on the Propounded MCPA's patch at 2.85, 10.76, and 14.81 GHz. At 2.85 GHz, maximum current is dispersed at the outer ring and the feed (Kaur, 2024). At the X-shaped pattern, current density is very low. At 10.76 GHz, there is homogenous current flow. The current density on the patch is excellent. At 14.81 GHz, largest flow of the current is at the bottom of the designed geometry. Very less current flow is at the upper and middle portion (Kaur & Sivia, 2019). For better understanding, the vector current distribution on the Propounded MCPA's patch at 2.85, 10.76, and 14.81 GHz is elucidated in Fig. 11. In all these plots, direction of vectors in downwards, i.e., towards the feed, where the current density is maximum (Kaur & Sivia, 2019). The gain and efficiency values of the Propounded MCPA at all the observed resonant points is delineated in Fig. 12. The antenna's performance is analyzed carefully, revealing gain of 3.15, 4.93, 3.56, 5.13, and 8.11 dB at the claimed resonant points. In the similar manner, the examined efficiency values at the considered bands are 83%, 85%, 84%, 89%, and 92%. Software produced radiation patterns of the Propounded MCPA at the above-mentioned frequency points (2.85, 10.76, and 14.81 GHz) are described in Fig. 13. At 2.85, 10.76, and 14.81 GHz, the radiation patterns exhibit distorted bidirectional, butterfly shape and asymmetrical omnidirectional behavior, respectively. Distortion occurs at higher frequency points because the electrical length becomes greater than the associated wavelength. As a proof of concept, this way of antenna designing helps in achieving better and stable radiation characteristics. Due to its superior characteristics, this light weight antenna could be a best choice for S, C, X, and Ku band applications

Figure 12. Gain and efficiency values of the Propounded MCPA at the observed resonant points

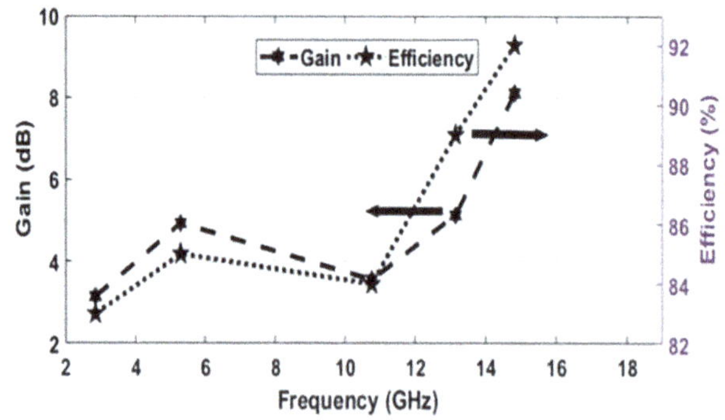

Figure 13. Software produced radiation patterns of the Propounded MCPA at (a) 2.85, (b) 10.76, and (c) 14.81 GHz

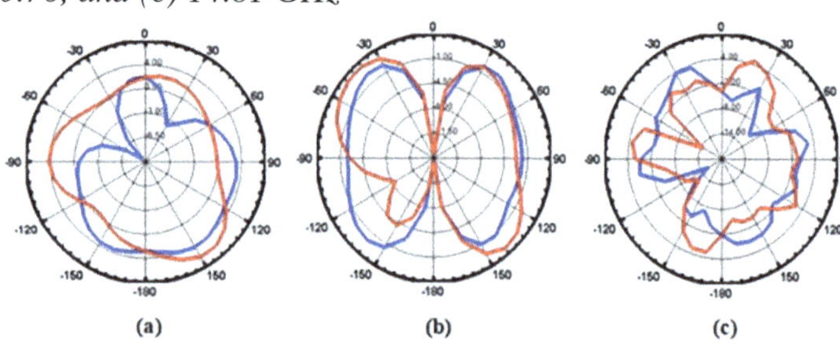

7. COMPARISON WITH FEW SELECTED ANTENNAS

The detailed comparative analysis of the Propounded MCPA with few reported antennas in given in Table 7. Several antenna designs bearing different sizes are studied and considered for comparison. Some of the considered antennas had large volume than the Propounded MCPA while others had less number of functional bands. After a rigorous analysis, the tabulated summary provides strong evidence that makes it clear that the suggested is highly competitive, taking into account size, fundamental bands and gain values.

Table 7. Comparison of Propounded MCPA with few reported antennas

Ref. no.	Area (mm^2)	Resonances (GHz)	Gain (dB)
(Bhatia et al., 2018)	38.92 x 40	4.4/6.8/9.6	7.50 (Max)
(Masroor et al., 2022)	32 x 37	3.5/7/9.7/12.7	8.37 (Max) 3.36 (Min)
(Kaur & Sivia, 2019)	38 x 38	1.42/2.43	3.63/7.80
(Kaur & Sivia, 2019)	34 x 34	2.42/5.80	10.19/5.74
(Sran & Sivia, 2023)	52.3 x 58.7	2.44/6.31	8.84/5.77
(Sran & Sivia, 2020)	52.3 x 58.7	2.53/4.91/7.6	8.61/3.32/5.38
(Akkole & Vasudevan, 2021)	39.7 x 47	2.10/2.91/4.32/ 5.83/7.88	1.36/-5.22/3.85/ 1.02/1.92
(Sharma & Sharma, 2018)	45 x 38.92	2.47/8.92	5.73/5.62
Propounded MCPA	**30 x 30**	**2.85/5.30/10.76/13.14/ 14.81**	**3.15/4.93/3.56/ 5.13/8.11**

8. CONCLUSION

This book chapter describes the operational characteristics of the Propounded MCPA which is constructed using FR4 based dielectric sheet of propositions of 30 x 30 x 1.6 mm^3, backed by defected ground plane. The structure follows circular ring radiator incorporated with slot loaded circular patterns and an X shape pattern. The maximum value of S11 (-29.85 dB) is achieved at 2.85 GHz, while at other resonances such as 5.30, 10.76, 13.14, and 14.81 GHz, the S11 parameter is -13.87, -12.81, -19.00, -20.85 dB, respectively. Analysis confirms that within the range of 2-19 GHz, the maximum and minimum realized gain is found to be 8.11 and 3.15 dB, respectively. At the claimed resonant points, the efficiency values are $\geq$ 83%. The proposed unique and novel design with significant flexibility exhibits assymetrical omnidirectional patterns at the above-mentioned frequency points. The feed position of the projected geometry is estimated with two different ANN types, namely FFBPN and GRNN. The detailed comparative analysis shows that the MCE values obtained using FFBPN and GRNN are 0.0111 and 0.2691, respectively. Based on these values, it has been stated that the FFBPN is preferred in this work. A comparative investigation in regard to vital antenna parameters shows that the presented antenna performs well for S, C, X, and Ku band applications.

REFERENCES

Aboualalaa, M., Abdel-Rahman, A. B., Allam, A., Elsadek, H., & Pokharel, R. K. (2017). Design of a dual-band microstrip antenna with enhanced gain for energy harvesting applications. *IEEE Antennas and Wireless Propagation Letters, 16*, 1622–1626.

Akkole, S., & Vasudevan, N. (2021). "Microstrip Fractal MultiBand Antenna Design and Optimization by using DGS Technique for Wireless Communication," 2021 6th International Conference on Inventive Computation Technologies (ICICT), pp. 72-77, Coimbatore, India DOI: 10.1109/ICICT50816.2021.9358712

Arif, A., Zubair, M., Ali, M., Khan, M. U., & Mehmood, M. Q. (2019, May). A Compact, Low-Profile Fractal Antenna for Wearable On-Body WBAN Applications. *IEEE Antennas and Wireless Propagation Letters, 18*(5), 981–985. DOI: 10.1109/LAWP.2019.2906829

Bhatia, S. S., Sahni, A., & Rana, S. B. (2018, January). A Novel Design of Compact Monopole Antenna with Defected Ground Plane for Wideband Applications. *Progress in Electromagnetics Research M. Pier M, 70*, 21–31. DOI: 10.2528/PIERM18050201

Bukhsh, Z. A., Molegraaf, H., & Jansen, N. (2023, April). A Maintenance Planning Framework Using Online and Offline Deep Reinforcement Learning. *Neural Computing & Applications*. PMID: 40584427

Dash, R. K., Saha, P. B., & Ghoshal, D. (2022, June). Slotted Patch Based Multiband Antenna with Multiple DGS Effect to Suppress Cross Polarized Radiation. *Progress in Electromagnetics Research C. Pier C, 120*, 179–193. DOI: 10.2528/PIERC22031707

Dhaliwal, B. S., Kaur, G., Kaur, S., & Pattnaik, S. (Feb 2020). Low-Cost Fractal MIMO Antenna Design for ISM Band Using ANN. In *Advances in Computational Intelligence Techniques* (pp. 219–226). Springer. DOI: 10.1007/978-981-15-2620-6_15

Dhaliwal, B. S., Kaur, G., Saini, N., Pattnaik, S. S., & Josan, S. K. (2019, August). A Modified Two-Step ANN Ensemble Approach to Improve Generalization and its Application in Fractal Antenna Design. *Journal of Circuits, Systems, and Computers, 29*(07), 2050106. DOI: 10.1142/S0218126620501066

Dhaliwal, B. S., & Pattnaik, S. S. (2017, September). Development of PSO-ANN Ensemble Hybrid Algorithm and Its Application in Compact Crown Circular Fractal Patch Antenna Design. *Wireless Personal Communications, 96*(1), 135–152. DOI: 10.1007/s11277-017-4157-8

Elabd, R. H., & Al-Gburi, A. J. A. (2024, June). Design and Optimization of a Circular-Ring Shaped UWB Fractal Antenna for Wireless Multiband Applications Using Particle Swarm Optimization. *Progress in Electromagnetics Research B. Pier B, 106*, 101–112. DOI: 10.2528/PIERB24033002

Elabd, R. H., & Al-Gburi, A. J. A. (2024, June). Design and Optimization of a Circular-Ring Shaped UWB Fractal Antenna for Wireless Multiband Applications Using Particle Swarm Optimization. *Progress in Electromagnetics Research B. Pier B, 106*, 101–112. DOI: 10.2528/PIERB24033002

Elmezughi, M. K., Salih, O., Afullo, T. J., & Duffy, K. J. (2023, September). Path Loss Modeling Based on Neural Networks and Ensemble Method for Future Wireless Networks. *Heliyon, 9*(9), e19685. DOI: 10.1016/j.heliyon.2023.e19685 PMID: 37809436

Gupta, A., Joshi, H. D., & Khanna, R. (2017, June). An X-shaped Fractal antenna with DGS for Multiband Applications. *International Journal of Microwave and Wireless Technologies*, *9*(5), 1075–1083. DOI: 10.1017/S1759078716000994

Gupta, M., & Mathur, V. (2017, September). Wheel Shaped Modified Fractal Antenna Realization for Wireless Communications. *AEÜ. International Journal of Electronics and Communications*, *79*, 257–266. DOI: 10.1016/j.aeue.2017.06.017

Jamil, A., Rauf, M., Sami, A., Ansari, A., & Idrees, M. D. (2022, February). "A Wideband Hybrid Fractal Ring Antenna for WLAN Applications," International Journal of Antennas and Propagation, Vol. 2022. *Article ID*, *6136916*, 1–8.

Kaur, A., & Malik, P. K. (2021, March). Multiband Elliptical Patch Fractal and Defected Ground Structures Microstrip Patch Antenna for Wireless Applications. *Progress in Electromagnetics Research B. Pier B*, *91*, 157–173. DOI: 10.2528/PIERB20102704

Kaur, M. (2024). Multifunctional and multistandard compact antenna design using hybridization of fractals and artificial neural network. In *Bio-Inspired Intelligence for Smart Decision-Making* (pp. 26–45). IGI Global Scientific Publishing.

Kaur, M., Krishan, R., Kaur, N., & Sivia, J. S. (2024). Invasive weed optimization based compact hybridized fractal antenna design for RF energy harvesting and multifunctional wireless applications. *AEÜ. International Journal of Electronics and Communications*, *184*, 155428.

Kaur, M., Krishan, R., Sivia, J. S., & Kaur, N. (2024). Shuffled frog Leaping algorithm based circular patterns loaded ring shaped fractal antenna for multistandard wireless applications. *AEÜ. International Journal of Electronics and Communications*, *176*, 155123.

Kaur, M., & Sivia, J. S. (2019, March). ANN-based Design of Hybrid Fractal Antenna for Biomedical Applications. *International Journal of Electronics*, *106*(8), 1184–1199. DOI: 10.1080/00207217.2019.1582712

Kaur, M., & Sivia, J. S. (2019, November). Giuseppe Peano and Cantor Set Fractals based Miniaturized Hybrid Fractal Antenna for Biomedical Applications using Artificial Neural Network and Firefly Algorithm. *International Journal of RF and Microwave Computer-Aided Engineering*, *30*(1), 1–11.

Kaur, M., & Sivia, J. S. (2019, February). Minkowski, Giuseppe Peano and Koch Curves based Design of Compact Hybrid Fractal Antenna for Biomedical Applications using ANN and PSO. *AEÜ. International Journal of Electronics and Communications*, *99*, 14–24. DOI: 10.1016/j.aeue.2018.11.005

Kaur, N., Singh, J., & Kumar, M. (2020, August). Hexagonal Ring Shaped Dual Band Antenna Using Staircase Fractal Geometry for Wireless Applications. *Wireless Personal Communications*, *113*(4), 2067–2078. DOI: 10.1007/s11277-020-07307-0

Kim, I.-K., Yook, J.-G., & Park, H.-K.II-K. Kim. (2002, July). J-G. Yook and H-K. Park, "Fractal-Shape Small Size Microstrip Patch Antenna,". *Microwave and Optical Technology Letters*, *34*(1), 15–17. DOI: 10.1002/mop.10359

Kumar, S. A., & Dileepan, D. (2017, June). Design and Development of CPW Fed Monopole Antenna at 2.45 GHz and 5.5 GHz for Wireless Applications. *Alexandria Engineering Journal*, *56*(2), 231–234. DOI: 10.1016/j.aej.2016.12.018

Mark, R., Mishra, N., Mandal, K., Sarkar, P. P., & Das, S. (2018, September). Hexagonal Ring Fractal Antenna with Dumb Bell Shaped Defected Ground Structure for Multiband Wireless Applications. *AEÜ. International Journal of Electronics and Communications*, *94*, 42–50. DOI: 10.1016/j.aeue.2018.06.039

Masroor, I., Ansari, J. A., Pandey, A., & Mishra, P. K. (2022, June). Design and Analysis of Multiband Fractal Reconfigurable Antenna Using PIN Diodes for Smart Wireless Communications. *Progress in Electromagnetics Research C. Pier C*, *120*, 209–222. DOI: 10.2528/PIERC22041402

Misala, S., & Mosa, S. A. (2023, December). A Novel Stacked Rectangular with Surface Mounted Short Rectangle Dielectric Resonator Antenna in C-Band Applications. *Progress in Electromagnetics Research Letters*, *115*, 81–89. DOI: 10.2528/PIERL23070902

Panda, P. K., & Ghosh, D. (2019, November). Wideband and High Gain Tuning Fork Shaped Monopole Antenna Using High Impedance Surface. *AEÜ. International Journal of Electronics and Communications*, *111*(1), 152920. DOI: 10.1016/j.aeue.2019.152920

Regulagadda, N. R., & Kumari, U. V. R. (2023, July). A Multiband Flexible Wideband CPW Wearable Slot Antenna for Biomedical and IoT Applications. *Progress in Electromagnetics Research C. Pier C*, *135*, 131–144. DOI: 10.2528/PIERC23052102

Rodriguez-Soto, M., Lopez-Sanchez, M., & Rodriguez-Aguilar, J. A. (2025). Multi-objective reinforcement learning for designing ethical multi-agent environments. *Neural Computing & Applications*, *37*(31), 25619–25644.

Sharma, N., & Bhatia, S. S. (2021, March). Ultra-wideband Fractal Antenna using Rhombus Shaped Patch with Stub loaded Defected Ground Plane: Design and Measurement. *AEÜ. International Journal of Electronics and Communications*, *131*, 153604. DOI: 10.1016/j.aeue.2021.153604

Sharma, N., & Sharma, V. (2018, December). A Design of Microstrip Patch Antenna using Hybrid Fractal Slot for Wideband Applications. *Ain Shams Engineering Journal*, *9*(4), 2491–2497. DOI: 10.1016/j.asej.2017.05.008

Sharma, N., Sharma, V., & Bhatia, S. S. (2018, September). A Novel Hybrid Fractal Antenna for Wireless Applications. *Progress in Electromagnetics Research M. Pier M*, *73*, 25–35. DOI: 10.2528/PIERM18052403

Sharma, S., Dudeja, R. K., Aujla, G. S., Bali, R. S., & Kumar, N. (2020, September). DeTrAs: Deep Learning-Based Healthcare Framework for IoT-Based Assistance of Alzheimer Patients. *Neural Computing & Applications*.

Shi, Y., & Wang, B. (2022, October). Cartoon Art Style Rendering Algorithm Based on Deep Learning. *Neural Computing & Applications*.

Singh, G., Kumar, A., Singh, M. J., & Singh, A. P. (2024, March). Modified Pentagonal Fractal Antenna for Ultra-Wideband Applications in Communication Systems. *International Journal of Electronics Letters*, *12*(4), 413–421. DOI: 10.1080/21681724.2024.2327091

Singh, U., Rathor, S., & Kumar, M. (2024, November). Hybrid Deep Learning and Machine Learning Approach for Detecting Spatial and Temporal Forgeries in Videos. *Neural Computing & Applications.*

Sivia, J. S., Pharwaha, A. P. S., & Kamal, T. S. (2013, November). Analysis and Design of Circular Fractal Antenna using Artificial Neural Networks. *Progress in Electromagnetics Research B. Pier B*, *56*, 251–267. DOI: 10.2528/PIERB13091611

Sran, S. S., & Sivia, J. S. (2020, August). ANN and IFS based Wearable Hybrid Fractal Antenna with DGS for S, C and X band Application. *AEÜ. International Journal of Electronics and Communications*, *127*, 153425. DOI: 10.1016/j.aeue.2020.153425

Sran, S. S., & Sivia, J. S. (2021, February). PSO and IFS Techniques for the Design of Wearable Hybrid Fractal Antenna. *International Journal of Electronics*, *108*(12), 2039–2057. DOI: 10.1080/00207217.2021.1885067

Sran, S. S., & Sivia, J. S. (2023, July). Design of a Novel Wearable Hybrid Fractal Antenna for Wi-Fi, Bluetooth, and WiMAX Applications. *Wireless Personal Communications*, *132*(4), 737–755. DOI: 10.1007/s11277-023-10635-6

Verulkar, S., Rochkari, A., Trimukhe, M., Bodade, V., & Gupta, R. (2024, February). High Gain Compact Dual Band Antenna Using Frequency Selective Surface for 5G and WLAN Applications. *Progress in Electromagnetics Research C. Pier C*, *142*, 1–11. DOI: 10.2528/PIERC24010101

Wang, E., Liu, X., & Chang, H. (2023, July). En. C. Wang, X. F. Liu and H. Chang, "Wideband Circular Polarized Fractal Antenna for RFID/WiMAX/WLAN Applications,". *Progress in Electromagnetics Research Letters*, *111*, 111–120. DOI: 10.2528/PIERL23032601

Wang, Z.-G., You, R., Yang, M., Zhou, J., & Wang, M. (2024, March). Design of a Monopole Antenna for WiFi-UWB Based on Characteristic Mode Theory. *Progress in Electromagnetics Research M. Pier M*, *125*, 107–116. DOI: 10.2528/PIERM24012611

Yu, Z., Yu, J., Ran, X., & Zhu, C. (2017, June). A Novel Ancient Coin-like Fractal Multiband Antenna for Wireless Applications. *International Journal of Antennas and Propagation*, *2017*(1), 1–10. DOI: 10.1155/2017/6459286

Zade, P. L., & Khade, S. S. (2024, July). Miniaturized Novel Multi Resonance Monopole Planar Antenna with Slots, Slits, Split Ring Resonator. *Progress in Electromagnetics Research C. Pier C*, *145*, 75–82. DOI: 10.2528/PIERC24051102

Chapter 3
Design and Performance Evaluation of Microstrip Patch Antenna for 5G Applications

S. Regina
http://orcid.org/0000-0003-4369-1211
Velammal College of Engineering and Technology, Madurai, India

S. Vasuki
Velammal College of Engineering and Technology, Madurai, India

S. Savita
http://orcid.org/0009-0003-5717-3287
Velammal College of Engineering and Technology, Madurai, India

ABSTRACT

In this study, a microstrip patch antenna is designed for 5G applications and is presented, with a resonance frequency of 3.33 GHz. The antenna, created with CST Microwave Studio and built on a FR4-lossy substrate with a dielectric constant of 4.3, is suitable for satellite and WiMAX communications. The square patch features three well-positioned slots, which enhance the antenna's electromagnetic performance. The slots measure 12mm x 30mm, 12mm x 8mm, and 12mm x 12mm, and are crucial for optimizing impedance matching, bandwidth, and radiation characteristics. The fully grounded patch ensures a broader bandwidth, and the antenna achieves a high gain of 7.63 dBi, a return loss of -28.457 dB, and a VSWR of 1.467. The simulation results confirm the antenna's potential for 5G applications, enabling fast data transfer and reliable communication. The proposed antenna is

DOI: 10.4018/979-8-3373-0852-4.ch003

a viable option for 5G-enabled devices due to its compact size, high gain, and low return loss, addressing the need for efficient and space-saving solutions in modern wireless communication systems.

I. THE EVOLUTION OF MOBILE NETWORK TECHNOLOGY FROM 1G TO 5G

The first mobile phone call was made by Motorola engineer Martin Cooper in the 1970s, marking the beginning of mobile communication history. With the ability to connect with one another at any time and from any location, this signalled the start of a new era in communication. Launched in the 1980s, the original generation of mobile phones, or 1G, was built on Analog technology. These phones were bulky, expensive, and had limited coverage, but they revolutionized the way people communicated. The Advanced Mobile Phone System (AMPS) technology, which transmitted data and speech over Analog signals, served as the foundation for the 1G network. The network's coverage was restricted to cities, and it was built to accommodate a small number of customers. The phones were also expensive, with prices ranging from $3,000 to $4,000, making them accessible only to a select few. The Nordic Mobile Telephone (NMT) system was introduced in 1981, marking the beginning of the first mobile phone networks in Europe throughout the 1980s. This method ushered in a new era of mobile communication by becoming the first to use digital technology. Future mobile phone networks were made possible by the NMT system, which was in use in a number of European nations, including Norway, Sweden, and Denmark.

The second generation, or 2G, of mobile phones was introduced in the 1990s. The digital technology that underpinned this generation of mobile phones was a major advancement over the Analog technology found in 1G phones. Text messaging and internet access were two additional capabilities added to the 2G network, which was built to accommodate more users. The Global System for Mobile Communications (GSM) technology, which transmitted data and voice via digital signals, served as the foundation for the 2G network. The GSM technology was more efficient than the AMPS technology used in 1G phones, and it allowed for more users to be supported on the same network. The 2G network also introduced the concept of roaming, which allowed users to use their phones in different countries.

The third generation, or 3G, of mobile phones was introduced in the 2000s. This generation of mobile phones was designed to support high-speed internet access, and it marked a significant improvement over the 2G network. The Universal Mobile Telecommunications System (UMTS) technology, which employed wideband code division multiple access (WCDMA) for data transmission, served as the foundation for the 3G network. Numerous applications, such as online gaming, mobile TV, and video conferencing, were intended to be supported by the 3G network. In addition to offering innovative capabilities like video calling and high-speed internet access, the network was built to accommodate a big user base. A major turning point in the evolution of mobile communication was reached with the deployment of the 3G network in a number of nations, including the US, South Korea, and Japan.

The fourth generation, or 4G, of mobile phones was introduced in the 2010s. This generation of mobile phones was designed to support even faster internet access, and it marked a significant improvement over the 3G network. The Long-Term Evolution (LTE) technology, upon which the 4G network was built, transmitted data using orthogonal frequency division multiple access (OFDMA). Numerous applications, such as cloud computing, video streaming, and online gaming, were intended to be supported by the 4G network. Along with introducing new capabilities like mobile payments and HD video calling, the network was built to accommodate a big number of customers. A major turning point in the evolution of mobile communication was reached with the debut of the 4G network in a number of nations, including the US, China, and India.

The fifth generation, or 5G, of mobile phones was introduced in the 2020s. This generation of mobile phones is designed to support even faster internet access, and it marks a significant improvement over the 4G network. The 5G network is based on the New Radio (NR) technology, which uses millimeter wave (mmWave) frequencies to transmit data. Virtual reality, video streaming, and online gaming are just a few of the many applications that the 5G network is intended to handle. In addition to offering innovative capabilities like mobile augmented reality and ultra-high resolution video calling, the network is built to accommodate a large number of users. Several nations, including the US, China, and South Korea, are launching the 5G network, which represents a major turning point in the evolution of mobile communication. Over the last few decades, there have been notable developments in mobile phone technology in addition to the creation of new mobile phone networks. The first mobile phones were bulky and expensive, but they have become smaller, cheaper, and more powerful over the years.

II. INTRODUCTION

The need for creative antenna designs that make use of cutting-edge materials and techniques to improve performance has been fuelled by the growing demand for high-speed data transmission in 5G networks, according to Sumit Kumar et al. (Kumar et al., 2020). In order to enable dependable and effective communication in a variety of applications, such as wearable technology and healthcare monitoring,

5G wideband antenna development for wireless body-centric networks (WBCNs) is essential. as shown by Mohammad Monirujjaman Khan and associates (Khan et al., 2022). The necessity for small, effective antennas that can operate at high frequencies has increased due to the growing demand for 5G technology. The construction of compact patch antennas operating at 3.5 GHz is a possible option for 5G applications, providing a potential answer for future wireless communication systems, as proven by Nayla Ferdous et al. (Ferdous et al., 2019). High-performance antennas that can facilitate satellite, mobile, and fixed communications are now required due to the advent of 5G technology. Microstrip patch antennas, as suggested by Rashmitha R et al. (Rashmitha et al., 2020), can be made to satisfy the demands of 5G communications, providing a small and effective solution for a range of uses. The need for high-performance antennas that can accommodate a variety of applications has significantly increased since the introduction of 5G technology. The design, development, and use of 5G antennas have advanced significantly, allowing for higher connectivity, lower latency, and faster data rates, as described by Pant and Malviya (Pant & Malviya, 2023). 5G technology, which offers high-speed connectivity, low latency, and increased reliability, has completely changed the mobile network scene. Significant technological improvements, such as the creation of MIMO antennas, Defected Ground Structure (DGS), and other methods to improve parameter performance, have characterized the progression of mobile networks from 1G to 5G, as discussed by Swati Patil et al. (Patil et al., 2022). However, 5G technology also faces challenges, such as mutual coupling reduction, which can be addressed through various techniques, including substrate selection, antenna element placement, and isolation methods. For 5G communication systems, the creation of small and effective Multiple-Input Multiple-Output (MIMO) antennas is essential. A compact two-element MIMO antenna design can offer a potential option for 5G communication, offering reduced size and enhanced performance, according to Ahmad et al. (Ahmad et al., 2022). A key component of 5G communication systems is the creation of Multiple-Input Multiple-Output (MIMO) antennas, which allow for faster data transfer and increased network capacity. Recent developments in MIMO antenna design have resulted in notable performance enhancements, such as higher gain, smaller size, and improved isolation between antenna elements, as reported by Tej Raj et al. (Raj et al., 2023). For 5G mobile communications, which demand low latency and high-speed data transfer, high-performance antenna development is crucial. It is possible to create a microstrip patch antenna to function in the 5G frequency spectrum, as suggested by Yitong Li (Li, 2023), providing a small and effective mobile communications solution. For 5G communications, which demand low latency and high-speed data transfer, multiband microstrip patch antenna development is essential. In order to provide a small and effective solution for 5G communications, Punith et al. (Punith et al., 2019) suggested designing a unique multiband microstrip patch antenna that can function in various frequency bands. For 5G wireless communication systems, which demand fast data transfer and minimal latency, broadband dual-polarized antenna development is crucial. A broadband dual-polarized antenna can be made to function in a large frequency range, as suggested by N. Sathishkumar et al. (Sathishkumar et al., 2023), providing a small and effective solution for 5G wireless communication systems. High-frequency microstrip patch antenna development is essential for 5G communication systems, which need minimal latency and fast data transfer. A microstrip patch antenna with a C-shaped slot can be made to function at 30 GHz, as suggested by Siddharth Kishore et al. (Kishore & Abdul Rajak, 2022), providing a small and effective option for 5G communication systems. 5G communication technologies, which demand fast data transfer and minimal latency, depend on the development of high-frequency antennas for on-body applications. An elliptical arc-shaped antenna can be made to function at 37 GHz, as suggested by Anupma Gupta et al. (Gupta et al., 2024), providing a small and effective solution for 5G on-body applications. For 5G applications, the creation of microstrip patch antennas is essential to attaining low latency and fast data transfer. A microstrip patch antenna can be made to function in the 5G frequency band, as suggested by Yogendra Kumar Patel (Patel, 2022), providing a small and effective solution for 5G applications. For 5G applications, the creation of microstrip patch antennas is essential to attaining low latency and fast data transfer. A microstrip rectangular patch antenna can be constructed and evaluated to function in the 5G frequency spectrum, as suggested by Rishitha Kyama et al. (Kyama et al., 2023), providing a small and effective solution for 5G applications. For 5G wireless applications, the creation of microstrip patch antennas is essential to attaining low latency and fast data transmission speeds. A microstrip patch antenna can be made to function in the 5G frequency spectrum, as suggested by T. Maheswari et al. (Maheswari et al., 2019), providing a small and effective solution for 5G wireless applications. For 5G applications, the creation of circular microstrip patch antennas is essential to attaining low latency and fast data transfer. A circular microstrip patch antenna can be made to function in the 5G frequency spectrum, as suggested by Mohammed Mahdi Salih Altufaili et al. (Altufaili et al., 2022), providing a small and effective solution for 5G applications. High-frequency microstrip patch antenna development is essential for 5G applications in order to achieve reduced latency and high-speed data transfer. According to Omar Darboe et al. (Darboe et al., 2019), a 28 GHz rectangular microstrip patch antenna can be made to function in the 5G frequency range, providing a small and effective solution for 5G applications. In order to achieve low latency and high-speed data transmission, microstrip patch antenna development is essential for wireless communication systems. A microstrip patch antenna can be developed and researched to function in several frequency bands, providing a small and effective solution for wireless communication systems, as suggested by Md.

Sohel Rana et al. (Rana & Rahman, 2022). For satellite communication systems to send and receive signals across great distances, high-gain and effective antennas are essential. A microstrip patch antenna, as suggested by A. Manimegalai and Dr. I. Kalphana (Manimegalai & Kalphana, 2024), can be planned and examined for satellite communication applications, providing a small and effective way to send and receive signals in space.

III. LITERATURE SURVEY

For 5G networks to achieve low latency and high-speed data transfer, wideband antenna development is essential. A broad and effective solution for 5G networks can be provided by designing a hexagonal type antenna to function in the 5G frequency band, as suggested by Orçun Yıldıran and Hüsnü Deniz Başdemir (Yıldıran & Başdemir, 2017). For 5G applications, the creation of small and effective antennas is essential to obtaining low latency and fast data transfer. A small and effective solution for 5G applications can be provided by designing a tiny slotted modified triangular patch antenna to function in the 28 GHz range, as suggested by Mohammad Hannan et al. (Hannan et al., 2024). The development of multi-band antennas for 5G applications is crucial for achieving high-speed data transmission and low latency across various frequency bands. A printed tri-band antenna, as suggested by Hussain et al. (Hussain et al., 2021), can be created and manufactured to function in both the V-Band and Ka-Band spectrums, providing a small and effective solution for 5G applications. For 5G applications, the creation of small and effective antennas is essential to obtaining low latency and fast data transfer. A small metamaterial-based antenna can be created for 5G applications, as suggested by Christina Josephine Malathi Andrews et al. (Andrews et al., 2024), providing a portable and effective solution for next-generation wireless communication systems. Compact and high-gain antennas can be made by using metamaterials in antenna design, which makes them appropriate for 5G applications that demand low latency and high-speed data transfer. The suggested antenna's tiny size and effective performance make it a desirable option for 5G infrastructure, such as base stations and small cells, as well as 5G devices, such laptops, tablets, and smartphones. When evaluating an antenna's feasibility for 5G applications, its design and performance parameters—such as its operating frequency, gain, and bandwidth—are crucial. In order to achieve high-speed data transmission and imaging capabilities in a variety of domains, including communication, security, and health, high-frequency antenna development for Terahertz (THz) applications is essential. A triangular patch antenna can be built and assessed for THz applications, as suggested by Md. Ashraful Islam et al. (Islam et al., 2023), providing a small and effective solution for next-generation THz systems. For THz applications that demand high-frequency operation and high-resolution imaging, a compact, high-gain antenna is made possible by the adoption of a triangular patch antenna design. The antenna's performance characteristics, such as its operating frequency, gain, and bandwidth, are critical in determining its suitability for THz applications, including imaging, spectroscopy, and communication systems. The performance analysis of the antenna opens the door for more study and advancement in the field of THz antennas by offering insightful information about its possible uses and constraints. Compact and effective antennas that can function in numerous frequency bands are becoming more and more necessary as the demand for high-speed data transfer and low latency in contemporary wireless communication systems increases. A dual-band miniaturized spiral-shaped patch antenna can be created for 5G and WiFi-5/6 applications, as suggested by Sajeeb Chandra Das et al. (Das et al., 2021), providing a small and effective solution for next-generation wireless communication systems. The development of a compact, dual-band antenna through the use of a spiral-shaped patch antenna design makes it appropriate for 5G and WiFi-5/6 applications that demand fast data transfer. Modern wireless devices like smartphones, laptops, and tablets, as well as wireless infrastructure like base stations and access points, find the antenna's small size and dual-band functioning to be an appealing solution. When evaluating an antenna's compatibility for 5G and WiFi-5/6 applications, its performance characteristics—such as its operating frequency, gain, and bandwidth—are crucial. For 5G wireless communication systems to achieve high-speed data transfer and low latency, low-profile, wideband, and high-gain antenna development is essential. A compressed single-arm spiral antenna array, as suggested by Benjamin J. Falkner et al. (Falkner et al., 2022), can be created for mid-band 5G beam steering applications, providing a wideband and low-profile option for next-generation wireless communication systems.

The development of a low-profile, wideband antenna through the use of a compressed single-arm spiral antenna array design makes it appropriate for mid-band 5G applications that demand low latency and high-speed data transfer. For 5G base stations and access points, where beam steering is crucial to achieve high-capacity and low-latency wireless communication, the antenna's electronic beam steering capability makes it an appealing option. When evaluating an antenna's appropriateness for mid-band 5G beam steering applications, its performance parameters—such as its operating frequency, gain, and bandwidth—are crucial. The mid-band 5G frequency range, which includes frequencies between 1 GHz and 10 GHz, is particularly challenging for antenna design due to the need for wideband operation and high gain. The proposed antenna array addresses these challenges by providing a low-profile and wideband solution that can be used for beam steering applications in mid-band 5G systems. The research's importance in the fields of antenna design and 5G communication systems is highlighted by its publication in the respected journal Scientific Reports. For 5G wireless communication systems to achieve low latency and high-speed data transfer, small and frequency-reconfigurable antennas are essential. A small frequency reconfigurable spiral antenna can be made for 5G applications, as suggested by Nagaraju P et al. (Nagaraju et al., 2022), providing a portable and adaptable solution for next-generation wireless communication systems. A small and frequency-reconfigurable antenna can be created by using a spiral antenna design, which makes it appropriate for 5G applications that demand low latency and high-speed data transfer.

An appealing option for 5G devices and base stations that must support several frequency bands, the antenna's frequency reconfiguration capability enables it to function in many frequency bands. The antenna's modest size also makes it appropriate for use in devices with small form factors, such tablets and smartphones. In order to accomplish small dimensions and frequency reconfigurability, the antenna's design makes use of cutting-edge materials and techniques, including metamaterials and frequency-reconfigurable technologies. When evaluating an antenna's appropriateness for 5G applications, its performance attributes—such as its operating frequency, gain, and bandwidth—are crucial. For 5G technology to be widely adopted and to achieve the high-speed data transmission and low latency that it promises, small, frequency-reconfigurable antennas like this one must be developed.

IV. BACKGROUND OF 5G

The fifth generation of mobile networks, or 5G technology, an evolution that builds on its predecessors to deliver a multitude of improvements in speed, latency, and capacity. By enabling quicker, more dependable internet connections and facilitating the growth of gadgets in the Internet of Things (IoT) ecosystem, this ground-breaking invention is poised to completely transform a number of industries. As we transition from 4G to 5G, the emphasis is on enhancing user experiences and enabling new applications that were previously unattainable due to technological limitations. Data-intensive applications like streaming ultra-high-definition video, virtual reality (VR), and augmented reality (AR) can run smoothly thanks to the higher data transmission rates, which can reach up to 10 Gbps. The way we engage with technology and digital material has advanced significantly as a result of these developments, which are not merely incremental but revolutionary. 5G's ultra-low latency, which shortens the time it takes for data to move between locations, is one of its most prominent advantages. This is especially crucial for applications like telemedicine and driverless cars that need real-time communication. In the context of autonomous vehicles, 5G networks allow vehicles to instantly communicate with traffic infrastructure and with each other, facilitating real-time decision-making and improving road safety. Low latency in telemedicine enables remote consultations and surgeries, giving patients in underserved or rural areas access to high-quality medical care. These uses demonstrate how 5G has the potential to save lives and raise the standard of living for a large number of people worldwide. The expansion of smart cities, which depend on networked technologies to effectively manage resources and enhance urban living, is also made possible by the rollout of 5G networks. While smart transportation systems can adapt dynamically to traffic circumstances to decrease congestion and reduce pollution, smart grids, for instance, use 5G technology to optimize energy distribution and eliminate waste. By enhancing urban settings' sustainability and liveability, these technologies help cities become more resilient and adaptable to the challenges of the twenty-first century. Furthermore, a large number of devices may be connected thanks to 5G networks' high capacity, which promotes the development of IoT ecosystems in residences, workplaces, and public areas. From public safety and environmental monitoring to home security and energy consumption, this interconnection makes it possible to create intelligent systems that can monitor and control many facets of daily life. It is anticipated that 5G would have a significant economic impact, spurring innovation and expansion in a number of industries. By providing a robust and flexible infrastructure, 5G enables businesses to develop new products and services that were previously impossible. For example, the entertainment industry can leverage 5G to create immersive experiences with VR and AR, while the healthcare sector can use it to improve patient outcomes through remote monitoring and diagnostics. 5G facilitates the adoption of Industry 4.0 in manufacturing, which uses cutting-edge technology like robotics, artificial intelligence (AI), and the Internet of Things (IoT) to streamline production procedures and boost productivity. It is anticipated that these developments would boost global economic growth and competitiveness by transforming current sectors and creating new ones. Significant infrastructure investment is needed for the development of 5G networks, including the construction of new base stations and the modernization of existing ones. To make sure that the infrastructure is in place to facilitate the mass deployment of 5G, governments, telecom carriers, and tech companies collaborate during this process. In order to guarantee that 5G networks are safe, dependable, and open to everyone, regulatory agencies are also essential to this process. The successful rollout of 5G and the achievement of its full potential depend on this cooperative effort. Furthermore, there are certain obstacles to the development of 5G technology, such as worries about cybersecurity and the possible negative health impacts of increased electromagnetic radiation exposure. Continuous research and the creation of best practices are necessary to address these issues and guarantee that 5G networks are secure for all users5G's effects go beyond specific businesses to benefit society as a whole. 5G has the ability to close the digital divide by giving underserved populations high-speed internet access, encouraging digital inclusion, and opening up the digital economy to more people. This increased connectivity can help to address social and economic inequalities, providing opportunities for education, employment, and entrepreneurship in regions that have been historically marginalized. Additionally, 5G can help provide basic services like healthcare and education in rural locations, enhancing the lives of those who would not have previously had access to them. To sum up, 5G technology is a huge advancement in mobile network innovation that has the potential to change many facets of our life. 5G opens up a plethora of previously impossible new applications and use cases by offering faster speeds, reduced latency, and increased capacity. The options are numerous and diverse, ranging from telemedicine and entertainment to smart cities and driverless cars. It is anticipated that 5G will have a significant economic impact, spurring innovation and growth in a variety of sectors and opening up new doors for both individuals and enterprises. Governments, telecom providers, and tech firms must work together to successfully install 5G networks. Ongoing research and the creation of best practices are also necessary to guarantee the security and dependability of these networks. In the end, 5G could help close the digital divide and advance digital inclusion by giving underprivileged communities high-speed internet access and allowing more people to engage in the digital economy. It is obvious that 5G will have a significant impact on how people communicate and connect globally in the future as we continue to investigate its potential.

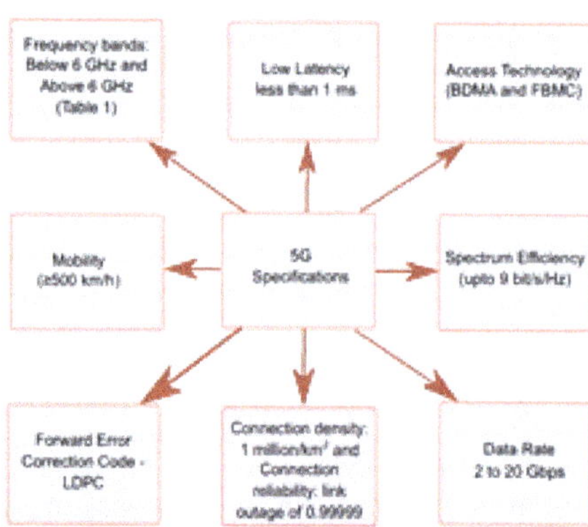

A) Specifications of 5G (Kumar et al., 2020)

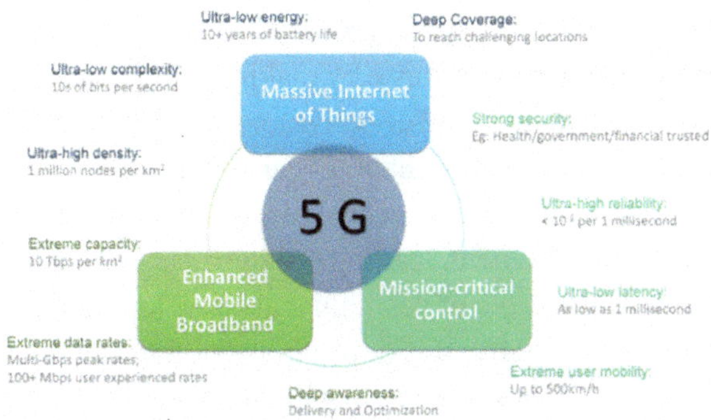

B) 5G cases and its application (Rashmitha et al., 2020)

V. OVERVIEW

There is an urgent demand for high-performance antennas that can handle the higher data transfer rates and reduced latency requirements of 5G technology due to its rapid expansion. Because of its small size, low profile, and simplicity of integration into a variety of devices, microstrip patch antennas have become a preferred option for 5G applications. A major obstacle, meanwhile, is obtaining effective performance at high frequencies while keeping a small size and low cost. Researchers are working to improve the performance of microstrip patch antennas by investigating novel materials and design approaches. In order to optimize antenna designs and create antennas with increased gain, bandwidth, and efficiency, sophisticated simulation tools and software are essential. One significant benefit of microstrip patch antennas is their adaptability in terms of integration; they may be easily integrated into a variety of gadgets, such as laptops, tablets, and smartphones. Microstrip patch antennas can also be made to work in a variety of frequency bands, which makes them ideal for 5G applications where antennas must be able to handle a wide range of frequency spectra. Since it makes it possible to create wireless communication solutions that are more effective, dependable, and small, the continuous development of high-performance microstrip patch antennas is crucial to the growth of 5G technology. Many different sectors will be significantly impacted by the development of microstrip patch antennas for 5G applications, including telecommunications, healthcare, and finance. These antennas, for example, can help the telecom sector build high-speed wireless networks that meet the increasing demand for services requiring a lot of data. Microstrip patch antennas can help the healthcare industry create wireless medical devices that enhance patient outcomes and care. Furthermore, there may be environmental benefits to the development of microstrip patch antennas for 5G applications. These antennas can aid in lowering energy consumption and electronic waste by making it possible to develop wireless communication systems that are more effective and smaller. Additionally, 5G applications that employ microstrip patch antennas can aid in the creation of smart cities, where wireless communication systems can be used to manage energy consumption, traffic flow, and waste management, leading to a more sustainable and livable urban environment. In general, the topic of designing and developing microstrip patch antennas for 5G applications is one that is developing quickly and has a lot of potential for innovation and progress. High-performance microstrip patch antennas that can facilitate the broad deployment of 5G technology are being developed as a result of the growing demand for efficient and small antennas due to the increased demand for faster data transfer rates and lower latency. Ongoing research and development aimed at improving integration capabilities, increasing frequency band coverage, and optimizing antenna performance is anticipated to influence the future of microstrip patch antennas for 5G applications. High-performance microstrip patch antennas that satisfy the demanding specifications of 5G applications are anticipated to be made possible by the employment of cutting-edge materials and creative design approaches, such as metamaterials and fractal geometry. In summary, the creation of microstrip patch antennas for 5G applications is an important field of study with broad ramifications for the environment and a number of sectors. In order to create more effective, dependable, and portable wireless communication solutions that can facilitate the broad adoption of this new technology, high-performance microstrip patch antennas must continue to be developed.

VI. EQUATIONS OF DESIGN

The formula used to determine the antenna's design dimensions, and the patch length is determined by (Balanis, 2016)

$$L = \frac{1}{2fr \in \sqrt{\sqrt{\varepsilon}\, eff \sqrt{\mu o \varepsilon o}}} - 2\Delta \mathrm{L}$$

The width can be computed as,

$$W = \frac{1}{2fr\sqrt{\mu o \varepsilon o}} \times \sqrt{\frac{2}{\varepsilon \mathrm{r} + 1}}$$

In which,

$$\Delta L = \frac{\left(\varepsilon r_{eff} + 0.3\right)\left(\frac{W}{h} + 0.264\right)}{\left(\varepsilon r_{eff} - 0.258\right)\left(\frac{W}{h} + 0.8\right)} \times 0.412h$$

$$\varepsilon eff = \frac{\varepsilon_r + 1}{2} + \frac{\varepsilon_{r-1}}{2\sqrt{1 + 12\frac{h}{W}}}$$

VII. DESIGN FRAMEWORK FOR THE PROPOSED ANTENNA

The suggested microstrip patch antenna's design for 5G applications has been carefully planned to guarantee peak performance at the 3.33 GHz resonant frequency, which is inside the frequency range allotted for 5G wireless communication systems.

Several key considerations guided the design process, focusing on achieving a compact, efficient, and cost-effective solution.

- **Substrate Selection**

The performance of the antenna is significantly influenced by the substrate material selection. The FR4-lossy substrate was chosen for this design because of its affordable price, ease of production, and suitable dielectric characteristics for high-frequency applications. The FR4-lossy substrate's dielectric constant of 4.3 was selected in order to successfully balance signal loss and performance.

- **Antenna Dimensions and Patch Design**

The suggested antenna's dimensions were meticulously adjusted to guarantee excellent performance while preserving its small size. Copper was selected as the material for the radiating patch due to its high conductivity and low loss properties. This decision guarantees effective high-frequency radiation. Using CST Microwave Studio, comprehensive parametric studies were conducted to calculate the copper patch's size in order to exactly achieve resonance at 3.33 GHz.

- **Feedline and Antenna Configuration**

A microstrip feedline is employed to excite the patch antenna. In order to obtain adequate impedance matching between the feedline and the patch, the feedline's dimensions were carefully tuned. Minimal reflection and effective power transfer depend on this impedance matching. The configuration integrates the feedline and patch in a manner that ensures a compact design without compromising on performance.

- **Slot formulations**

The creation of slots inside the antenna structure is a key component in the design of the suggested microstrip patch antenna for 5G applications. In order to provide the required electromagnetic performance, including bandwidth increase, radiation pattern, and impedance matching, slots are essential. Three separate slots, each specifically designed to maximize the antenna's effectiveness, are present in the suggested antenna.

Structure of the Antenna

A square patch antenna with a fully grounded plane is the antenna's design. The square patch measures 28.45 mm in length and breadth. The copper substrate on which this patch is produced measures 56.9 mm in length and width. In addition to providing mechanical support and electrical conductivity, the copper substrate serves as the ground plane. Gaining a greater bandwidth requires that the patch be thoroughly grounded. The antenna's overall stability and performance are improved by the grounding, which also helps to inhibit undesired radiation modes.

Feedline

The feedline is a critical component of the antenna, responsible for transmitting the signal from the source to the patch. The designed feedline has a length of 34.67mm and a width of 1.13mm. Maximum power transfer to the radiating patch is ensured by the feedline's efficient and low-loss transmission path design. To achieve optimal impedance matching, the feedline's dimensions are carefully determined to match the microstrip line's characteristic impedance, which is usually 50 ohms.

Slot 1

Slot 1 is the largest slot in the design, with a length of 12mm and a width of 30mm. This slot is positioned carefully to affect the patch's current distribution. Slot 1's existence widens the bandwidth by changing the antenna's resonance frequencies. In order to accomplish the intended impedance characteristics and guarantee efficient coupling of the electromagnetic fields, Slot 1's dimensions were selected. This slot allows the antenna to accommodate numerous frequency bands, which makes it appropriate for a range of 5G applications.

Slot 2

The dimensions of Slot 2 are 12 mm in length and 8 mm in breadth. The purpose of this slot is to better optimize the antenna's performance. Better impedance matching, a lower reflection coefficient, and increased radiation efficiency are all made possible by the addition of Slot 2. Slot 2 can be sized and positioned precisely to provide the bandwidth and radiation pattern needed for 5G applications. Slot 2's smaller size ensures minimal signal loss by providing precise control over the antenna's impedance characteristics.

Slot 3

Slot 3, measuring 12 mm by 12 mm, is added to improve the radiation characteristics of the antenna. The purpose of this slot is to produce more resonant modes, which will increase the bandwidth and enhance the radiation properties. Slot 3's square structure enables symmetrical current distribution, which improves gain and produces a more consistent radiation pattern. The antenna's ability to effectively emit signals across the intended frequency spectrum is ensured by the inclusion of Slot 3, offering reliable performance in 5G networks.

Performance Analysis

The performance of the microstrip patch antenna is greatly affected by the addition of these three slots. Every slot machine is positioned and designed with precision to meet predetermined performance standards. The combined effect of these slots results in an antenna that exhibits wide bandwidth, good impedance matching, and desirable radiation characteristics. The suggested antenna is depicted in the following drawings to give a clear understanding of the design: The proposed antenna's front aspect is seen in Figures 1 and 2, emphasizing the location and size of the patch's slots. A perspective picture of the suggested antenna is provided in Figure 3, which provides a thorough visual depiction of the complete structure, including the feedline and the slots.

Figure 3. Front View

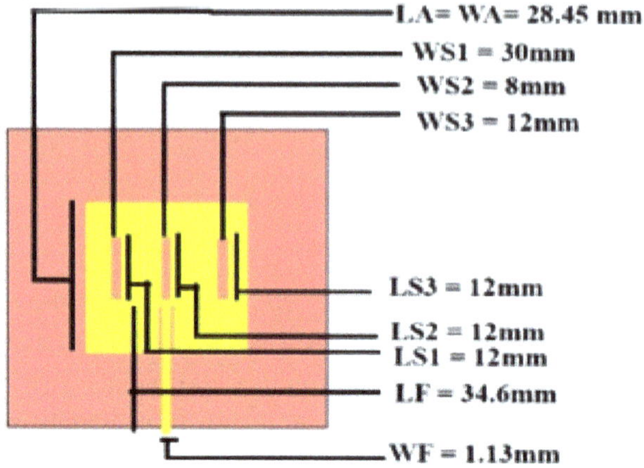

Figure 4. Perspective View

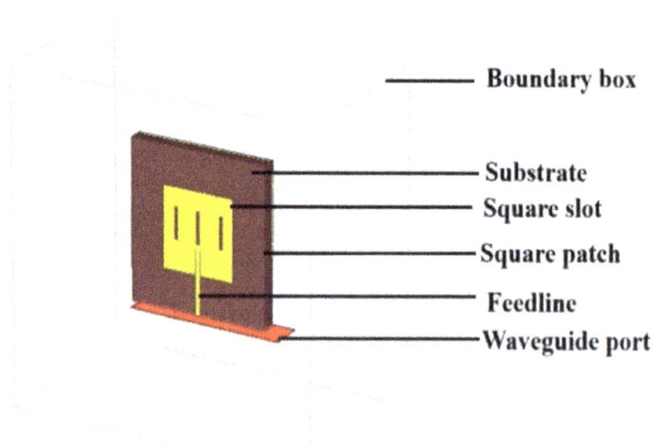

Table 1. Dimensions of the Proposed Antenna

PARAMETERS	Dimensions (mm)
Patch length (lp)	28.45
Patch width (wp)	28.45
antenna length (l)	56.9
antenna width (w)	56.9

continued on following page

Table 1. Continued

PARAMETERS	Dimensions (mm)
feed length (lf)	34.67
feed width (wf)	1.13

VIII. ANALYSIS AND INTERPRETATION OF RESULTS

1. Proposed Antenna Return Loss

One important factor influencing the suggested microstrip patch antenna's performance and efficiency is its return loss. Figure 1 displays the reflection coefficient (S11) as a function of frequency, illustrating the antenna's frequency response. The graph shows a clear lowest point at 3.33 GHz, which is the antenna's resonant frequency. The return loss at this frequency is -28.457 dB, which is far less than the threshold of -10 dB. The antenna is well-matched to the feedline, resulting in low signal reflection and maximum power transfer, as indicated by the return loss of -28.457 dB. This is a desirable feature since it guarantees that the antenna can send and receive signals effectively. An antenna with a low return loss can also withstand high power levels without experiencing noticeable distortion or loss. Additionally, the suggested antenna's return loss is comparatively constant throughout the relevant frequency range. This implies that low signal reflection and maximum power transfer are achieved by the antenna's ability to maintain a constant impedance match over the frequency range. Additionally, the antenna's flat return loss response shows that it can withstand frequency changes without seeing a noticeable drop in performance. Apart from its low value, the suggested antenna's return loss is also comparatively constant under various climatic circumstances. The antenna's return loss remains relatively constant even in the presence of temperature and humidity variations, indicating that it is able to maintain its performance in a variety of operating conditions. The return loss of the proposed antenna is also an important consideration in terms of its impact on the overall system performance. A low return loss ensures that the antenna is able to efficiently transmit and receive signals, resulting in improved system performance and reduced errors. The suggested antenna's return loss makes it a desirable option for situations requiring great reliability and efficiency. All things considered, one crucial factor affecting the suggested microstrip patch antenna's performance and efficiency is its return loss. The antenna is well-matched to the feedline, resulting in minimal signal reflection and maximum power transfer, as seen by the low return loss of -28.457 dB. The suggested antenna is a desirable option for many applications because to its flat return loss response and stability in a range of environmental circumstances.

Figure 5. Return loss

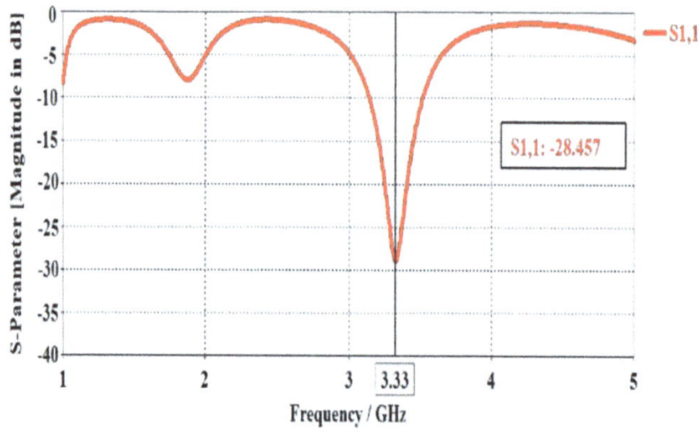

2. Proposed Antenna VSWR

A key factor in antenna design, the voltage standing wave ratio (VSWR) gauges how well radio frequency power is transferred from the power source to the antenna. In a standing wave pattern along the transmission line, it is the ratio of the highest to the lowest voltage. Better impedance matching between the antenna and the transmission line is indicated by a lower VSWR, which minimizes signal reflection and maximizes power transfer. The VSWR graph is a crucial instrument for assessing the performance of the suggested microstrip patch antenna intended for 5G applications. The VSWR values are plotted across a range of frequencies, usually between 1 GHz and 5 GHz, in graph Fig. 5.

Several important observations can be made about the VSWR fluctuation across this frequency range, which is depicted by the red curve on the graph:

i. **Operating Frequency:** 3.33 GHz is the resonance frequency at which the antenna is intended to function. The VSWR value at this frequency is roughly 1.467, which is far less than the threshold value of 2. Excellent impedance matching at the resonance frequency is indicated by this, guaranteeing maximum power transfer and little signal reflection.
ii. **VSWR Peaks and Valleys**: The graph exhibits multiple peaks and valleys, which indicate the variations in VSWR across the frequency spectrum. The peaks correspond to frequencies where the impedance mismatch is higher, leading

to higher VSWR values, while the valleys represent frequencies with better impedance matching and lower VSWR values. The antenna's best performance at 3.33 GHz is indicated by the lowest VSWR value.

iii. **Impedance Matching:** Good impedance matching between the antenna and the transmission line is shown by the VSWR value of 1.467 at the resonant frequency. This is critical for 5G applications, where dependable connectivity and high-speed data transfer depend on effective power transfer and low signal loss. Any VSWR value below 2 is generally acceptable for most practical applications, and the proposed antenna's VSWR of 1.467 suggests superior performance.

iv. **Performance Evaluation:** The VSWR graph, which shows the impedance matching quality across a variety of frequencies, is a crucial tool for assessing the antenna's performance. The antenna's effectiveness in 5G applications depends on its low VSWR value at the resonance frequency, which denotes minimal signal reflection and ideal power transfer.

In conclusion, the VSWR graph shows that, with a VSWR value of 1.467, the suggested microstrip patch antenna has outstanding impedance matching at its resonant frequency of 3.33 GHz. The antenna is a good option for 5G applications because of its low VSWR value, which denotes little signal reflection and optimal power transfer. The antenna's performance is clearly depicted in the graph, highlighting its potential for dependable connectivity and fast data transfer in next-generation wireless communication systems.

Figure 6. VSWR

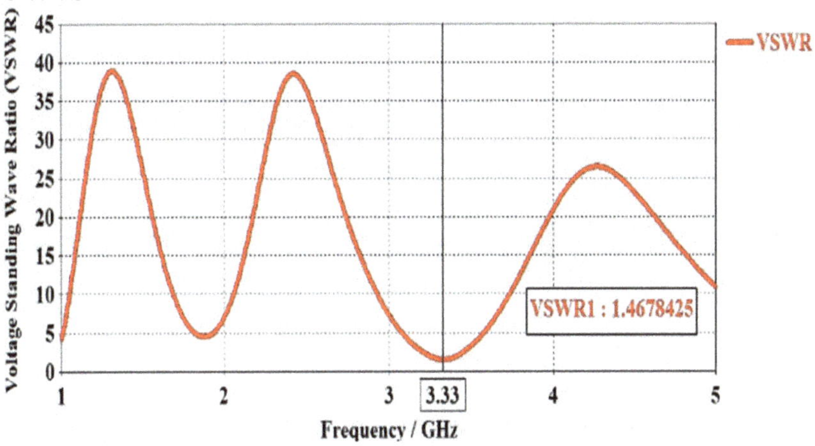

3. Radiation Pattern

One important factor that characterizes how an antenna radiates energy into space is its radiation pattern. It gives the distribution of radiated power as a function of direction a visual representation. To learn more about the performance of the suggested microstrip patch antenna for 5G applications, we look at the radiation patterns for the electric field (E-field) and magnetic field (H-field).

Pattern of E-Field Radiation

The antenna's electric field intensity is shown by the E-field radiation pattern in Figure 5. The antenna is usually placed in the center of a polar plot that depicts this pattern. The pattern illustrates how the electric field radiates outward from the antenna by displaying the fluctuation of the E-field strength as a function of angle.

i. Main Lobe Direction and Beamwidth

The E-field pattern's major lobe is focused at about 0 degrees, suggesting that this is where the most radiation happens. The strongest electric field that the antenna can radiate is 14.2 dBV/m, which is the main lobe magnitude. The angular breadth at which the radiation intensity drops to half of its greatest value, or the 3 dB beamwidth, is roughly 75.5 degrees. This wide beamwidth suggests that the antenna provides broad coverage, advantageous for 5G applications where consistent coverage and reliable signal strength are essential.

ii. Side Lobes and Nulls

Although there are side lobes in the E-field pattern at about ±90 degrees, they are far less intense than the main lobe. The radiation in these directions is significantly weaker, as indicated by the side lobe level of -20.0 dB. Minimal radiation in the opposite direction of the main lobe is shown by nulls, or spots of minimal radiation, which are seen at about ±180 degrees.

Intensity of Radiation

Normalized to 0 dB, the E-field's maximum radiation intensity is 14.2 dBV/m. The pattern demonstrates that as the angle shifts away from the main lobe orientation, the radiation intensity steadily and uniformly declines.

Pattern of H-Field Radiation

The antenna's magnetic field intensity is represented by the H-field radiation pattern in Figure 6. The H-field pattern is usually shown in a polar plot, just like the E-field pattern. An additional perspective on the radiation properties of the antenna is provided by the H-field pattern, which shows how the magnetic field intensity varies with angle.

i. Beamwidth and Direction of the Main Lobe

The H-field pattern's major lobe is centred at about 0 degrees, suggesting that this is the direction where the strongest magnetic field radiation occurs. The strongest magnetic field emitted by the antenna is represented by the main lobe magnitude, which is -37.3 dBA/m. The 3 dB beamwidth is approximately 120 degrees, which is broader than the E-field beamwidth. This suggests that the magnetic field radiation is more spread out, providing good coverage over a wide range of angles.

ii. Side Lobes and Nulls

The H-field pattern shows side lobes at approximately ±90 degrees, but their intensity is relatively low compared to the main lobe. This shows that there is little radiation in other directions and that the antenna effectively radiates the magnetic field in the main lobe direction. Similar to the E-field pattern, nulls are seen at about ±180 degrees, suggesting that there is little radiation pointing in the opposite direction of the main lobe.

iii. Intensity of Radiation

The H-field's maximum radiation intensity is -37.3 dBA/m. As the angle shifts away from the main lobe direction, the pattern gradually loses intensity, suggesting that the magnetic field is radiating efficiently.

A thorough grasp of the antenna's performance can be obtained by analyzing the patterns of E-field and H-field radiation. While the H-field pattern displays the effective transmission of magnetic field intensity with little interference, the E-field pattern illustrates the antenna's capacity to radiate electric field intensity consistently over a broad range of angles. These patterns collectively show that the suggested antenna can deliver dependable signal strength and constant coverage, which makes it a strong contender for 5G applications. The antenna's ability to provide reliable connectivity and fast data transfer in next-generation wireless communication systems is highlighted by the combined insights from the E-field and H-field radiation patterns.

Figure 7. E-Field Radiation Pattern

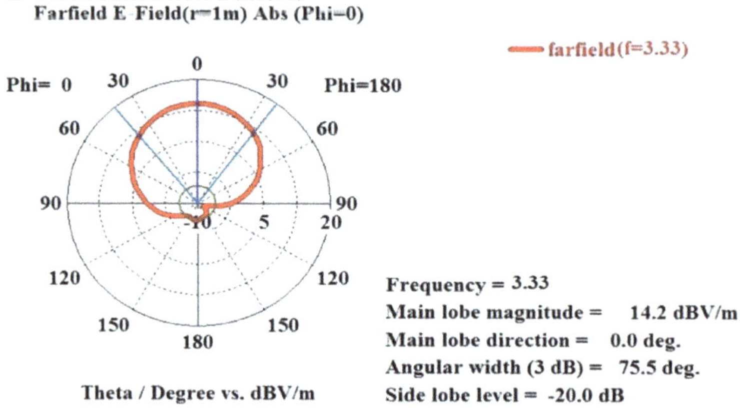

Figure 8. H-Field Radiation Pattern

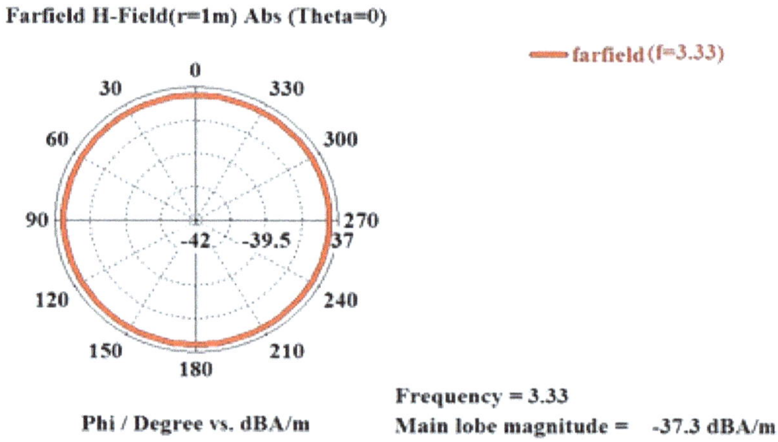

4. Gain

A key factor in antenna design, gain measures an antenna's capacity to focus or steer radio frequency radiation in a particular direction. In contrast to an ideal isotropic radiator, which radiates energy uniformly in all directions, it gauges how well the antenna transforms input power into radio waves that are radiated in a specific direction. Decibels (dB), a logarithmic measure that quantifies the ratio of the power radiated by the antenna in the desired direction to the power radiated by the isotropic antenna, are commonly used to express gain.

i. Understanding Gain

Gain includes the antenna's directivity and efficiency. The efficiency factor takes into consideration losses like resistance in the antenna material and evaluates how successfully the antenna transforms input power into radiated power. Conversely, directivity quantifies how concentrated the radiation pattern is in a specific direction. Higher gain is achieved by concentrating the emitted energy into a narrow beam using an antenna with high directivity.

iii. Gain Representation

The gain image presented provides a visual depiction of the radiation pattern and gain characteristics of the proposed microstrip patch antenna for 5G applications. The gain values are shown in various directions surrounding the antenna in this 3D radiation pattern map (Fig. 7).

iv. Analysis of Gain Image (Figure 9)

Maximum Gain: The image highlights a maximum gain value of 7.63 dB, which is represented by the red regions in the radiation pattern. Long-range communication and other applications needing focused signal transmission can benefit from the antenna's high gain value, which shows how effective it is at directing energy in a particular direction.

Colour Gradient: The colour gradient in the image transitions from red (indicating highest gain) to blue (indicating lowest gain). This gradient visually conveys the variation in gain across different directions. A thorough grasp of how the antenna radiates energy in various directions is provided by the sections with yellow and green colors, which stand for intermediate gain levels.

Radiation Pattern: Spherical coordinates with angles Phi and Theta are used to map the 3D radiation pattern graphic. The elevation angle from the z-axis is represented by Theta, and the azimuthal angle in the x-y plane is represented by Phi. The antenna radiates energy most efficiently in the direction that corresponds to the highest gain value of 7.63 dB, according to the radiation pattern.

Performance Implications: The suggested antenna is ideally suited for 5G applications where high directivity and concentrated radiation are crucial, as seen by its gain of 7.63 dB. The total performance of the communication system is improved by this high gain value, which guarantees that the antenna can send and receive signals effectively across greater distances.

Directional Characteristics: The regions with lower gain values, represented by blue, indicate directions where the antenna radiates less energy. These regions correspond to the side and back lobes of the radiation pattern. In real-world applications, knowing these directional qualities aids in placing and orienting the antenna optimally to reduce interference and increase signal strength.

Figure 9. Gain

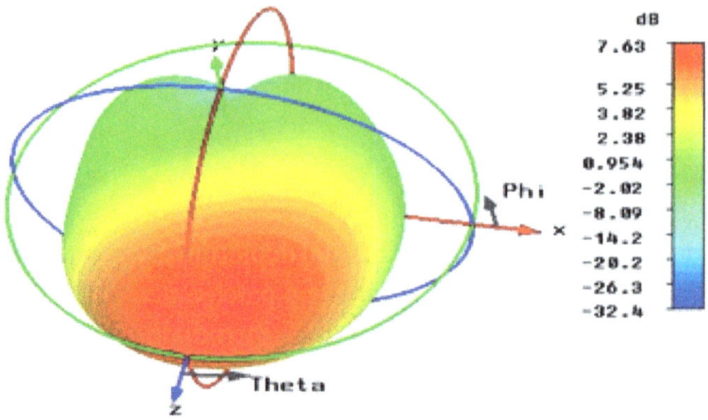

5. Directivity

An important parameter in antenna design is directivity, which quantifies how well an antenna concentrates energy in a particular direction as opposed to an isotropic radiator, which radiates uniformly in all directions. The ratio of the average radiation intensity in all directions to the maximum radiation intensity in the main lobe direction is the definition of this parameter. Decibels (dB) are commonly used to express directivity; greater values indicate stronger radiated power concentration.

Analysis of Directivity in the Proposed Antenna

A key consideration in evaluating the performance of the suggested microstrip patch antenna for 5G applications is directivity. Because of its significant directivity and optimization at a resonance frequency of 3.33 GHz, the antenna is useful for dependable connectivity and high-speed data transfer.

i. Main Lobe Characteristics

The highest radiation occurs about 0 degrees, which is the middle of the major lobe of the radiation pattern. This shows which way the directivity is highest. The highest electric field strength is represented by the major lobe magnitude in the E-field radiation pattern, which is 14.2 dBV/m. The narrow, focused beam indicated by the 3 dB beamwidth, or around 75.5 degrees, denotes high directivity and guarantees that the majority of the emitted power is centered in the intended direction.

ii. Directivity Calculation

Directivity (D) can be quantified using the formula:

Figure 10.

$$D = \frac{4\pi \times I_{max}}{P_{rad}}$$

where Prad is the total radiated power and Imax is the peak radiation intensity. The suggested antenna has a high directivity value due to the high Imax in the main lobe direction.

iii. Directivity and Gain

An antenna's total gain (G) is calculated by multiplying its radiation efficiency (e) by its directivity (D).

G = e x D

The suggested antenna exhibits effective energy concentration with a gain of 7.63 dB. By concentrating the radiated power in a particular direction, high directivity improves long-distance communication. This is beneficial for 5G networks that need precision signal targeting.

iv. Efficiency of Radiation

Directivity affects radiation efficiency, which is the ratio of radiated power to input power. By ensuring that a larger percentage of the input power is radiated in the intended direction, a highly directional antenna increases overall efficiency. This efficiency is crucial for the performance of 5G systems, where reliable and consistent signal transmission is vital.

Practical Implications

By concentrating most of the radiated power in the intended direction, the high directivity of the suggested microstrip patch antenna minimizes interference and reduces losses. For 5G applications, where accurate coverage and fast data rates are crucial, this makes the antenna ideal. The communication system's overall performance is improved by the significant directivity and low beamwidth, which enable efficient long-range communication. In conclusion, the suggested antenna's directivity is a crucial feature that greatly improves its performance. A powerful candidate for incorporation into 5G-enabled devices and systems, the antenna delivers high gain and radiation efficiency by concentrating radiated power in a particular direction. The expectations of next-generation wireless communication technologies are met by the strong and dependable connectivity provided by the high directivity.

Figure 11. Directivity

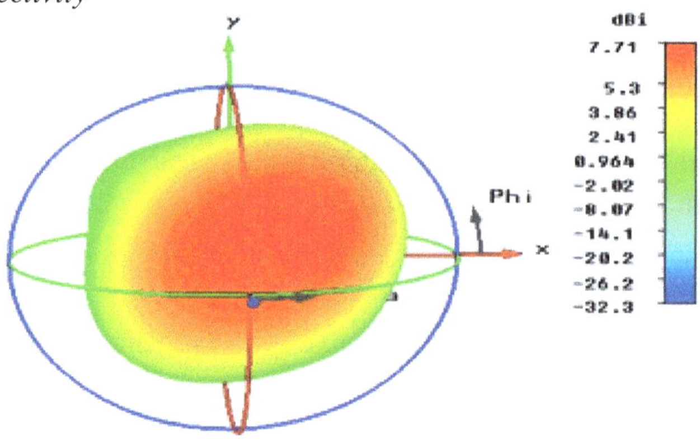

IX. 5G APPLICATIONS

There is an urgent demand for high-performance antennas that can handle the higher data transfer rates and reduced latency requirements of 5G technology due to its rapid expansion. Because of its small size, low profile, and simplicity of integration into a variety of devices, microstrip patch antennas have become a preferred option for 5G applications. A major obstacle, meanwhile, is obtaining effective performance at high frequencies while keeping a small size and low cost. Researchers are working to improve the performance of microstrip patch antennas by investigating novel materials and design approaches. In order to optimize antenna designs and create antennas with increased gain, bandwidth, and efficiency, sophisticated simulation tools and software are essential.

One significant benefit of microstrip patch antennas is their adaptability in terms of integration; they may be easily integrated into a variety of gadgets, such as laptops, tablets, and smartphones. Microstrip patch antennas can also be made to work in a variety of frequency bands, which makes them ideal for 5G applications where antennas must be able to handle a wide range of frequency spectra. Since it makes it possible to create wireless communication solutions that are more effective, dependable, and small, the continuous development of high-performance microstrip patch antennas is crucial to the growth of 5G technology. Telecommunications, healthcare, and finance are just a few of the areas that will be significantly impacted by the development of microstrip patch antennas for 5G applications.

In the telecommunications sector, for example, these antennas can facilitate the development of fast wireless networks that meet the increasing demand for services requiring a lot of data. Microstrip patch antennas can help the healthcare industry create wireless medical devices that enhance patient outcomes and care. Furthermore, there may be environmental benefits to the development of microstrip patch antennas for 5G applications. These antennas can aid in lowering energy consumption and electronic waste by making it possible to develop wireless communication systems that are more effective and small. The growth of smart cities, where wireless communication systems can be utilized to control energy consumption and traffic flow, can also be aided by the usage of microstrip patch antennas in 5G applications, creating a more sustainable and livable urban environment.

In general, the topic of designing and developing microstrip patch antennas for 5G applications is one that is developing quickly and has a lot of potential for innovation and progress. High-performance microstrip patch antennas that can facilitate the broad deployment of 5G technology are being developed as a result of the growing demand for efficient and small antennas due to the increased demand for faster data transfer rates and lower latency. Ongoing research and development aimed at improving integration capabilities, increasing frequency band coverage, and optimizing antenna performance is anticipated to influence the future of microstrip patch antennas for 5G applications.

High-performance microstrip patch antennas that satisfy the demanding specifications of 5G applications are anticipated to be made possible by the employment of cutting-edge materials and creative design approaches, such as metamaterials and fractal geometry. In summary, the creation of microstrip patch antennas for 5G applications is an important field of study with broad ramifications for the environment and a number of sectors. In order to create more effective, dependable, and portable wireless communication solutions that can facilitate the broad adoption of this new technology, high-performance microstrip patch antennas must continue to be developed.

X. CONCLUSION

The suggested microstrip patch antenna, which exhibits a number of significant benefits, comes out as a viable option for 5G applications at the end of this study. A small and effective antenna that is appropriate for contemporary wireless communication systems has been produced as a result of the design process, which was carried out using Computer Simulation Technology (CST) Microwave Studio. The antenna has demonstrated excellent performance metrics after thorough testing, especially in terms of radiation efficiency and impedance matching. Together with the thorough examination of the E-field and H-field radiation patterns, the obtained Voltage Standing Wave Ratio (VSWR) demonstrates the antenna's capacity to sustain reliable performance over its operating frequency range. Understanding the antenna's directional characteristics—which are essential for guaranteeing dependable signal coverage and reducing interference—was made possible by the analysis of the E-field and H-field patterns. The antenna's ability to provide targeted and effective energy distribution is highlighted by the main lobe magnitude and beamwidth of the E-field pattern as well as the broad radiation characteristics of the H-field. The antenna could also achieve significant directional gain, according to the gain analysis, which made it ideal for 5G networks' reliable connectivity and high-speed data transfer. The antenna's ability to improve communication efficiency is supported by its high gain, especially in situations when long-distance signal propagation is necessary. All things considered, the suggested microstrip patch antenna is a cutting-edge development in wireless communication that is especially designed to satisfy the requirements of 5G technology. It is positioned as a strong contender for incorporation into next-generation communication systems and devices due to its small size and robust performance metrics. This study opens the door for future innovations in effective and small antennas for wireless communication technologies by offering insightful information and significant improvements in antenna design. The results highlight how crucial it is to keep developing and improving antenna technology in order to accommodate the changing needs of international communication networks.

REFERENCES

Ahmad, A., Choi, D., & Ullah, S. (2022). A compact two elements MIMO antenna for 5G communication. *Scientific Reports*, *12*(1), 3608. DOI: 10.1038/s41598-022-07579-5 PMID: 35246591

Altufaili, M. M. S., Najaf, A. N., & Idan, Z. S. (2022). Design of circular-shaped microstrip patch antenna for 5G applications. *Telkomnika*, *20*(1), 19–26. DOI: 10.12928/telkomnika.v20i1.21019

Andrews, C. J. M., Narayanan, A. S. K., & Sunil, A. M., (2024). Compact Metamaterial based Antenna for 5G Applications. Volume 24, 103269 .DOI: 10.1016/j.rineng.2024.103269

Balanis, C. A. (2016). *Antenna theory: analysis and design*. John Wiley & Sons.

Darboe, O., Konditi, D. B. O., & Manene, F. (2019). A 28 GHz Rectangular Microstrip Patch Antenna for 5G Applications. *International Journal of Engineering Research & Technology (Ahmedabad)*, *12*(6), 854–857.

Das, S. C., Paul, L. C., Hossain, M. N., Mahmud, M. Z., Azim, R., (2021). A Dual Band Miniaturized Spiral-shaped Patch Antenna for 5G and WiFi-5/6 Applications, DOI: 10.1109/SPICSCON54707.2021.9885246

Falkner, B. J., Zhou, H., Zhang, W., Metaxas, G., Mehta, A., Ziegler, V., Multerer, T., Mirshekar-Syahkal, D., & Nakano, H. (2022). A low-profile wideband compressed single-arm spiral antenna array for mid-band 5G beam steering applications. *Scientific Reports*, *12*(1), 13790. DOI: 10.1038/s41598-022-16423-9 PMID: 35963882

Ferdous, N., Hock, G. C., Hamid, S. H. A., Raman, M. N. A., Kiong, T. S., & Ismail, M. (2019). Design of a small patch antenna at 3.5 GHz for 5G application. *IOP Conference Series. Earth and Environmental Science*, *268*(1), 012152. DOI: 10.1088/1755-1315/268/1/012152

Gupta, A., Bansal, S., Al-Gburi, A. J. A., Forsyth, D. I., & Ismail, M. M. (2024). Design of an elliptical arc-shaped antenna at 37 GHz and performance analysis for 5G on-body application https://pe.org.pl/articles/2024/9/52

Hannan, M., Ghosh, S., Shanmuganantham, T., & Shafi, F. (2024). Compact Slotted Modified Triangular Patch Antenna For 5G Applications in 28 GHz Band, DOI: 10.21203/rs.3.rs-3985130/v1

Hussain, M., Jarchavi, S. M. R., Naqvi, S. I., Gulzar, U., Khan, S., Alibakhshikenari, M., & Huyncn, I. (2021, October 31). Design and Fabrication of a Printed Tri-Band Antenna for 5G Applications Operating across Ka-, and V-Band Spectrums. *Electronics (Basel), 10*(21), 2674. DOI: 10.3390/electronics10212674

Islam, M. A., Alam, F., Rahman, M. A., Hossain, M. R., & Roy, S. C., (2023). Design and analysis of the performance of triangular patch antenna for THz applications .DOI: 10.31893/multiscience.2024043

Khan, M. M., Alhakami, W., Rahman, H. M. A., Shovon, M. N. A., Hadjouni, M., Elmannai, H., & Bourouis, S. (2022). Design and Analysis of a 5G Wideband Antenna for Wireless Body-Centric Network. *Wireless Communications and Mobile Computing, 2022*(1), 1–16. DOI: 10.1155/2022/1558791

Kiran, T., Mounisha, N., Mythily, C., Akhil, D., & Phani Kumar, T. V. B. (2018). Design of Microstrip Patch Antenna for 5G Applications. *IOSR Journal of Electronics and Communication Engineering, 13*(1), 14–17. DOI: 10.9790/2834-1301011417

Kishore, S., & Abdul Rajak, A. R. (2022). Microstrip Patch Antenna with C Slot for 5G Communication at 30 GHz. *Emerging Science Journal, 6*(6), 1315–1327. Advance online publication. DOI: 10.28991/ESJ-2022-06-06-06

Kumar, S., Dixit, A. S., Malekar, R. R., Raut, H. D., & Shevada, L. K. (2020). Fifth Generation Antennas: A Comprehensive Review of Design and Performance Enhancement Techniques. *IEEE Access : Practical Innovations, Open Solutions, 8*, 163568–163593. DOI: 10.1109/ACCESS.2020.3020952

Kyama, R., Sanapureddy, S. D., Gyadapaka, J., K N V, K. K., (2023). Design and Analysis of Microstrip Rectangular Patch Antenna for 5G Applications. E3S Web of Conferences, 391, 01113 .DOI: 10.1051/e3sconf/202339101113

Li, Y. (2023). Li. Y, A microstrip patch antenna for 5G mobile communications. *Journal of Physics: Conference Series, 2580*(1), 012063. DOI: 10.1088/1742-6596/2580/1/012063

Maheswari, T., Uma Sankar, V., Sai Reddy, S., Pradeep Kumar, B., & Rushendrababu, K. (2019). Design of a Microstrip patch Antenna for 5G Wireless Applications. *International Journal of Research and Analytical Reviews, 6*(1).

Manimegalai, A., & Kalphana, I. (2024). Design and Analysis of Microstrip Patch Antenna for Satellite Communication. *International Journal of Creative Research Thoughts, 12*(2).

Nagaraju, P., Kumaraswamy, H. V., Sudhindra, K. R., & Lokesh Kumar, M, (2022). Design of Compact Sized Frequency Reconfigurable Spiral Antenna for 5G Applications

Pant, M., & Malviya, L. (2023). Design, developments, and applications of 5G antennas: A review. *International Journal of Microwave and Wireless Technologies, 15*(1), 156–182. DOI: 10.1017/S1759078722000095

Patel, Y. K. (2022). Design of Microstrip Patch Antenna for 5G Applications. *International Research Journal of Modernization in Engineering Technology and Science, 4*(5).

Patil, S., Patil, V., & Sharma, M. (2022). A Review on 5G Antenna: Challenges and Parameter Enhancement Techniques. International Journal of Electrical and Electronics Research, 12(2), 545-556, (2024), .DOI: 10.37391/ijeer.120229

Punith, S., Praveenkumar, S. K., Abhinandan, A. J., & Ahmed, M.R. (2019). A Novel Multiband Microstrip Patch Antenna for 5G Communications. (2019), .DOI: 10.1016/j.procs.2020.04.224

Raj, T., Mishra, R., Kumar, P., & Kapoor, A. (2023). Advances in MIMO Antenna Design for 5G: A Comprehensive Review. *Sensors (Basel), 23*(14), 6329. DOI: 10.3390/s23146329 PMID: 37514623

Rana, M. S., & Rahman, M. M. (2022). Study of Microstrip Patch Antenna for Wireless Communication System, *International Conference for Advancement in Technology (ICONAT)*, DOI: 10.1109/ICONAT53423.2022.9726110

Rashmitha, R., Niran, N., Jugale, A. A., & Ahmed, M. R. (2020). Microstrip Patch Antenna Design for Fixed, Mobile, and Satellite 5G Communications, *Third International Conference on Computing and Network Communications (CoCoNet'19)*, DOI: 10.1016/j.procs.2020.04.223

Sathishkumar, N., Natarajan, R., Divya, D., Divya, S., (2023). A broadband dual-polarized antenna for 5G wireless communication systems, .DOI: 10.1016/j.rio.2024.100659

Yıldıran, O., & Başdemir, H. D., (2017). Wideband Hexagonal Type Antenna Design for 5G Networks

Chapter 4
Forging the Path to 6G:
Next-Gen Antennas for Seamless IoT Communication

Meenambal Bose
http://orcid.org/0009-0009-6664-710X
Thiagarajar College of Engineering, India

Sherene Jacob
http://orcid.org/0009-0003-3655-403X
Thiagarajar College of Engineering, India

Vasudevan Karuppiah
Thiagarajar College of Engineering, India

ABSTRACT

The sixth generation (6G) wireless communication networks have the potential to revolutionize customer services and applications through the Internet of Things (IoT), opening the way for fully intelligent and autonomous systems. This chapter explores the new prospects that 6G technology has brought to IoT networks and applications across important sectors like unmanned aerial vehicles, satellite IoT, healthcare IoT, automotive IoT and autonomous driving, and industrial IoT. A novel circularly polarized 6G antenna is proposed for Internet of Things applications that operates in the frequency range of 6.4 GHz to 8.4 GHz with a bandwidth of 2 GHz. To accomplish miniaturization, metamaterial approaches guarantee compactness without sacrificing performance. The antenna's parameters are fine-tuned using the Chameleon Swarm Algorithm, a bio-inspired optimization method that achieves the required bandwidth of 2 GHz, return loss S11 <-25 dB, high gain, and an excellent radiation pattern. Simulation results validate the antenna's potential for next-generation 6G IoT applications.

DOI: 10.4018/979-8-3373-0852-4.ch004

1. INTRODUCTION

1.1 Overview

Wireless communication technologies have rapid advancement. It has paved the way for the evolution of the sixth generation (6G) networks. It has envisioned to revolutionize customer services and applications through the Internet of Things (IoT). These networks aim to enable a future of fully intelligent and autonomous systems. It addresses the demands for higher data rates. It gives improved connectivity and seamless integration of diverse devices. The potential of 6G spans a wide range of domains. It is including Healthcare IoT, Vehicular IoT and Autonomous Driving, Unmanned Aerial Vehicles (UAVs), Satellite IoT, and Industrial IoT, making it a transformative technology for global communication infrastructure.

In this chapter, the design and development of antenna is explained. It is capable of meeting the stringent requirements of 6G IoT applications. Antennas serve as the main component of wireless communication systems. By efficiently transmitting and receiving electromagnetic waves. With the age of 6G, the challenges in antenna design have become stringent. It encompasses the need for compactness, wide bandwidth, high gain, minimal return loss and radiation patterns. These features must be realized while ensuring seamless integration with IoT devices. It prioritizes miniaturization and energy efficiency. Addressing these challenges requires innovative approaches. It combines advanced materials such as metamaterials, with optimization techniques like the Chameleon Swarm Algorithm (CSA). These solutions make the way for antennas that should meet and exceed the performance benchmarks. These are necessary for the next generation of wireless communication.

This chapter introduces a novel antenna design. It is tailored for 6G IoT applications which is circularly polarized. It is operating in the frequency range of 6.4 GHz to 8.4 GHz with a bandwidth of 2 GHz. A main highlight of this work is the implementation of metamaterial. It helps to achieve miniaturization. It is essential for integrating antennas into modern IoT devices. Metamaterials are known for their ability to manipulate EM waves in unconventional ways. It offers a promising solution to the challenge of size reduction with the maintenance of high performance. To further enhance the antenna's performance, the Chameleon Swarm Algorithm (CSA) is employed for optimization. CSA is a bio-inspired optimization technique that mimics the adaptive behavior of chameleons in nature, enabling efficient exploration and exploitation of the solution space. By optimizing critical parameters such as slot radius and position, CSA ensures that the antenna meets the desired performance metrics, including the required bandwidth, return loss ($S11 < -10$ dB), high gain, circular polarization and optimal radiation pattern. The metamaterial techniques and the CSA optimization are integrated together. It represents a significant advancement in antenna design for 6G. The proposed approach addresses the technical challenges. but also aligns with the overarching goals of 6G networks to support a wide array of IoT applications. Simulation and experimental results validate the effectiveness of the proposed antenna, demonstrating its suitability for next-generation wireless communication systems.

This chapter aims to provide a comprehensive understanding of the proposed antenna design. It has relevance to 6G IoT applications. It begins with an overview of 6G and IoT technologies, followed by a detailed discussion of metamaterial techniques and the CSA optimization process. The results and analysis section highlights the antenna performance. It is showing its potential to meet the demands of 6G IoT networks. The chapter concludes with plans for future research. It discusses the broader spectrum of this work in advance of 6G communication.

1.2 Evolution From 1G to 6G: The Need for Next-Gen Communication Systems

The evolution of wireless communication has been driven by the needs. It needs higher data rates, lower latency, good connectivity. The 5G has significantly enhanced network capabilities with speeds up to 10 Gbps and latency as low as 1 millisecond, it primarily focuses on enabling massive machine-type communications and enhanced mobile broadband. The applications like autonomous systems, virtual reality (VR), industrial automation, and the Internet of Things (IoT) are recent trends. It demands ability beyond 5G.

6G represents the next step in this growth trajectory. It addresses the limitations of 5G. It also paves the way for transformative communication systems. It is required to operate in frequency ranges in the terahertz (THz) spectrum. 6G aims to provide data rates exceeding 1 Tbps. The latency is on the order of microseconds. The connectivity is supported for trillions of devices. Such capabilities will support the integration of intelligent and autonomous systems into daily life, transforming industries ranging from healthcare and transportation to space exploration. One of the basic drivers for 6G is the integration of IoT devices into a communication system. IoT applications, such as smart cities, UAVs, satellite communications, and remote healthcare, require ultra-reliable low-latency communication (URLLC), massive connectivity, and high energy efficiency. These demands necessitate the development of next-generation antennas. This is capable of operating efficiently in the higher frequency bands required for 6G. It addresses sustainability challenges. This is by incorporating energy-efficient technologies and supporting green communication systems. The network infrastructure will leverage AI and machine learning to optimize resource allocation, enhance performance, and dynamically adapt to changing environments.

Antenna is an important aspect of wireless communication. The design technologies must evolve in parallel with 6G. They have to meet the requirements of this revolutionary system. Innovations such as compact, high-performance antennas with wide bandwidth, high gain, and low return loss are crucial for supporting the enhanced capabilities of 6G. By addressing these needs, next-generation communication systems will redefine global connectivity and usher in an era of unprecedented technological advancement. Figure 1 shows the evolution of generations with future 6G IOT applications (Nguyen et al., 2021).

Figure 1. Evolution from 1G to 6G: The need for next-gen communication systems

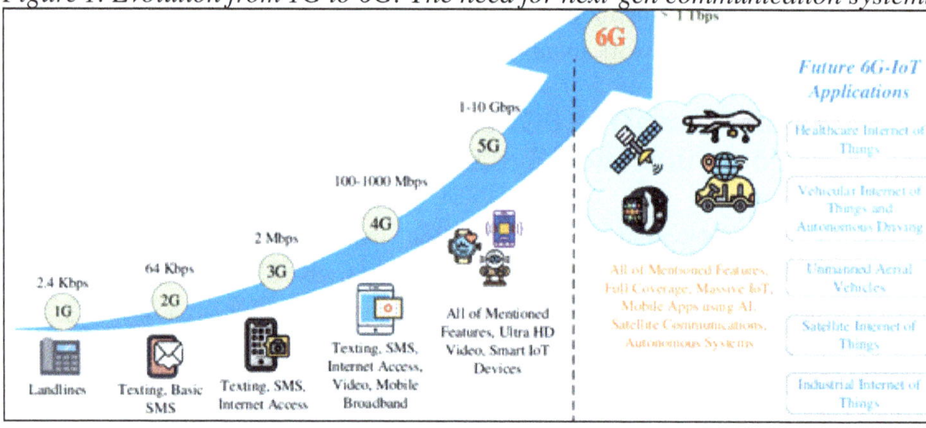

1.3 Impact of IoT on Shaping Future Intelligent and Autonomous Systems

The Internet of Things (IoT) has grown into a transformative technology. It is revolutionizing how devices interact and communicate. They form an interconnected ecosystem. It is enabling continuous data exchange between sensors and systems. IoT has paved the way for intelligent and autonomous solutions across various domains. This novel connectivity adopts real-time decision making. This improved operational efficiency and enhanced user experiences. This is setting the stage for future growth in intelligent systems.

IoT plays a pivotal role in shaping the landscape of smart cities, healthcare, transportation, and industrial automation. In the smart cities, sensors monitor and checks various parameters. It manages traffic flow, and public safety. It is creating more greener urban environments. Autonomous vehicles include self-driving cars and drones. It relies heavily on IoT to gather and process data for navigation and decision-making. In healthcare, IoT enables remote monitoring. It also helps in predictive diagnostics. This is helping monitor the patients.

As 6G communication networks evolve, the role of IoT becomes even more significant. The high data rates, ultra-low latency, and massive device connectivity promised by 6G will empower IoT to support complex applications, such as unmanned aerial systems, autonomous industrial robots, and satellite-based IoT networks. These growths will help the development of complete intelligent and reliant systems. These systems are capable of adapting to dynamic environments and user needs. Achieving this vision poses challenges. It is including the need for robust connectivity, security, and scalability. IoT creates an impact on future systems too. It is completely connected to advancements in technologies. This includes edge computing, AI, and antenna designs. These innovations ensure that IoT will continue to play a central role in realizing intelligent and autonomous systems. It helps in shaping a future characterized by efficiency, intelligence, and connectivity.

1.4 Key Domains for 6G IoT Application

The age of 6G networks is set to evolve IoT applications across various sectors. It is enabling unimaginable connectivity, efficiency, and innovation. The areas where 6G IoT will have a great impact are mentioned as follows:

1.4.1 Healthcare Internet of Things (HIoT)

6G-powered IoT will change healthcare systems. This can be done by enabling real-time remote patient monitoring. It can provide advanced diagnostics, and predictive healthcare solutions. High-speed data transmission and ultra-low latency will support wearable medical devices and implantable sensors, ensuring seamless communication between patients, doctors, and healthcare systems. The applications include telemedicine, remote surgeries, etc.,

1.4.2 Vehicular Internet of Things (V2X) and Autonomous Driving

The vehicular IoT domain will benefit significantly from 6G's ultra-reliable low-latency communication (URLLC) and massive machine-type communication (mMTC). These features will support autonomous vehicles, smart traffic systems, and vehicle-to-everything (V2X) communication, enhancing road safety, reducing congestion, and enabling self-driving cars to operate efficiently in complex urban environments.

1.4.3 Unmanned Aerial Vehicles (UAVs)

IoT-enabled UAVs includes drones. It will leverage 6G networks for applications in delivery, surveillance, disaster management, and agriculture. High-frequency 6G bands will provide UAVs with long-range connectivity and robust data links, ensuring precise navigation and real-time data processing, even in challenging environments.

1.4.4 Satellite Internet of Things

The 6G IoT will include satellite communication systems. It has to provide coverage throughout the globe. This has to achieved in remote and isolated regions. Satellite IoT will enable applications like environmental monitoring, maritime and aviation tracking, and disaster response. Thus, ensures good connectivity in areas that are beyond reach of land-based networks.

1.4.5 Industrial Internet of Things (IIoT)

The 6G will revolutionize industrial automation. This is done by connecting machines, sensors, and control systems. The connection is carried out in real-time. Smart factories are powered by 6G IoT. It will feature predictive maintenance. It carries out automated workflows, and advanced robotics. This transformation will lead to increased productivity, reduced downtime, and enhanced safety in industries such as manufacturing, energy, and logistics.

1.4.6 Smart Cities

6G IoT will benefit smart city initiatives. It is providing the infrastructure needed for intelligent urban systems. The applications include smart grids, waste management, water supply optimization, and traffic monitoring. The 6G IoT in smart cities will improve resource management and enhance the quality of life for urban residents.

1.4.7 Consumer IoT

6G IoT will benefit personal devices like smart home systems. It includes wearables, and virtual reality (VR)/augmented reality (AR) devices. Enhanced connectivity and data processing capabilities will available. It enables great user experiences, energy-efficient devices, and integration across consumer IoT ecosystems.

1.5 Role of Antennas in Enabling 6G IoT Communication

Antenna play a key role in the realizing of 6G wireless communication networks. It is particular in facilitating seamless Internet of Things (IoT) integration. It is the primary interface between the wireless system and the physical environment. The antennas are critical for transmitting and receiving electromagnetic waves with efficiency. It provides reliability and precision. In the context of 6G, the expectations from antenna systems are more demanding than ever. It is thus envisioned as ultra-high data rates, low latency, massive device connectivity, and energy efficiency goals.

1.5.1 Miniaturization and Integration

With IoT devices becoming increasingly compact, antennas must achieve significant size reduction without compromising performance. Advanced techniques such as metamaterials and metasurfaces, are used to miniaturize antenna structures. Also done while maintaining desirable characteristics such as wide bandwidth and high gain. These materials enable the manipulation of electromagnetic waves in unconventional ways, offering a path to overcoming size constraints in IoT applications.

1.5.2 Performance Metrics: Bandwidth, Gain, Polarization and Efficiency

6G antennas are designed to operate in circularly polarized manner. It works in the higher frequency bands. It includes millimeter-wave and sub-terahertz range. This is to accommodate the immense data transfer requirements of IoT networks. Ensuring wide bandwidth is crucial to support diverse IoT applications. The high gain and radiation efficiency ensure reliable communication over extended distances. Return loss (S11 < -10 dB) and well-optimized. The radiation patterns are also essential for effective signal transmission and reception.

1.5.3 Adaptability Through Optimization Algorithms

The complex of 6G antenna design increases. It requires the use of higher optimization algorithms. This makes use of algorithms such as the Chameleon Swarm Algorithm (CSA). These algorithms tune parameters like slot radius, element position, and material properties. Thus, helps to achieve the optimal performance. Bio-inspired techniques are used for this purpose. These algorithms ensure easy exploration of the solution space. Thus, leading to antennas that meet the strict requirements of 6G IoT network.

1.5.4 Support for Diverse IoT Domains

6G antennas are expected to make up to a broad spectrum of IoT applications. This includes healthcare, autonomous vehicles, industrial automation, and satellite communication. Each domain presents different challenges. This may include challenges such as frequency bands, energy efficiency, and environmental adaptability. The antennas must address these problems. The wearable devices demand ultra-compact, low-power antennas, The vehicular IoT systems prioritize high-gain, robust designs for reliable connectivity in dynamic environments.

1.5.5 Energy Efficiency and Sustainability

As 6G emphasizes energy-efficient communication, antenna designs must align with these sustainability goals. Low-power consumption and materials with minimal environmental impact are integrated into antenna systems to ensure green communication technologies for IoT networks.

2. LITERATURE STUDY

The rapid evolution of wireless communication technologies has increased demands on antenna design for 5G, 6G, and IoT applications. Antennas in these systems must meet rigorous requirements. It includes high efficiency, miniaturization, high gain, and polarization diversity. The advent of metamaterials and nature-inspired optimization algorithms, such as the Chameleon Swarm Algorithm (CSA), has opened new avenues for achieving these goals. 6G antennas are envisioned to support a wide range of applications. This includes IoT, autonomous systems, smart cities, and tactile internet (Nguyen et al., 2021). The 5G primarily focuses on high-speed data transmission and low latency. The 6G aims to integrate a broader spectrum of applications, such as terahertz communication (Rasilainen et al., 2023), artificial intelligence-driven networks (Mwang'amba et al., 2024), and ubiquitous connectivity. The design faces several challenges for 6G antennas. This includes compact, high radiation efficiency, gain, and circular polarization. These parameters enable robust communication in dynamic environment. The IoT devices form the backbone of smart systems (Zhang et al., 2024). It requires antennas that are not only efficient and compact but also cost-effective and compatible with diverse deployment scenarios.

The leap from 5G to 6G requires significant advancement in antenna technologies. 5G systems widely adopts massive multiple-input multiple-output (MIMO) antenna arrays (Wong et al., 2024). This enhances spectral efficiency and achieve beamforming capabilities. These antennas operate at millimeter-wave frequencies

(Azari et al., 2024) (Sim et al., 2022). They range from 24–40 GHz. These frequencies allow high data rates and low latency. It also faces challenges such as increased path loss and limited coverage. To address these issues certain methods are followed. The 5G antenna designs have incorporated advanced materials (Meenambal, Vasudevan, Maheshwari et al, 2024) (Meenambal, Jacob, & Vasudevan, 2024). They have innovative geometries and uses various optimization techniques. Metamaterials are useful in reducing the antenna size and gain improvement. Metamaterials are artificially engineered structures. These materials have unique EM properties not found in natural materials. They have the ability to manipulate electromagnetic waves (Esmail & Koziel, 2022). This enables the design of compact and efficient antennas. Metamaterials leverage size reduction by their negative refractive index properties (Pallavi et al., 2024). This property creates subwavelength resonators. These resonators allow antennas to operate at lower frequencies with smaller physical size. This characteristic is particularly beneficial for IoT devices, which often have stringent space constraints. Using metamaterial structures size is reduced with the required antenna characteristics. Metamaterials have been used to enhance the antenna (Bose & Karuppiah, 2024). High gain is essential for long-range communication and overcoming signal attenuation, especially at higher frequencies. Metamaterial-based structures, such as electromagnetic bandgap (EBG) materials and high-impedance surfaces (HIS), have been employed to achieve gain enhancement. EBG materials mitigates surface wave propagation. This reduces radiation loss and improve the overall antenna efficiency (Sherene et al., 2024) (Jacob et al., 2024). HIS structures provide control of reflection phase. It enables the design of antennas with improved directivity and gain.

Circular polarization (CP) is an important critical requirement for modern antennas. This is required for IoT and 6G applications. CP antenna offer several advantages. It offers robustness to polarization mismatches. It is also used to reduced multipath fading, and improved signal integrity. Metamaterials are widely used to achieve circular polarization (Hui et al., 2023) (Lin et al., 2022). For instance, split-ring resonators (SRRs) and complementary split-ring resonators (CSRRs) are commonly integrated into antenna designs to create circularly polarized radiation. These structures introduce phase differences between orthogonal components of the electric field. This results in CP radiation. Metamaterial-based CP antennas are useful in applications where the orientation of the transmitting and receiving antennas is not fixed. This scenario includes mobile or satellite communication systems (Zhao et al., 2023). The optimization of antenna parameters is important. It helps in meeting the requirements of the multifaceted 6G and IoT applications. Usual trial-and-error methods are not viable due to the complexity of antenna designs. Hence, optimization algorithms are employed to systematically refine antenna parameters, such as slot dimensions, feed positions, and substrate properties. The nature-inspired techniques have significant attention recently. This is due to their ability to handle multiple objectives.

The Chameleon Swarm Algorithm (CSA) is a recent addition to the nature-inspired optimization. This is inspired by the adaptive behavior of chameleons. CSA mimics their ability to change colour based on environmental stimuli. This adaptability is modelled in the algorithm. This algorithm is designed as a dynamic exploration and exploitation mechanism. Hence, enabling CSA to maintain diversity in the solution space while converging toward the global optimum (Cai et al., 2022). In antenna design, CSA has been successfully applied to optimize slot dimensions and positions, achieving improvements in bandwidth, gain, and polarization characteristics (Uluslu, 2022). CSA offers many advantages over other optimization techniques. This includes Genetic Algorithms (GA), Particle Swarm Optimization (PSO), and Differential Evolution (DE). The GA relies on crossover and mutation operations to evolve solutions. It often requires extensive parameter tuning. Hence it may suffer from premature convergence. PSO is based on the social behavior of flocking birds. Its implementation is faster and easier. The disadvantage is it may but struggle with maintaining diversity in many dimensional design spaces. DE, known for its robustness in handling nonlinear and multimodal problems, can be computationally intensive and sensitive to parameter selection. CSA overcomes these limitations by dynamically adjusting its exploration and exploitation capabilities. This adaptation is based on the current state of the search process. This ensures a more balanced and efficient search, making it particularly suitable for complex antenna design problems.

The CSA is applied in slot optimization of antenna. This has yielded promising results. Slot antennas are widely used in modern communication systems. It requires precise tuning of their dimensions and positions. Hence to achieve the desired performance. With CSA it can efficiently explore the design space and identify the optimal configuration that meets multiple objectives. This includes maximizing gain, bandwidth, and achieving CP. The adaptability of CSA allows it to handle the trade-offs. The main trade-offs are between size reduction and performance enhancement. These are common challenges in 6G and IoT antenna design. Several studies have demonstrated the effectiveness of CSA in optimizing antenna designs. For MTM based antenna, CSA was used to optimize parameters of the metamaterial unit cells and the feed structure. The results had significant improvements in gain, axial ratio bandwidth, and efficiency. Compared traditional algorithms, CSA had better performance. CSA has been applied to optimize the design of compact slot antennas for IoT devices. Hence achieve substantial miniaturization and maintaining high performance. Metamaterials and CSA complement each other in addressing the challenges design challenges. The metamaterials provide the means to achieve miniaturization and performance enhancement (Swetha et al., 2024). The optimization algorithms make sure that these capabilities are utilized to their full potential. The integration of these technologies is important for 6G and IoT applications. The requirements are diverse and often conflicting. By leveraging the unique properties of metamaterials and the adaptability of CSA, designers can create antennas that meet the stringent demands of next-generation communication systems (Akyildiz et al., 2020). The change to 6G proliferated the IoT devices. It will continue to drive advancements in antenna technologies. Demand for smaller, more efficient, and higher-performing antennas grows. The role of metamaterials and nature-inspired optimization algorithms are increasingly important. This offers a powerful approach to tackling the challenges of modern antenna design. This enables the development of innovative solutions that support the next generation of wireless communication.

3. DESIGN REQUIREMENTS FOR 6G IOT ANTENNAS

The design of antennas for 6G IoT applications plays a key role. It helps in achieving the advanced capabilities. It is required for next-generation wireless communication systems. These antennas must meet stringent performance metrics. It has to be done while addressing the specific requirements of IoT devices. The key design requirements for 6G IoT antennas are given below.

3.1 Operating Frequency and Bandwidth

The 6G IoT antennas must support high speed communication. The communication must be for wide frequency range to support various applications of next-gen networks. The proposed antenna operates within the frequency range of 6.4 GHz to 8.4 GHz. It is delivering a bandwidth of 2 GHz. This wide bandwidth ensures efficient data transmission, meeting the low-latency and high-capacity needs of 6G applications, such as real-time healthcare monitoring, autonomous vehicles, and industrial automation.

3.2 Return Loss (S11)

Return loss is an important value for the antenna performance. This reflects the amount of electric field reflected due to impedance mismatch. The antenna has a S11 of -25 dB. This value shows good impedance matching. This is resulting in minimal power loss. This has improved efficiency across the operating frequency range.

3.3 High Gain

Gain is vital for signal strength and coverage in high-frequency bands where propagation losses are more pronounced. The proposed antenna ensures high gain. It is enabling reliable communication in complex IoT environments. Urban areas and industrial settings have reliable communication. High-gain antennas also improve energy efficiency by reducing the need for additional power amplification.

3.4 Radiation Pattern

The antenna's radiation pattern is a key factor in determining its suitability for different IoT applications. Some applications, such as satellite IoT and UAVs, require highly directional patterns. This is for focused communication. Other applications like healthcare IoT, benefit from omnidirectional coverage. The proposed antenna demonstrates an excellent radiation pattern. It is adaptable to these diverse application scenarios.

3.5 Circular Polarization

CP is defined as the rotation of the Electric field. It rotates in a circular motion as it travels. It turns in a right or left-hand direction. This type of rotation makes sure that the signal remains intact. Even with changes in orientation, it is efficient. It makes the antenna ideal for satellite communications and IoT networks. CP enhances signal robustness in dynamic environments. CP does this by reduction in being prone to polarization mismatches and multi-path fading. It is important for IoT devices with different orientations. The axial ratio is the measure of the circular polarization. The AR should be below 3 dB across the bandwidth.

3.6 Miniaturization Through Metamaterial Techniques

IoT devices require compact and lightweight antennas. It is due to their limited size factors. The metamaterial techniques in the proposed design enables miniaturization. It is done without compromising performance. Metamaterials are known to manipulate electromagnetic waves in unconventional ways. It allows the antenna to maintain its high performance while achieving the compactness required for modern IoT devices.

3.7 Optimization Using Chameleon Swarm Algorithm (CSA)

To achieve the desired performance metrics requires optimization of antenna. The CSA is a bio-inspired optimization technique. It is used to tune the critical parameters. The parameters include slot radius and position. This makes sure that the antenna meets the required bandwidth of 2 GHz. Also, the return loss (S11 < -25 dB). To make sure high gain and maintaining a compact size. CSA's adaptive and explorative utility make it an effective method for optimizing antenna for 6G applications.

4. METAMATERIALS FOR ANTENNA

Metamaterials (MTMs) are artificially tailored materials. These are designed to exhibit EM properties not found in natural substances. These properties are achieved through the arrangement of subwavelength structural elements. Thus, enabling metamaterials to manipulate EM waves based on their structure. It does not depend on their material nature. A feature of metamaterials is their ability to show unique behaviours. This includes negative refractive indices and perfect lensing. These are not observed in conventional materials. Depending on their electromagnetic properties, metamaterials can be classified into categories such as epsilon-negative (ENG), mu-negative (MNG), and double-negative (DNG) materials, where permittivity, permeability, or both are negative. Metamaterials enable significant size reduction in antennas. This is by using their ability to manipulate EM waves at subwavelength scales. This is crucial for modern applications like IoT devices and 6G communication systems. Metamaterial antenna can achieve high performance with smaller size. This makes them good for compact environments. These properties make metamaterials for innovation. It is used in applications like antenna design.

4.1 Classification of Metamaterials

Metamaterials are divided based on the values of their making parameters as shown in Figure 2. They are namely permittivity (ε) and permeability (μ). These are classified as epsilon-negative (ENG) and mu-negative (MNG) materials. These are which exhibit negative permittivity ($\varepsilon<0$) or negative permeability ($\mu<0$). Such materials are designed to achieve unique electromagnetic responses within specific frequency ranges. Double-negative (DNG) metamaterials possess negative permittivity and permeability ($\varepsilon<0$and $\mu<0$) simultaneously. This is enabling artificial phenomena like negative refraction and backward wave propagation. Also, epsilon-near-zero (ENZ) and mu-near-zero (MNZ) materials exist. They exhibit permittivity or permeability values close to zero. They are corresponding to frequencies near their electric or magnetic plasma frequencies.

Figure 2. Classification of Metamaterial

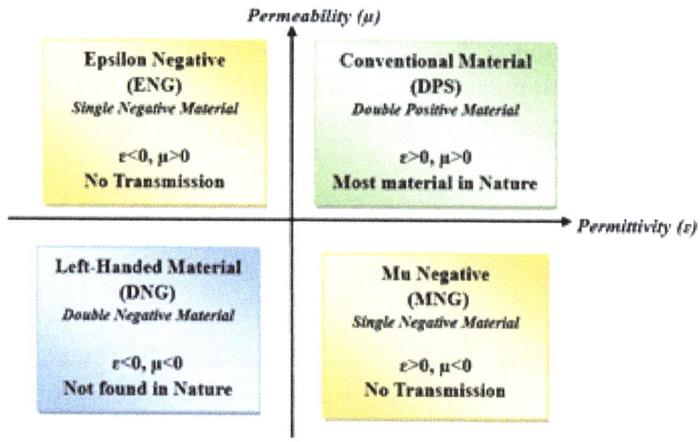

4.1.1 Single Negative Metamaterial

Single-Negative (SNG) metamaterials are a class of designed materials. They exhibit either negative permittivity (ε) or negative permeability (μ). They are not both simultaneously within a specific frequency range. These are further divided into two types:

a) *Epsilon Negative (ENG)* materials are a type of single-negative (SNG) metamaterial characterized by a negative permittivity ($\varepsilon<0$) over a specific frequency range. These materials are designed to exhibit unique electromagnetic properties. These are not found in natural materials. This is enabling novel applications in communication and sensing systems. The negative permittivity is typically achieved using periodic arrangements of thin metallic wires or other subwavelength structures as shown in Figure 3. These structures interact with the electric component of EM waves. These materials are useful in controlling the propagation of electromagnetic waves. They are used as they can support evanescent waves and inhibit the transmission of certain frequency bands. Their ability to create unconventional electromagnetic responses, such as backward wave propagation, makes ENG materials indispensable in advancing modern antenna technologies and other components for next-generation communication systems.

Figure 3. Schematic and Equivalent Circuit Model of Epsilon-Negative Metamaterial

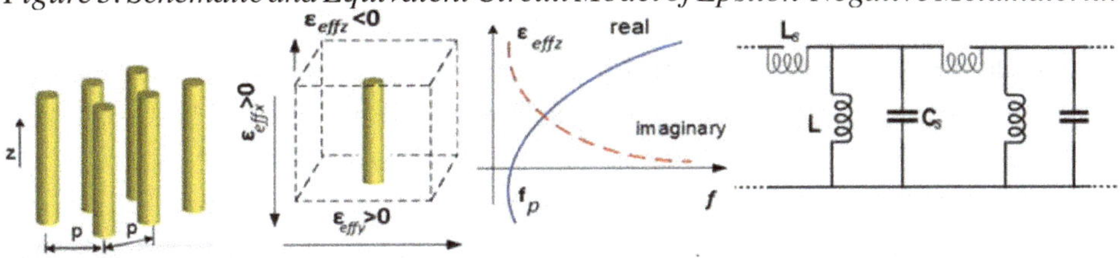

b) *Mu Negative (MNG)* materials are a type of single-negative (SNG) metamaterial. They are characterized by a negative magnetic permeability (μ<0) over a specific frequency range. Natural materials have positive magnetic permeability. MNG materials are designed to exhibit negative magnetic properties. This is achieved by using periodic structures. This includes structures such as split-ring resonators or resonant elements as shown in Figure 4. These structures react with the magnetic component of EM waves. These materials enable unique electromagnetic behaviours, including backward wave propagation, where the energy flow is opposite to the wave vector, and negative refraction, which can be exploited for applications like super lenses. The MNG materials are used to design compact waveguides and antennas. This helps in improving radiation efficiency and miniaturization. These properties make MNG materials in next-gen communication systems. This may include 5G and 6G, and for innovations in radar, microwave devices, and cloaking technologies, where precise control of electromagnetic wave propagation is crucial.

Figure 4. Schematic and Equivalent Circuit Model of Mu Negative Metamaterial

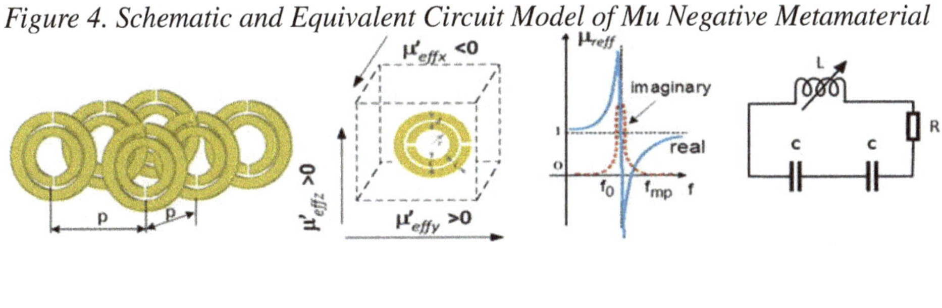

4.1.2 Double Negative Metamaterial

Double-negative (DNG) metamaterials are not natural materials. These materials have negative permittivity ($\varepsilon<0$) and negative permeability ($\mu<0$). These materials are known as negative index metamaterials. This is because their refractive index ($n=\sqrt{\varepsilon\mu}$) is negative. This property not found in natural substances. These metamaterials are designed using periodic structures. This includes split-ring resonators. This interacts with both the electric and magnetic components of electromagnetic waves. Key attributes of DNG materials include reverse Snell's law. The refraction angle is on the same side as the incident angle. The ability to focus EM waves is with high precision. They also exhibit negative phase velocity. This is leading to show reverse Doppler shifts and inverse Cherenkov radiation. These properties are exploited in developing devices for controlling wave propagation.

Double-negative (DNG) metamaterials are achieved by combining the individual properties of mu-negative (MNG) and epsilon-negative (ENG) materials as shown in Figure 5. MNG materials exhibit a negative permeability ($\mu<0$) and a positive permittivity ($\varepsilon>0$). The ENG materials display negative permittivity ($\varepsilon<0$) and positive permeability ($\mu>0$). These characteristics are brought together in a single material. Thus, both permittivity and permeability become negative simultaneously ($\varepsilon<0$, $\mu<0$). This is resulting in a DNG material. While MNG materials influence the magnetic response of electromagnetic waves and ENG materials affect the electric response, their integration leads to unique wave propagation phenomena. For instance, the simultaneous negativity creates a negative refractive index, enabling behaviours like reverse Snell's law and backward wave propagation, which neither MNG nor ENG can achieve alone

Figure 5. Structural Development of Double-Negative Metamaterials

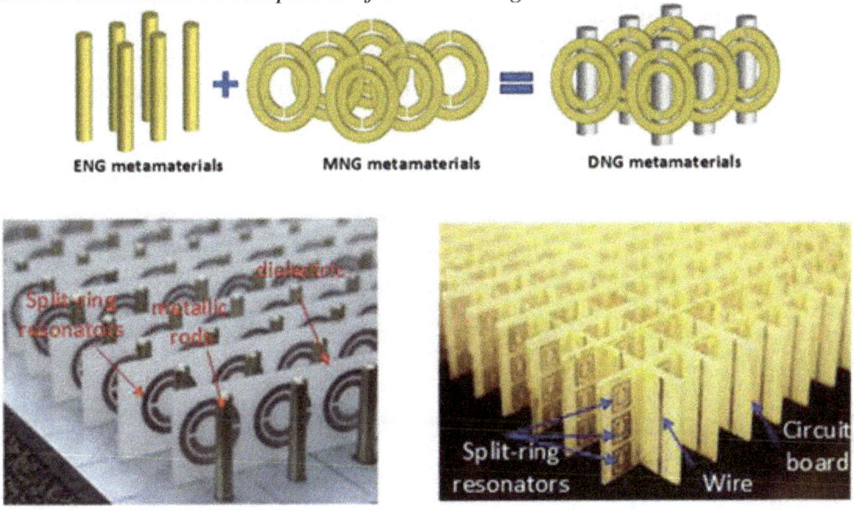

5. PROPOSED ANTENNA DESIGN AND METHODOLOGY

The proposed antenna design prioritizes enhanced bandwidth, higher efficiency, and reduced size to meet the growing needs of the upcoming 5G and 6G IoT networks. This design incorporates cutting-edge techniques such as metamaterials to achieve miniaturization, circular polarization, high gain, high efficiency, and stable performance over a wide frequency range.

5.1 Evolution and Antenna Configurations

Figure 6 shows the evolution of antenna at different stages. Figure 7 shows the proposed antenna design with compact dimensions of 16 x 15 x 0.64 mm^3, making it suitable for modern wireless communication systems. The substrate material used is Rogers RO3210 with thickness of 0.64mm. It has the relative permittivity (ε_r) of 10.2 allows for size reduction by lowering the resonant wavelength. The dielectric loss tangent (tan δ) of 0.003 increases the antenna's efficiency by providing minimal power loss. Table 1 gives the parameters and dimensions of the proposed antenna.

The methodical development of the antenna design has been divided into three parts. Each one concentrates on specific performance standards and challenges. The objective was to create a compact, circularly polarized antenna that will operate at the desired frequency with high efficiency. The patch dimensions were determined based on the resonant frequency, with the microstrip feed ensuring efficient power transfer and proper impedance matching. This design performed well in terms of radiation characteristics and bandwidth. However, it had size restrictions because the patch's dimensions were directly related to the operational wavelength.

Figure 6. Evolution of antenna at different stages

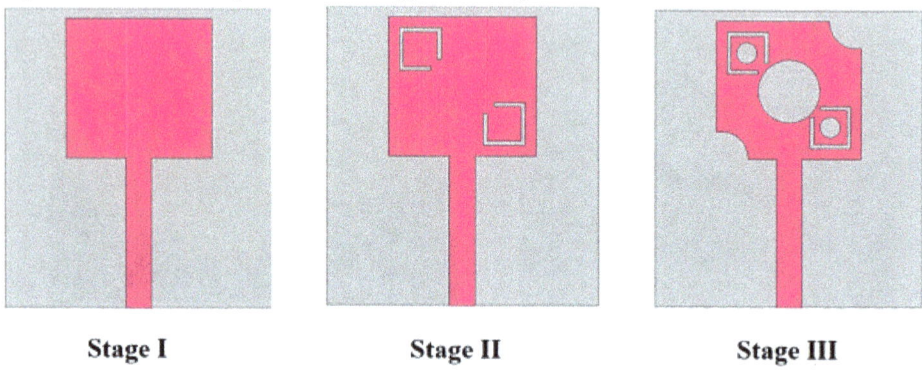

In the second stage, metamaterial techniques were incorporated in order to expand beyond the square patch design's limitation on dimension. The patch included two complementary square split-ring resonators (SRRs). Because of their special electromagnetic characteristics, metamaterials facilitate size reduction without sacrificing performance.

Figure 7. Antenna Geometry a) Front view b) Back view

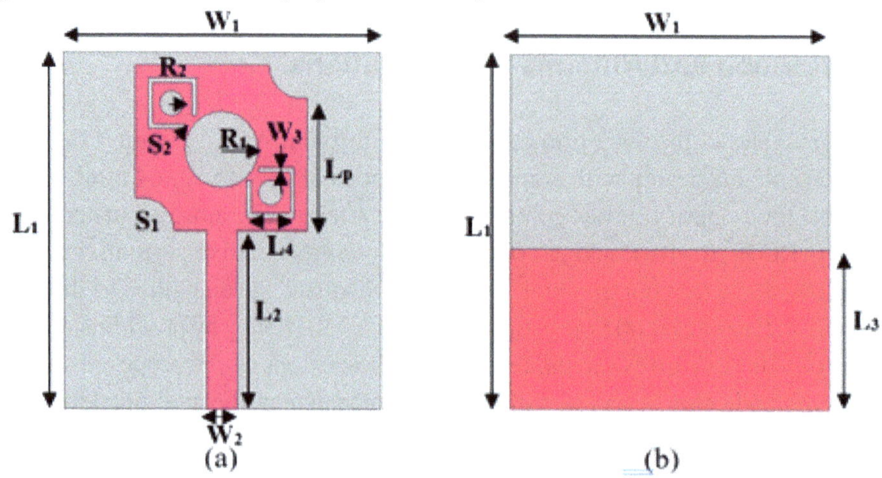

Table 1. Antenna parameters and Dimensions

Parameter	*Dimension(mm)*	*Parameter*	*Dimension(mm)*
L_1	16	W_3	0.24
L_2	8	S_1	2.4
L_3	7.56	S_2	0.5
L_4	2.4	R_1	1.8
W_1	15	R_2	0.6
W_2	1.44	L_P	6

Complementary SRR reduce capacitive effects by introducing discontinuities in the patch, which improve impedance matching. They also increase inductive effects by elongating current paths, lowering the resonant frequency for size reduction. This has made it feasible for the antenna to have an improved form factor while continuing to operate at the desired frequency. The concerns of circular polarization and impedance matching were resolved in the design's endure phase. To ensure effective power transfer and minimal reflection loss, slots have been introduced to the patch to minimize capacitive effects and achieve optimal impedance matching. Also, additional slots were strategically positioned at the patch's corners to provide circular polarization.

Using the Chameleon Search Algorithm (CSA), an optimization method inspired by nature, the slot placements and dimensions were optimized. In order to achieve an axial ratio < 3dB for circular polarization while retaining impedance matching and reducing return loss, the Chameleon Swarm Algorithm is used by tuning the slots position and dimension. This optimization technique ensured that the antenna could perform satisfactorily spanning its operational bandwidth.

5.2 Current Distribution and Radiation Mechanism

Figure 8 illustrates the patch antenna's radiating behavior at various phases based on the surface current distribution at center frequency 7.4 GHz. Phase cancellation lowers radiation efficiency for phases 0° and 180° because the current density is relatively small and fluctuates significantly over the patch. And, phases 45° and 225° exhibit substantial increases in current density, especially in particular patch locations. These high-current-density areas shows, where the antenna exhibits strong radiation characteristics.

Figure 8. Surface current distribution at 7.4 GHz

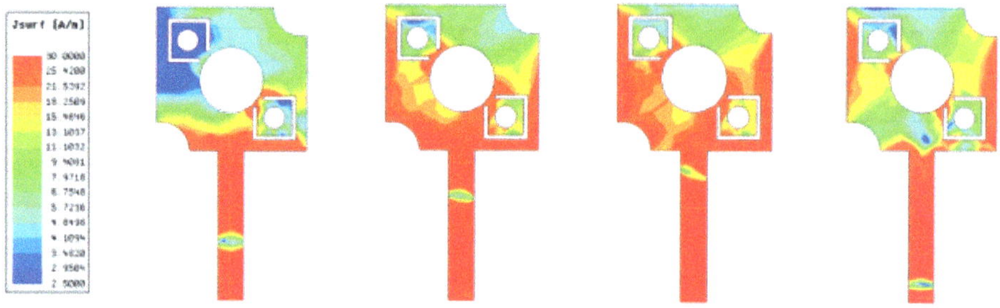

a) Phase=0⁰,180⁰　　b) Phase=45⁰,225⁰　　c) Phase=90⁰,270⁰　　d) Phase=135⁰,315⁰

Likewise, the current remains concentrated in the radiating regions at phases 90° and 270°, reaching maximum density and providing the strongest radiation performance. The current density continued to be considerable at phases 135° and 315°, suggesting an acceptable level of radiation efficiency. The appropriate current density distribution remains intact in order to achieve optimal impedance matching, thereby decreasing signal reflection. Concentrated high current densities also result in elevated resistive losses. Owing of the antenna's more uniform current density distribution at the desired frequency, the return loss of *S*11 has become lower, at around -25 dB.

5.3 Metamaterial Characteristics Extraction

In order to examine the metamaterial characteristics, parameters such as refractive index (η), permeability (μ), permittivity (ε), and impedance (Z) must be extracted. These properties helps in understanding of how electromagnetic waves interact with metamaterial structures and functions. S-parameters, or scattering parameters, are often derived from electromagnetic properties. When electromagnetic waves interact with a material, its transmission coefficient and reflection coefficient are characterized by the S-parameters (S11 and S21). S21 indicates the behavior of the transmitted wave, and S11 indicates the size and phase of the reflected wave. These parameters are used to calculate the metamaterial characteristics.

The refractive index (n) describes the phase velocity of the electromagnetic wave within the material. It is derived from the phase of the transmission coefficient S21 using the following relation (1),

$$n = \frac{1}{k_0 d}\left\{ [[\ln(e^{ink_0 d})]'' + 2m\pi] - i[\ln(e^{ink_0 d})]' \right\} \tag{1}$$

where k_0 is the wave number and d is thickness of the material. The metamaterial's impedance (z) characterizes the ratio of the medium's magnetic and electric fields. Using the following formula (2), it is computed from the S-parameters,

$$z = \pm\sqrt{\frac{(1+S_{11})^2 - S_{21}^2}{(1-S_{11})^2 - S_{21}^2}} \tag{2}$$

This normalized impedance is needed to produce materials with particular impedance-matching desires and also for understanding the metamaterial's transmission and reflection behavior. The effective permittivity (ε) and permeability (μ) can be calculated using the following equations (3) and (4) after the refractive index (n) and impedance (z) have been determined,

$$\varepsilon = \frac{n}{z} \tag{3}$$

$$\mu = nz \tag{4}$$

As a characteristic of negative index materials, metamaterials often exhibit negative permittivity and permeability over specific frequency ranges. When the electric polarization of a material resists the electric field that is being applied, which frequently occurs in metallic structures due to resonance, negative permittivity emerges. Negative permeability, which is commonly observed in split-ring resonators, occurs when magnetic polarization resists the applied magnetic field. These unique characteristics transform wave manipulation applications by enabling phenomena like reverse Snell's law.

Figure 9 displaying the retrieved metamaterial parameters such as permeability (μ), permittivity (ε), impedance (z), and refractive index (n) which provides significant information describing the characteristics and performance of the proposed structure between 6.4 GHz and 8.4 GHz. Using every parameter, the metamaterial's effectiveness in achieving the desired characteristics for the proposed antenna design is verified. The metamaterial can effectively control electromagnetic waves if the refractive index behaves uniformly over the frequency range. The impedance is matched properly within the bandwidth, so there is minimal reflection in the operating bandwidth. In 6G and Internet of Things applications, impedance matching is a crucial need for efficient antenna performance, since it optimizes power transfer and reduces signal losses.

Figure 9. Extracted Metamaterial Parameters a) Impedance b) Refractive index c) Permeability

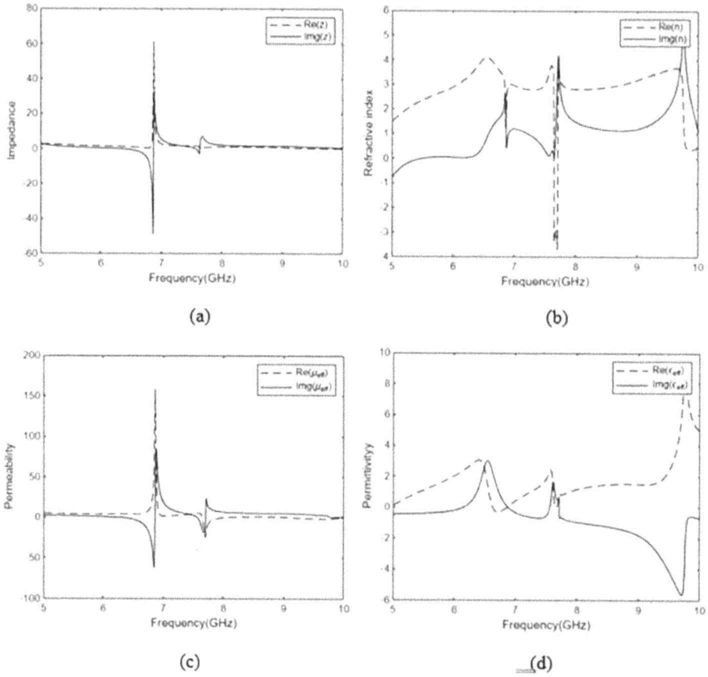

The negative permeability (μ) inside the intended frequency range is a notable property. This shows that the antenna is a single-negative (SNG) metamaterial, in which the permeability is negative. It is widely known that certain SNG materials allow for distinct wave manipulation capabilities, like size reduction and the support for certain resonance modes that enhance antenna performance. The electromagnetic wave propagation characteristics are supported by the permittivity (ε), which remains constant. When taken as a whole, the parameters confirm that the metamaterial might assist with circular polarization, enhanced gain, and compact antenna development. The bandwidth performance further demonstrates this metamaterial's effectiveness for applications that need long operational ranges, as those in next-generation wireless communication systems and the Internet of Things.

5.4 Chameleon Swarm Algorithm for Optimization

The Chameleon Swarm Algorithm (CSA) is a nature-inspired. It is a meta-heuristic optimization method. It is introduced to address high-dimensional, nonlinear, and complex optimization problems effectively. Inspired by the adaptive hunting behaviors of chameleons, CSA mimics their strategies for finding, tracking, and capturing prey. This algorithm is utilized for the context of optimizing the parameters. These are the radius and position of slot in the antenna. It is providing a study of its efficiency.

5.4.1 Mechanism of CSA

The CSA is inspired by the behaviors of chameleons. These reptiles possess unique abilities that include independently rotating their eyes to locate prey and using their long, sticky tongues to rapidly capture it. These biological traits are mathematically modelled in CSA to achieve a balance between exploration, searching for global optima across the design space and exploitation, fine-tuning solutions near the identified optima. The chameleons look focusing on its prey. This is converted into an algorithm. It maintains the diversity applied in optimization.

5.4.2 Phases

The algorithm of CSA is classified based on the hunting strategies of the chameleon

- *Initialize*: The algorithm starts by initializing a solution population. Each represents a chameleon within a multi-dimensional search space. Initial positions are generated randomly. Thus, a broad distribution across the solution space. This is to encourage exploration.
- *Search for Prey*: Chameleons update their positions by considering the global best solution (Gbest). Their personal best solutions (Pbest) is recorded. Randomness and exploration probabilities are considered. This makes sure during this phase to prevent premature convergence and ensure avoidance of local optima.
- *Eye Rotation*: The ability of chameleons to rotate their eyes independently. This is modelled mathematically to enhance the search capabilities of the algorithm. This allows CSA to scan the solution space more effectively. This helps to detect regions that might have been overlooked.
- *Prey Hunting*: When a solution approaches the vicinity of an optimal point, a rapid "attack" mechanism is triggered. This mechanism refines the position of the solution. This is done with high accuracy. This makes sure the convergence to the global optimum with minimal iterations.

5.4.3 Antenna Slot Optimization Using CSA

The proposed 6G IoT antenna operates in the frequency range of 6.4 GHz to 8.4 GHz with a 2 GHz bandwidth. The design incorporates slot structure, with key dimensions and placement parameters selected for optimization. Miniaturization is achieved using metamaterial techniques, ensuring compactness without sacrificing performance.

The optimization problem targets three objectives:

1. Minimization of return loss (S11)
2. Maximization of directivity gain
3. Desired Bandwidth

The **objective function** represents the key performance criteria of the antenna, such as return loss, gain, and bandwidth. This function captures the relationships between the slot radius and positions, correlating them with performance metrics like minimal return loss ($S_{11}<-25$ dB), maximum bandwidth, and high directivity gain. For instance, the objective function might take the form of a weighted sum that balances conflicting criteria, such as maximizing gain while minimizing return loss as,

$$F = \alpha . R + \beta . G \tag{5}$$

where R represents return loss, G denotes gain, and α, β are weights. Defining the objective function is crucial because it provides a measurable benchmark for evaluating candidate solutions. The critical values governing the CMA are initialized to ensure the algorithm operates effectively within the multidimensional search space. These parameters include the population size, which determines the number of candidate solutions or "chameleons" in the search space, and the maximum number of iterations, which limits the computational effort while ensuring convergence. Additionally, the mutation rate introduces diversity by probabilistically altering a candidate's position, preventing stagnation in local optima. The exploration and exploitation rates are also calibrated to balance the algorithm's ability to search for new solutions and refine existing ones.

An initial population of candidate solutions is generated randomly within the defined search space. Each solution corresponds to a specific configuration of the antenna's slot radius and positions. For example, one candidate might represent a slot radius of r_1=2.5 mm and positions (x_1, y_1), while another could represent r_2=3 mm and positions (x_2, y_2).

The optimization of fitness is evaluated by normalizing the objectives as,

$$Fitness = w_1 \frac{f_1 - f_1^{min}}{f_1^{max} - f_1^{min}} + w_2 \frac{f_2 - f_2^{min}}{f_2^{max} - f_2^{min}} \tag{6}$$

where f_1 and f_2 are the normalized objectives, and w_1, w_2 are weights balancing their significance. It evaluates the fitness of each candidate solution by assessing how well it meets the objective function. This involves simulating the antenna's performance for the given slot configuration and calculating key metrics such as return loss, gain and bandwidth. To compute the fitness, the CSA first determines the antenna's performance metrics based on its current configuration by using electromagnetic simulations tools.

To validate the bandwidth of the antenna, the -10dB return loss can be expressed as,

$$BW = f_{high} - f_{low} \tag{7}$$

where $S_{11}(f) \leq$ -10dB for $f \in |f_{low}, f_{high}|$. The gain of the antenna at desired frequency f_d is derived from,

$$G(f_d) = 10 . log_{10}\left(\frac{4\pi V(f_d)}{P_{in}}\right) \tag{8}$$

where $V(f_d)$ is the radiation intensity and P_{in} is the input power. The far field radiation pattern $E(\theta, \phi)$ for optimized antenna slot can be determined as,

$$E(\theta, \phi) = F(\theta, \phi). |I . e^{-jkr}| \tag{9}$$

where $F(\theta, \phi)$ is element factor that includes contribution from individual antenna element and *kr* is the wave phase term. Once the fitness values are calculated, the algorithm updates the positions of the candidate solutions. This involves simulating the chameleons' behavior, where each candidate adapts its position based on global best solution (G_{best})and personal best solution (P_{best}). The position update equation is typically expressed as,

$$X_i(t + 1) = X_i(t) + r_1 . \left(G_{best} - X_i(t)\right) + r_2 . \left(P_{best} - X_i(t)\right) + r_3 . \Delta_{rand} \tag{10}$$

Where $X_i(t)$ is the position of the i_{th} candidate at iteration t, r_1, r_2, r_3 are random coefficients and Δ_{rand} introduces randomness for exploration. The position is updated based on the iterations using the random coefficients as,

$$r_1 = r_1^{init}\left(1 - \frac{t}{T_{max}}\right)$$

,

$$r_2 = r_2^{init}\left(\frac{t}{T_{max}}\right) \tag{11}$$

Figure 10. Flowchart illustrating the Chameleon Swarm Algorithm (CSA) applied for optimizing antenna parameters.

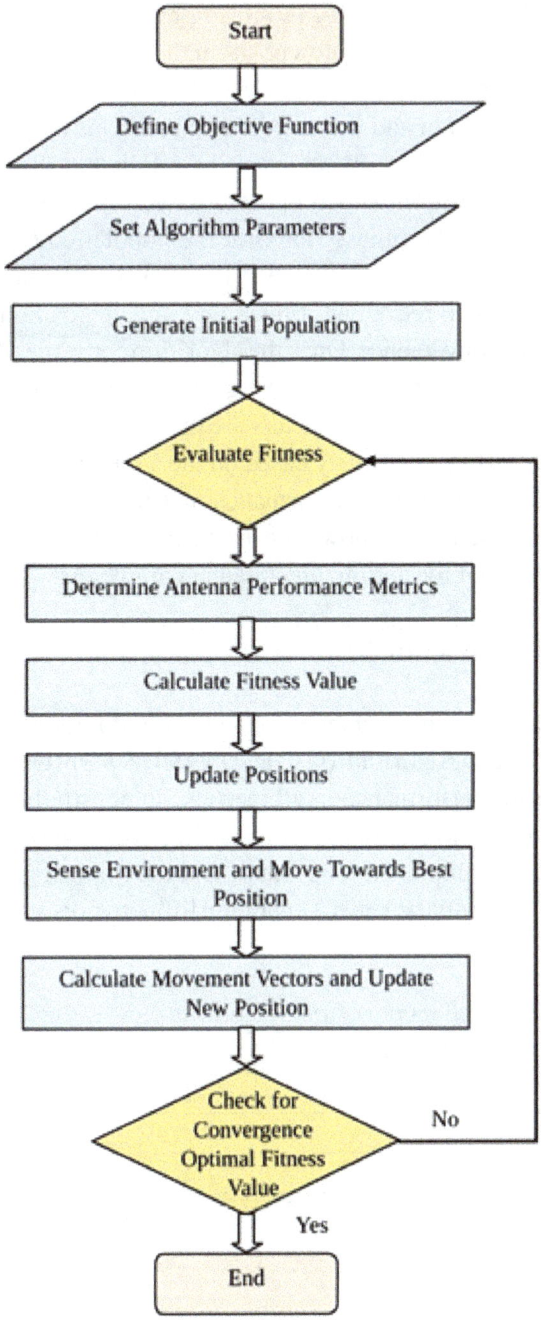

where t is the current iteration, T_{max} is the maximum iterations, and r_1^{init}, r_2^{init} are initial coefficients. This CSA algorithm incorporates a "sense-and-move" mechanism inspired by chameleon behavior. This involves sensing the environment by evaluating the search space to identify promising regions and moving toward the best position by adjusting the candidate's position to approach the global best solution. Now the movement vectors are calculated based on the difference between a candidate's current position and the best-known solutions (G_{best} and P_{best}). These vectors guide the candidate's movement, ensuring efficient navigation through the search space. The new position is then updated based on these vectors. The CSA algorithm checks whether the fitness values of the population have stabilized and the maximum number of iterations has been reached. If the criteria are not met, the algorithm loops back to reevaluate fitness and update positions. This iterative process continues until convergence. Once the algorithm converges, the optimization process ends. The best-performing solution, representing the optimal slot radius and positions, is outputted. These parameters are then used to fabricate the antenna, which is expected to perform optimally based on the simulation results and ensures that the antenna meets the desired performance metrics, making it suitable for its intended application. Figure 10 shows the flowchart illustrating the Chameleon Swarm Algorithm applied for optimizing antenna parameters.

5.4.4 Performance Evaluation

Table 2 compares the results achieved using CSA with traditional optimization algorithms such as Genetic Algorithm (GA) and Particle Swarm Optimization (PSO). CSA achieved superior performance in all metrics, demonstrating its capability to handle high-dimensional optimization problems effectively. The algorithm achieved the desired bandwidth, minimized return loss, and maximized gain. These results validate the antenna's potential for next-generation IoT networks and 6G applications.

Table 2. Comparative analysis of various algorithms

Algorithm	Bandwidth (GHz)	S11 (dB)	Gain (dB)
CSA	2.0	-24.8	7
GA	1.8	-22.3	6.2
PSO	1.9	-24.7	5.6

6. FINDINGS AND DISCUSSIONS

The performance of the proposed circularly polarized antenna has been evaluated based on its return loss, gain, radiation pattern, and axial ratio results.

6.1. Return Loss (S11)

The return loss (S11) is an indicator of the transmission power. It indicates impedance mismatch between the antenna and its feeding network. For the proposed design, the return loss was measured across the operational frequency range, with a minimum value of −25 dB at the center frequency as shown in Figure 11. This indicates excellent impedance matching, with less than 0.3% of the power being reflected back to the source. Such a low S11 value underscores the capability of the antenna to minimize reflection losses. It tells how efficiently transmit signals across the operating band.

Figure 11. Simulated S11 results over Frequency

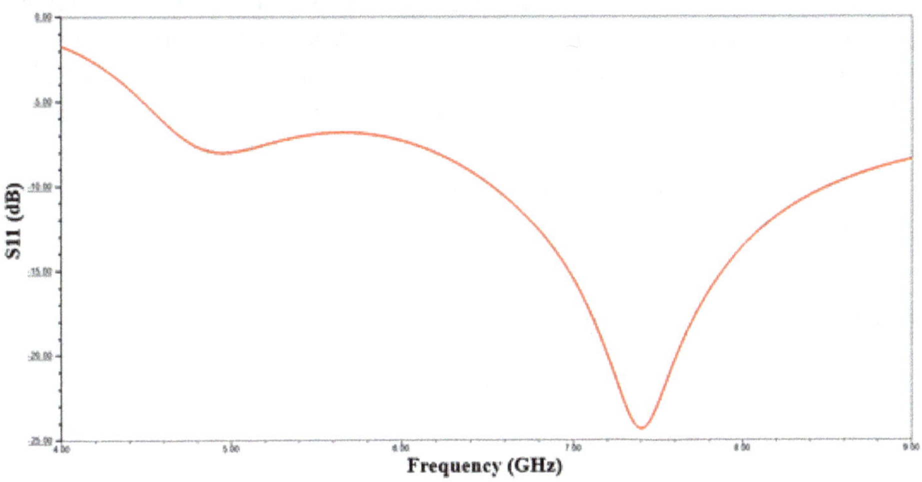

The frequency range for which S11is below the commonly accepted threshold of −10dB spans 2 GHz, from 6.4 GHz to 8.4 GHz, highlighting a wide operational bandwidth. This wide bandwidth makes sure of reliable performance. It is required in applications requiring high data rates and robust signal propagation. The obtained bandwidth of 2 GHz represents a significant achievement. It is accounting for 27% fractional bandwidth relative to the center frequency of 7.4 GHz. This result is indicative of the antenna's ability to support multi-channel communication systems and adapt to dynamic frequency allocations, which are essential features for modern wireless standards. The wide bandwidth facilitates seamless operation across various applications, including satellite communications, IoT networks, and radar systems. The wide bandwidth and low return loss ensures consistent and reliable performance. This is observed under varying environmental conditions. Also, Roger RO3210 substrate is used. It is with its low dielectric loss tangent. It contributed to minimizing energy dissipation and preserving signal integrity. Compared to existing designs operating within a similar frequency range, the proposed antenna demonstrates superior performance in terms of both bandwidth and return loss. Many conventional designs achieve bandwidths ranging from 1 GHz to 1.5 GHz with return losses of approximately −15dB. The proposed design has wider bandwidth and lower S11.

6.2. Radiation Pattern

The radiation characteristics of the proposed circularly polarized antenna were analysed. It is done using 2D and 3D far-field radiation patterns at the design frequency of 7.4 GHz. The 2D radiation patterns, evaluated in the E-plane and H-plane, reveal excellent directional behavior. The dipole like pattern exhibits smooth and symmetrical energy distribution in the E-plane. It makes sure a uniform coverage across the elevation axis. Similarly, the H-plane demonstrates consistent omnidirectional radiation around the azimuthal plane, providing stable signal coverage in all directions. Both patterns are free from side lobes. It is highlighting the antenna's efficiency. The need in focusing energy within the desired direction. while avoiding energy leakage or interference.

Figure 12. Radiation Pattern a) 3D plot b) 2D plot with E-Plane and H-Plane

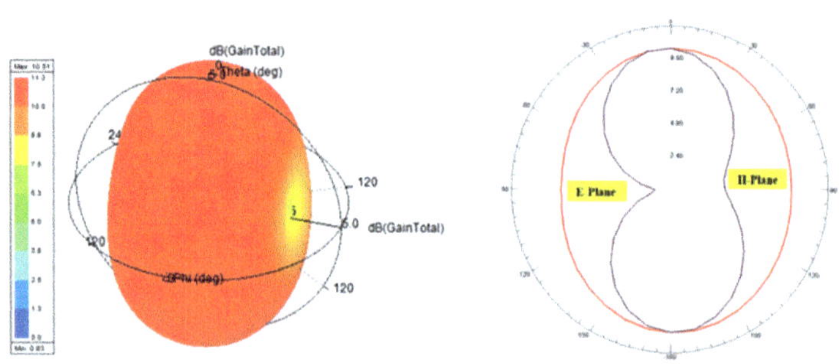

(a) (b)

The 3D far-field radiation pattern at 7.4 GHz corroborates the omnidirectional performance, showing a spherical energy distribution with no significant nulls or distortions. This consistency in 3D space enhances the antenna's suitability. It is required for applications requiring comprehensive spatial coverage. This might be suitable for application such as IoT networks and wireless communication systems. Across the operational band, the radiation patterns remain stable, with omnidirectional behavior and the absence of side lobes observed at 6.4 GHz, 7.4 GHz, and 8.4 GHz. This stability ensures the antenna's reliability across the entire frequency range, meeting the demands of dynamic and multi-channel communication systems. Additionally, the compact structural design and optimized placement of radiating elements eliminate side lobes and achieve symmetrical patterns. These design choices ensure high performance while maintaining a compact and efficient form factor. The proposed design exhibits superior radiation performance. Traditional designs often show side lobes or uneven energy distribution. This can degrade performance in practical scenarios. The clean and consistent patterns of the proposed antenna set it apart. It establishes its competitiveness in advanced communication systems. Figure 12 gives the visual representation of the radiation characteristics through 2D and 3D plots.

6.3. Gain

The gain of an antenna is a critical parameter. It reflects its ability to efficiently direct energy in a specific direction. High gain is desirable in communication systems. It makes sure a stronger signals and improved coverage. Here, the gain characteristics of the designed antenna were evaluated for frequency range of 6.4 GHz to 8.4 GHz. It has a particular focus on the desired frequency of 7.4 GHz. The results show that antenna achieves a peak gain of 7 dB at 7.4 GHz as shown in Figure 13. It is demonstrating its optimized performance for this specific operating point. Additionally, throughout the entire operating frequency range, the gain remains consistently above 6 dB. This consistent performance highlights the efficiency and robustness of the antenna design. It makes sure reliable signal transmission and reception across the intended bandwidth.

Figure 13. Obtained Gain over Frequency

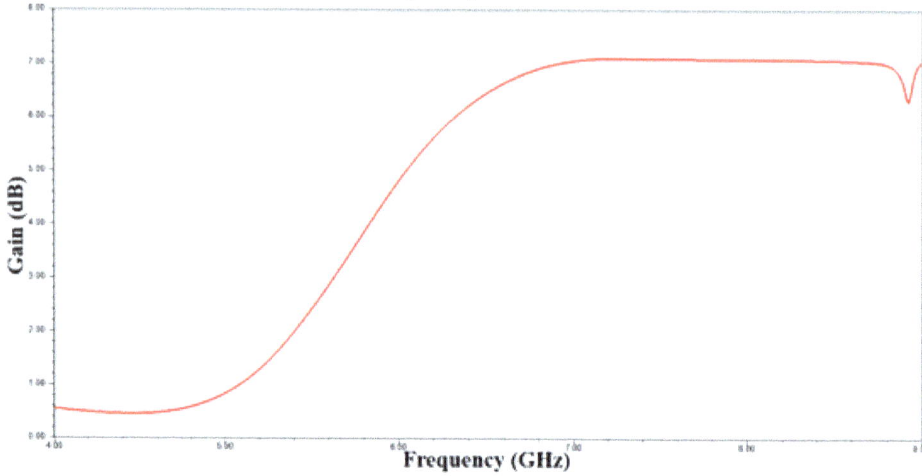

The peak gain of 7 dB at 7.4 GHz is particularly significant as it aligns with the desired frequency of operation, where the antenna is expected to deliver optimal performance. This indicates that the design has been effectively tuned. It is tuned to maximize radiation efficiency and directional performance at this frequency. Achieving a high peak gain at the target frequency is essential for applications requiring focused and efficient energy transmission, such as in satellite communication, radar systems, and emerging wireless technologies like 5G and 6G. The peak gain value is ability of the antenna to concentrate power in the intended direction. It makes sure minimal energy loss and robust communication links.

The observation that the gain remains above 6 dB at 6.4 GHz to 8.4 GHz. It further underscores the versatility and suitability for broadband applications. A gain above 6 dB ensures that the antenna delivers reliable performance across the operating bandwidth. It is accommodating the varying frequency requirements of modern communication systems. This is especially important in systems where frequency agility is needed to adapt to changing conditions, such as in dynamic environments or multi-channel communication setups. Maintains a high gain over a broad frequency range is a significant challenge in antenna design. It is particularly for compact structures. The results of this study demonstrate that the proposed antenna effectively addresses this challenge, offering stable and efficient performance throughout the desired frequency band. The gain characteristics indicate that the antenna design successfully minimizes power losses while maximizing the effective radiated power. which is critical for ensuring consistent signal strength and quality.

The high gain values observed in this study are also indicative of the antenna's efficient design in terms of impedance matching and radiation efficiency. Efficient impedance matching ensures minimal signal reflection. It is allowing maximum power transfer from the source to the antenna. Meanwhile, high radiation efficiency ensures that most of the input power is effectively radiated, with minimal losses to heat or other forms of energy dissipation. The combined effect of these factors contributes to the consistently high gain performance for the operating frequency range. Compared to similar designs, the proposed antenna exhibits competitive gain characteristics, making it suitable for a wide range of applications. Its ability to deliver high gain over a broad frequency range positions for use in broadband communication systems. Both performance and flexibility are critical. The peak gain of 7 dB at 7.4 GHz coupled with the gain above 6 dB across the operating bandwidth. It demonstrates the effectiveness of the design in meeting the stringent requirements of modern wireless communication technologies.

6.4. Axial Ratio

The performance of the designed antenna with respect to its axial ratio, an essential parameter for determining circular polarization. The axial ratio measures the relationship between the major and minor axes of the polarization ellipse, and is a key indicator of the circular polarization quality. For a true circularly polarized antenna, the ideal axial ratio should approach 1 (0 dB), but in practical terms, a value below 3 dB is generally considered acceptable for efficient circular polarization.

The axial ratio of our antenna was tested over its specified operating frequency range of 6.4 GHz to 8.4 GHz. The results show that the axial ratio remains under 3 dB. The range between 6.6 GHz and 6.9 GHz and again between 7.3 GHz and 8.6 GHz as shown in Figure 14. Both of which are fully within the operating frequency range. This confirms that the antenna is successfully maintaining good circular polarization within the target frequency band. The axial ratio stays below 3 dB from 6.4 GHz to 8.4 GHz. It is proving that the antenna performs efficiently with circular polarization throughout the design range.

Circular polarization is crucial for many applications, such as satellite communications, radar, and advanced wireless systems like 5G and 6G, where stable signal reception is needed regardless of the orientation of the transmitting or receiving antennas. The low axial ratio observed in our measurements. It suggests that the antenna is well-suited for these applications. It makes sure reliable performance. The specific frequency ranges where the axial ratio is less than 3 dB (6.6 GHz to 6.9 GHz and 7.3 GHz to 8.6 GHz) demonstrate that the antenna can radiate circularly polarized signals with minimal polarization mismatch, even in real-world scenarios where the orientations of antennas may vary.

Figure 14. Axial Ratio over Frequency

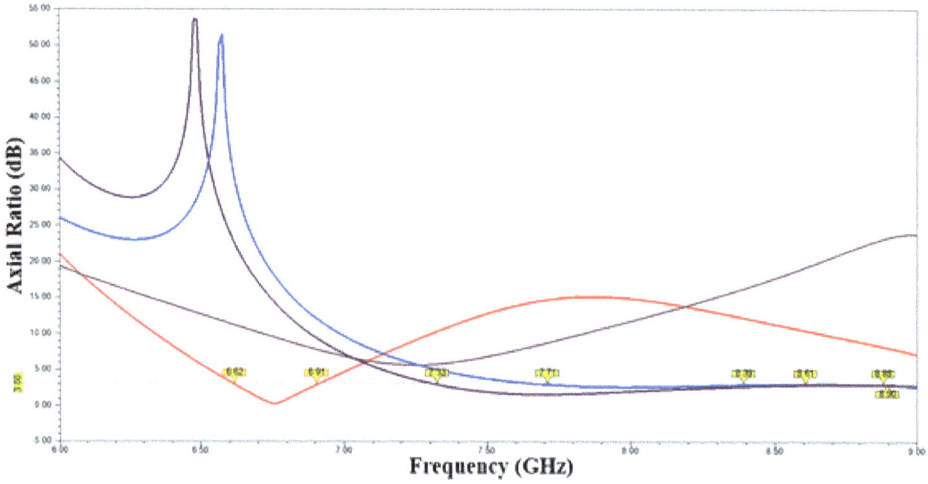

The ability to maintain a low axial ratio across a broad frequency range is especially significant for communication systems. The fact that the antenna achieves a low axial ratio over a substantial portion of the operating bandwidth. It highlights its potential for use in systems requiring consistent polarization and frequency performance. This ensures that the antenna continues to perform reliably as the operating frequency fluctuates within the design range. This minimizing the risk of signal loss or degradation due to polarization mismatches.

The axial ratio behavior indicates that the antenna design successfully supports circular polarization across the desired bandwidth. It is aligning well with theoretical predictions for such designs. The overlap between the frequency ranges where the axial ratio remains below 3 dB and the operating frequency range of 6.4 GHz to 8.4 GHz further confirms the antenna's ability to maintain efficient circular polarization. This wide bandwidth for CP adds to the reliability. It is making the antenna a strong candidate for modern communication systems, where stable frequency and polarization performance are essential.

When compared to other circularly polarized antenna designs, our antenna stands out for its ability to maintain a low axial ratio over a wide frequency range. Many antennas design struggle to achieve circular polarization across broad bandwidths. This design has effectively overcome this challenge. By positioning it as a viable option for use in real-world applications. The consistently low axial ratio across operating range. It means the antenna can deliver high-quality signals with minimal distortion, improved performance and reliability in practical applications.

Table 3. Comparison of State of art

Ref	Antenna Dimensions (mm^2)	Resonating frequency (GHz)	Bandwidth (MHz)	Gain (dB)	Polarization	Verification of Metamaterial property
22	65 x 65	4.4	580	2	circular	Verified
23	40 x 10	2.4	300	1.35	linear	-
24	50 x 50	0.87	200	- 1.49	linear	-
25	25 x 21	5.8	167.6	2.54	linear	-
26	92 x 121.5	0.919	210	4.2	linear	-
Proposed Antenna	**16 x 15**	**7.4**	**2000**	**7**	**circular**	**Verified**

6.5 Future Directions

As 5G and 6G networks evolve, the demands for antennas will continue to grow, especially in terms of miniaturization, multi-band support, and higher efficiency. Machine learning and artificial intelligence (AI) with optimization algorithms like CSA will be integrated. For antennas, metamaterials will remain crucial for reducing size and improving performance. Future research efforts are most likely to be focused on novel metamaterial forms and their combinations, including hybrid metamaterials, which combine several functions into one structure. Antennas in dynamic and mobile environments will also require the creation of meta-surfaces with dynamic property tuning capabilities.

Future of designing IOT antennas is the development of ultra-low-power, multi-functional, and compact antennas that can support a wide range of communication protocols ensuring long-range connectivity

7. CONCLUSION

In this chapter, the circularly polarized 6G antenna offers significant advancements for Internet of Things (IoT) applications within the frequency range of 6.4 GHz to 8.4 GHz, achieving a wide operational bandwidth of 2 GHz. With the metamaterial techniques, the antenna achieves remarkable miniaturization. Also, maintaining superior performance metrics such as high gain and compact size. This addresses important challenges associated with next-generation IoT devices. Challenges include space constraints and enhanced efficiency. The implementation of the Chameleon Swarm Algorithm (CSA) for antenna parameter optimization further enhances the design. As a bio-inspired optimization method, CSA ensures precision in achieving the desired bandwidth. An exceptional return loss (S11 < -25 dB), and a well-defined radiation pattern is achieved. The adaptability of the algorithm and ability navigates complex design spaces. This enables the realization of a high-performance antenna for the specific requirements of 6G IoT networks. The capability of the antenna to support circular polarization adds robustness to its communication performance. It is done by eliminating polarization mismatches and multipath fading. This feature coupled with its wide bandwidth and high efficiency. This positions the antenna as a reliable solution for various IoT applications across sectors. These are such as healthcare, satellite IoT, and UAV. These applications require connectivity and compact designs. Thus, making the proposed antenna a crucial enabler for future 6G networks. Simulation results validate the feasibility of the design. It highlights its potential to meet the stringent demands of next-generation communication systems. The proposed antenna addresses current technological challenges. It also sets a benchmark for future antenna designs in 6G and IoT applications. It paves the way for fully intelligent and autonomous systems.

REFERENCES

Akyildiz, I. F., Kak, A., & Nie, S. (2020). 6G and Beyond: The Future of Wireless Communications Systems. *IEEE Access : Practical Innovations, Open Solutions, 8*, 133995–134030. DOI: 10.1109/ACCESS.2020.3010896

Al-Sehemi, A., Al-Ghamdi, A., Dishovsky, N., Atanasova, G., & Atanasov, N. (2020). A flexible broadband antenna for IoT applications. *International Journal of Microwave and Wireless Technologies, 12*(6), 531–540. DOI: 10.1017/S1759078720000161

Anchidin, L., Lavric, A., Mutescu, P., Petrariu, A. I., & Popa, V. (2023). The design and development of a microstrip antenna for internet of things applications. *Sensors (Basel), 23*(3), 1062. DOI: 10.3390/s23031062 PMID: 36772099

Azari, A., Skrivervik, A., & Aliakbarian, H. (2024). A mmWave Low complexity and Low-Cost Super Wideband Dual-Polarized Aperture Coupled Antenna for 5G Applications. *IEEE Access : Practical Innovations, Open Solutions, 12*, 85601–85607. DOI: 10.1109/ACCESS.2024.3415167

Bose, M., & Karuppiah, V. (2024). Metamaterial inspired superstrate loaded miniaturized quad port MIMO antenna for 5G C-Band applications. *Optics Communications, 131123*. Advance online publication.

Cai, P., Zhang, Y., Tang, Z., & Gao, S. (2022). Chameleon Swarm algorithm enhanced by differential evolution. *2022 Joint 12th International Conference on Soft Computing and Intelligent Systems and 23rd International Symposium on Advanced Intelligent Systems (SCIS&Amp;ISIS)*, 1–6.

Elijah, A. A., & Mokayef, M. (2020). Miniature microstrip antenna for IoT application. *Materials Today: Proceedings, 29*, 43–47. DOI: 10.1016/j.matpr.2020.05.678

Esmail, B. F., & Koziel, S. (2022). Design and optimization of Metamaterial-Based Dual-Band 28/38 GHz 5G MIMO antenna with modified ground for isolation and bandwidth improvement. *IEEE Antennas and Wireless Propagation Letters, 22*(5), 1069–1073. DOI: 10.1109/LAWP.2022.3232622

Hui, W., Guo, Y., Xie, K., & Zhao, X. (2023). Investigation of circularly polarized microstrip antenna with Tri-Band polarization conversion metamaterials. *IEEE Antennas and Wireless Propagation Letters, 22*(12), 2831–2835. DOI: 10.1109/LAWP.2023.3300298

Islam, M. S., Islam, M. T., Ullah, M. A., Beng, G. K., Amin, N., & Misran, N. (2019). A modified Meander Line microstrip patch antenna with enhanced bandwidth for 2.4 GHz ISM-Band internet of things (IoT) applications. *IEEE Access : Practical Innovations, Open Solutions*, *7*, 127850–127861. DOI: 10.1109/ACCESS.2019.2940049

Jacob, S., Bose, M., & Karuppiah, V. (2024). Artificial Magnetic Conductor-Backed Microstrip-Fed Dual-Band antenna for 5G wearable sensor nodes and C Band application in WBAN network. In *Advances in information security, privacy, and ethics book series* (pp. 221–241).

Lin, Q., Alkaraki, S., Wong, H., & Kelly, J. R. (2022). A wideband circularly polarized antenna based on anisotropic metamaterial. *IEEE Transactions on Antennas and Propagation*, *71*(2), 1254–1262. DOI: 10.1109/TAP.2022.3225157

Liu, L., Liu, T., Zheng, Y., Chernogor, L. F., Jin, Z., & Sun, Z. (2022). Archimedean spiral antenna based on metamaterial structure with wideband circular polarization. *AEÜ. International Journal of Electronics and Communications*, *152*, 154257. DOI: 10.1016/j.aeue.2022.154257

Meenambal, B., Jacob, S., & Vasudevan, K. (2024). Metamaterial-Inspired concatenated dual ring antenna design for 5G NR C-Band applications. In *Advances in information security, privacy, and ethics book series* (pp. 200–220).

Meenambal, B., Vasudevan, K., Maheshwari, G. U., & Sherene, J. (2024). Offset-Fed Defected Ground Structure (DGS)-Based Metamaterial-Inspired Asymmetric SRR for 5G terminal applications. In *Lecture notes in networks and systems* (pp. 163–178).

Mwang'amba, R., Mei, P., Akinsolu, M. O., Liu, B., & Zhang, S. (2024). Gain bandwidth enhancement and sidelobe level stabilization of MM-Wave lens antennas using AI-driven optimization. *IEEE Antennas and Wireless Propagation Letters*, *1–5*. Advance online publication.

Nguyen, D. C., Ding, M., Pathirana, P. N., Seneviratne, A., Li, J., Niyato, D., Dobre, O., & Poor, H. V. (2021). 6G Internet of Things: A Comprehensive survey. *IEEE Internet of Things Journal*, *9*(1), 359–383. DOI: 10.1109/JIOT.2021.3103320

Pallavi, M., Kumar, P., Ali, T., Shenoy, S. B., Mane, P. R., & Prabhu, G. (2024). Design and experimental study of a coupled U and L-Shape negative index metamaterial for aircraft navigation applications. *IEEE Access : Practical Innovations, Open Solutions*, *12*, 16127–16150. DOI: 10.1109/ACCESS.2024.3358898

Rasilainen, K., Phan, T. D., Berg, M., Pärssinen, A., & Soh, P. J. (2023). Hardware aspects of Sub-THZ antennas and reconfigurable intelligent surfaces for 6G communications. *IEEE Journal on Selected Areas in Communications*, *41*(8), 2530–2546. DOI: 10.1109/JSAC.2023.3288250

Swetha, S., Vasudevan, K., Meenambal, B., & Sherene, J. (2024, April). Compact Wideband CPW Fed Metamaterial Antenna for V2X Communication. In *2024 International Conference on Recent Advances in Electrical, Electronics, Ubiquitous Communication, and Computational Intelligence (RAEEUCCI)* (pp. 1-6). IEEE.

Sherene, J., Vasudevan, K., Maheshwari, G. U., & Meenambal, B. (2024). Design and analysis of AMC-Based anti-symmetric Dual L-Shaped antenna for WBAN Application. In *Lecture notes in networks and systems* (pp. 179–194).

Sim, C., Lo, J., & Chen, Z. N. (2022). Design of a broadband Millimeter-Wave array antenna for 5G applications. *IEEE Antennas and Wireless Propagation Letters*, *22*(5), 1030–1034. DOI: 10.1109/LAWP.2022.3231358

Uluslu, A. (2022). Chameleon Swarm algorithm assisted optimization of U-Slot patch antenna for Quad-Band applications. *IEEE Access : Practical Innovations, Open Solutions*, *10*, 74152–74163. DOI: 10.1109/ACCESS.2022.3190378

Wong, K., Huang, Y., & Li, W. (2024). Compact 8-Port 2 × 2 Array Based on Dual-Polarized Patch Antennas with Modified Cavity Field Distribution for Enhanced Port Isolation for 5G IoT Device MIMO Antennas. *IEEE Access : Practical Innovations, Open Solutions*, *12*, 79311–79326. DOI: 10.1109/ACCESS.2024.3409355

Zhang, H., Chang, L., & Zhang, A. (2024). Extremely Low-Profile wideband patch antennas using electric- and Magnetic-Coupling for 5G mobile IoT devices application. *IEEE Antennas and Wireless Propagation Letters*, *23*(9), 2698–2702. DOI: 10.1109/LAWP.2024.3404779

Zhao, H., Gu, P. F., Fan, Z. H., He, Z., Cao, J., Tao, X., Wang, J. X., & Ding, D. Z. (2023). Ultrawideband dual circularly polarized waveguide array antenna for K- and Ka-Bands satellite communications. *IEEE Antennas and Wireless Propagation Letters*, *22*(8), 1808–1812. DOI: 10.1109/LAWP.2023.3266003

Chapter 5
ML-Optimized Biotelemetry Antennas Redefining Next-Gen Healthcare Connectivity

Sherene Jacob
http://orcid.org/0009-0003-3655-403X
Thiagarajar College of Engineering, India

Meenambal Bose
http://orcid.org/0009-0009-6664-710X
Thiagarajar College of Engineering, India

Vasudevan Karuppiah
Thiagarajar College of Engineering, India

ABSTRACT

This chapter discuss the regression-based machine learning approach to achieve the resonance by computing the line length, impedance matching width and substrate thickness of the Implantable Microstrip Antennas in the frequency band of 2.4–2.5 GHz. In the design, a meandered antenna with microstrip feed is loaded. The resonant frequencies of 1,000 IMAs were simulated using HFSS by varying line length, width, substrate thickness. These variations adhered to the specifications required for implantable medical device (IMD) for biotelemetry. Various regression-based ML algorithms, like Artificial Neural Networks were studies. The study found that the Gaussian Process Regression (GPR) model has the most accurate predictions parameters at resonance. The proposed GPR model was validated by simulating and characterizing an implantable microstrip antenna. The antenna has achieved a

DOI: 10.4018/979-8-3373-0852-4.ch005

size of the antenna achieved is 3.5x3.5 mm2. The designed antenna's performance, including its bandwidth characteristics, closely matched both the simulation results and the GPR predictions.

1. INTRODUCTION

The need to continuously monitor patients is satisfied with the advent of Implantable Medical Devices (IMD) (Abbas et al., 2024). They are designed to monitor various health conditions. At the center of this technology are implantable antennas that enable wireless communication between these devices and external systems. The design of such antenna is challenging due to the constraints. This includes the EM wave absorption, miniaturization, and safety compliance (Iqbal et al., 2024; Khattak et al., 2024).

Specific Absorption Rate (SAR) is a critical parameter in implantable antenna design, representing the rate at which human tissue absorbs EM energy (Sherene et al., 2024). The SAR must be limited to the safe values. It prevents tissue heating and even burn to the patient. To achieve this, the structure, material and the frequency of the antenna are optimized. In this context, innovative techniques and materials have been explored to enhance antenna performance while adhering to strict SAR limits.

Metamaterials has gained attention due to its ability to manipulate EM properties. Traditional materials do not have the ability to do so. Metamaterials consist of engineered structures of sub-wavelength sizes. They exhibit negative permittivity and permeability (Meenambal et al., 2024). This enhances the wave propagation, miniaturization and even SAR reduction. Metamaterials with implantable antennas also improves the gain of the antenna (Bose & Karuppiah, 2024). The energy is focused on the desired direction helps mitigate the signal loss. This contributes to increased efficiency.

Another emerging technology is the use of Artificial Magnetic Conductors (AMCs), which act as EM shields and reflective surfaces. AMC play a valuable role in SAR reduction. They direct the EM waves away from the body. This also improves the antenna efficiency. With AMC along with implantable antenna designs, researchers can improve the antenna performance like gain, bandwidth etc., without compromising the safety (Jacob et al., 2024).

Advanced structural techniques such as meandered lining or fractals offer solution to miniaturization of the antenna. This helps to meet the size constraint of the antenna for IMDs. Meandered line is folding the antenna to compact form. The overall size of the antenna reduces by reducing the electrical length (Fei et al., 2015). This is beneficial for miniaturization. But, the intricate designs of the meander line introduce complexities. It becomes tedious to analyze and optimize their performance. Also, more complex on interaction with tissues.

Machine learning algorithms are introduced to address the complexity of the implantable antenna design (Sarker et al., 2023). It provides a data driven approach to predict the performance of the antenna. It also predicts various input parameters, material properties, geometry and frequency of operation. Gaussian Process Regression (GPR) is one of the Machine Language techniques available (Cao et al., 2015). It is promising in modelling the relationship between design parameters and performance metrics. It is suitable for optimization of implantable antenna.

This chapter discuss the integration of advanced materials structural techniques and how machine language is used in the development of implantable antennas. It focuses on the usage of meandered line structure to address the problem of miniaturization, SAR Reduction, and performance enhancement. Also, it demonstrates how GPR is applied in this work to optimize the meander line antenna for biotelemetry application. Using the predictive power of GPR, the antenna design process is efficient. It enables faster convergence to optimal solutions. It reduces the computational cost. The combination of these approaches creates a significant leap in the field of biotelemetry involving implantable antennas. It paves way for safer, smaller and efficient devices. This chapter aims in detailed understanding of how these innovations work together to meet the demand of the next generation healthcare connectivity. It ultimately enhances the capabilities of IMDs and improve the patient outcome.

1.1 Literature Survey

In recent years, implantable meandered antenna plays a key role in bio-telemetry. The application of machine learning algorithms in design and optimization of is more common. This is driven by their potential applications in fields such as healthcare (Frontoni et al., 2020), biomedical telemetry (Liu et al., 2016), and wireless communication. Meandered antenna are space efficient designs. They are designed to be compact. Adding folded structure to the geometry achieve size reduction. By incorporating folded and serpentine structures into their geometry, meandered antennas achieve significant reductions in size without compromising performance. Several studies have explored the trade-offs associated with these designs, including bandwidth, gain, and efficiency. This is ideal for tough environment like inside of human body.

Although size reduction often leads to increased quality factor and narrower bandwidth, researchers have employed innovative techniques such as substrate selection, reactive loading, and advanced feeding mechanisms to mitigate these limitations. The development of implantable antennas, for operation within human or animal tissue.

This represents a challenging and essential domain of research. Implantable antennas are central to wireless biotelemetry systems, which enable the seamless transfer of physiological data from within the body to external monitoring systems. They find their applications in glucose monitoring (Malik et al., 2022), cardiac pacemakers (Anwar et al., 2022), and neural stimulation (Son & Cha, 2020) to more advanced systems for drug delivery and cancer detection. Meandered structures are particularly suited to resonate at frequencies, making them effective for applications in industrial, scientific, and medical (ISM) bands. Both wearable and implantable devices predominantly operate. A primary challenge in designing implantable antennas lies in achieving biocompatibility, stable performance in the lossy tissue environment, and adherence to strict safety regulations related to specific absorption rate (SAR). Human tissues are lossy dielectric mediums, and the absorption of electromagnetic waves affects both the antenna's efficiency and its operating frequency. Addressing these challenges, innovative geometries, materials and simulation techniques. For example, coating antennas with biocompatible materials like Parylene-C or PDMS not only ensures safe implantation but also helps in preserving antenna performance by providing a stable dielectric interface between the antenna and surrounding tissues. Furthermore, geometrical adaptations, such as meandered and spiral designs, have been employed to achieve size reduction and improve bandwidth while maintaining resonance frequencies suitable for operation in ISM bands such as 2.4 GHz or Medical Implant Communication Service (MICS) bands (402–405 MHz).

Meandered structures are relevant for space efficiency. These designs enable the development of miniaturized systems that are minimally invasive and suitable for long-term deployment. For instance, compact meandered antennas have been successfully used in intracranial pressure monitoring and glucose sensing, where minimizing the antenna's footprint is critical. Each layer with unique dielectric properties are considered. Multi-layered nature of human tissues, including skin, fat, muscle, and bone are considered. Testing in phantom models validates these designs. This ensures their performance in real-world scenarios. However, ensuring reliable performance over time remains a challenge due to tissue inflammation, encapsulation, and potential changes in tissue properties.

To overcome challenges in implantable antenna various methods are adopted. Reconfigurable antenna, dynamic impedance matching are employed. Machine Learning has emerged as a game-changer in design, analyze, and optimizing meandered and implantable antennas. Traditional approaches, which rely on trial-and-error simulations and repeated prototyping, can be slow and resource-intensive. ML offers a faster, smarter way by building predictive models. It is based on limited datasets. Techniques like regression models, neural networks, support vector machines, and genetic algorithms are widely used. These methods are helpful in managing the complex factors influencing antenna performance.

For IMAs, ML has been particularly useful in fine-tuning parameters. The predictive models like regression and neural networks (Peng et al., 2008), trained on data from EM simulations. This allows to estimate performance metrics quickly and accurately. This reduces the need for endless trial-and-error simulation cycles. Optimization techniques like genetic algorithms (Heo et al., 2025; Kerkhoff & Ling, 2007) and particle swarm optimization help tune antenna dimensions. Achieves a compact design meeting performance goal. ML helps analyze the variations in tissue properties affect antenna behavior. It creates a model that predict these interactions and design antennas that work consistently across different patients and adapt to natural changes in tissues over time.

The inverse design of antennas is done using Machine learning. Usually, design start with a specific shape and test how well it performs. Inverse design flips the process. It uses machine learning to suggest an ideal geometry based on the desired outcome. Advanced techniques, such as generative models like VAEs and GANs are used. It makes it possible to explore creative and unconventional designs. Novel shapes are achieved

Combining machine learning with physics-based models has proven especially effective in tackling the complexities of implantable antennas (Wang et al., 2024). Machine learning excels at finding patterns and relationships in large datasets. The physics-based models ensure that designs respect the fundamental rules of electromagnetics. Together, these hybrid approaches deliver more accurate predictions. It balances computational speed with physical realism. It allows for better generalization to new scenarios. For example, ML models trained with simulation data account for how antennas interact with realistic tissues. This integration makes it easier to prototype and evaluate antenna designs across a wide range of conditions, ultimately streamlining the entire development process (Mohammed et al., 2019).

Challenges remain in applying ML to meandered IMAs. The limitation is the availability of high-quality datasets for training ML models. Generating labelled data requires extensive simulations. This can be resource intensive (Pucci et al., 2025). Furthermore, ensuring that ML models generalize well to new design scenarios or operating conditions is a critical challenge. Overfitting datasets or neglecting important physical constraints lead to models that perform poorly in real-world applications. These issues are addressed by developing data efficient algorithms. This transfers learning and active learning, which reduce the amount of data required for training.

Ethical considerations play a role in the ML in biomedical applications. The domain data into the design of ML models with constraints ensures that the models produce reliable results. For instance, the use of ML-driven antenna designs in implantable medical devices concerns about safety, reliability, and accountability. Ensuring that these systems meet stringent regulatory standards and perform reliably over the long term. Collaborative efforts between engineers, medical professionals, and regulatory bodies are necessary to address these challenges. Promote the safe deployment of ML enabled antennas in healthcare. The meandered IMA design, and machine learning algorithms are rapidly evolving field with significant implications for biomedical communication systems. ML has emerged as a transformative tool in this domain (Zhang et al., 2023). It is enabling predictive modelling, optimization, and the discovery of novel antenna designs. Challenges remain in terms of data availability, model generalization, and regulatory compliance. The integration of machine learning with traditional design methodologies holds immense potential for accelerating innovation and enhancing the performance of meandered and implantable antennas. Future research will likely focus on developing hybrid approaches that combine the strengths of physics-based models and machine learning, as well as addressing the ethical and practical challenges associated with deploying these technologies in real-world scenarios. By fostering interdisciplinary collaboration and leveraging advances in artificial intelligence, researchers can drive the next generation of antenna systems that are smarter, smaller, and more adaptable to the needs of modern biomedical applications.

2. BIO-TELEMETRY ANTENNAS

Biotelemetry antennas are the backbone of wireless communication systems in biomedical applications. It facilitates the transfer of data between medical devices and external monitoring equipment. It is designed to operate within the human body or near them. These antennas enable real-time, wireless communication without physical connections. This makes them essential for modern healthcare solutions, particularly in continuous health monitoring and medical diagnostics. They are integral to implantable and wearable devices, which are transforming healthcare by offering personalized, non-invasive, and real-time health monitoring capabilities. The figure 1 represents the various applications of the biotelemetry antennas.

Figure 1. Applications of Biotelemetry Antenna

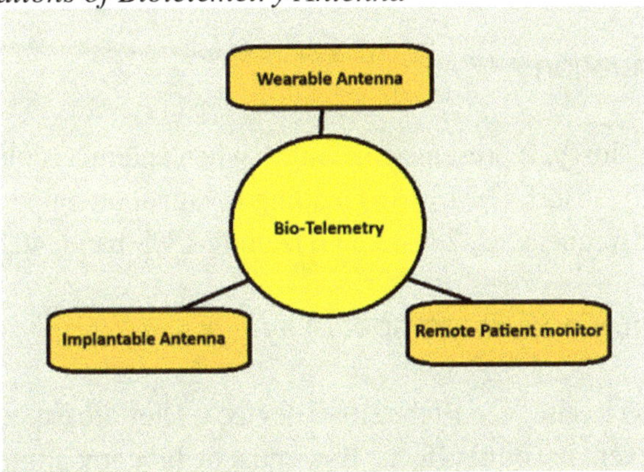

2.1 Applications of Biotelemetry Antennas

1. Implantable Medical Devices (IMD)

Biotelemetry devices within human body are IMD. These are devices such as pacemakers, and glucose monitors. They sense the bio-signals and transmit it real-time. these IMD has antenna to wirelessly transmit the cardiac data to external devices for remote monitoring efficiently.

2. Wearable Devices

Biotelemetry is also the main concern of wearable devices. This includes smart-watches, medical sensors etc, they monitor real-time physiological signals. They are famous as preventive healthcare The antenna must be flexible and efficiently perform near human body.

3. Remote Patient Monitoring

Remote monitoring of patients revolutionizes biotelemetry. Individuals with chronic diseases are helped by it. They are constantly monitored by the health care providers. hence frequency hospital visits can be minimized. This also helps in early diagnosis of diseases.

2.2 Design Challenges

A. Signal Attenuation

Human body is lossy. It presents challenges when antenna is placed near human body. Human body absorbs EM waves leading to signal attenuation. Antennas are tuned for specific frequencies, 2.4 - 2.5 GHz and MICS band, 402-405 MHz.

B. Miniaturization and Biocompatibility

Miniaturization is a issue to fit the small devices. They are expected to have good efficiency. To prevent harmful effects, they must be biocompatible. Materials such as high-permittivity substrates are often used to meet these requirements.

C. Power Efficiency and Battery Life

Implantable devices rely on limited battery power, making energy efficiency a critical aspect of antenna design. Frequent surgical replacements for recharging batteries are impractical, so efficient antenna operation and advancements in energy harvesting, such as wireless power transfer, are crucial.

D. Specific Absorption Rate (SAR) Compliance

Since biotelemetry antennas operate near or within the human body, they must comply with strict safety standards for electromagnetic exposure. Devices are supposed to be within the regulated values.

2.3 Advancements in Biotelemetry Antennas

Biotelemetry is booming. Limitations are overcome and capabilities are expanded. Advanced materials, miniaturization techniques enabling more efficient and reliable designs. AI-based optimization is further enhancing the functionality of these antennas. As healthcare increasingly adopts personalized, remote, and non-invasive solutions, biotelemetry antennas will continue to revolutionize the delivery of medical care.

3. MACHINE LEARNING

Machine Learning (ML) is a part of Artificial Intelligence (AI). It helps the algorithms to learn from the given data and make suitable decisions. The ML builds a mathematical representation of the input-output-relationship. Thus, it predicts new data points within the existing range. ML is categorized as:

Supervised Learning:

Input and its output are known. Then the model is trained to predict the outcomes for the new data given as the input.

Unsupervised Learning:

The labelled outputs are not available for the inputs. The main need is to discover a pattern or structure and the relation between the input and output without a guidance.

Semi-Supervised Learning:

This method is a combination of. supervised down on supervised learning. It's a mix of labelled and unlabelled data during the training process. It uses labelled data to guide the interpretation of unlabelled data.

Reinforcement Learning

The model interacts with the environment. It receives feedbacks as a reward and penalty for. The positive and negative outcomes respectively. It enhances the decision-making overtime.

The problems handled by ML is classified as follows:

1.**Regression**: helps to predict continuous outputs.
2.**Classification**: helps assigning labels to data points.
3.**Clustering**: Groups data points into clusters based on similarities without predefined labels.

3.1 Linear Regression

It is a tool in statistics to model the relationship between the independent and dependent variables. A best fitting line or hyperplane is found. It is such that it minimizes the difference between the predicted and actual values. The simplest form of linear regression, known as simple linear regression, involves one predictor variable and can be expressed as:

$$y = \beta_0 + \beta_1 + \in \quad (1)$$

Where y is the response variable, x is the predictor variable, β_0 is the intercept, β_1 is the slope of the line and ϵ is the error term.

3.2 Application of Regression in IMA

Regression lines are used to study IMAs. They predict behaviour performance such as material properties, geometrical dimensions or operating frequencies. Hence, it creates a predictive model that estimates the outcomes of the antenna parameters.

4. PROGRAMMING THE GPR

GPR is a sophisticated non parametric statistical method. It gives the relationship between the input and output variables in a complex relationship. GPR works in probabilistic framework. Programming GPR involves several steps. It begins from defining the kernel (covariance function) to training the model and making predictions. GPR predicts antenna parameters like frequency, length width based on the input variables. The input variables are sample index of the data and dielectric thickness. The heart of the GPR is the kernel function. It defines the relationship between the input points and understands the similarity between them.

This chapter uses squared exponential kernel. It is chosen to model smooth and continuous functions. The similarity is plotted by calculating exponential decay based on distance between them. it ensures model smoothly interpolates between the points. The flow chart of the program process is represented in figure 2.

The first step in implementing GPR is to train the model using a dataset. It requires fitting the Gaussian process to the observed data. The observed data is the frequency, length and width. It is achieved by optimizing the kernel's parameters. These parameters control the smoothness and variance of the predicted function. During training, the program uses the data samples and dielectric thickness as the input features to learn how the parameters behave across different conditions.

GPR does not only provide a single predicted value for each input but also a measure of uncertainty in the form of confidence intervals, which is one of its key advantages. The model's predictions are smooth and probabilistic, meaning they can be adjusted as more data is collected, and the uncertainty decreases with additional information. Hence, GPR is effective for predicting real-world data. The relationship is non-linear and noisy. MATLAB has fitgrp function. Using this GPR model can be trained quickly. Hence this becomes a valuable tool.

Figure 2. Flow chart representing the program process of GPR

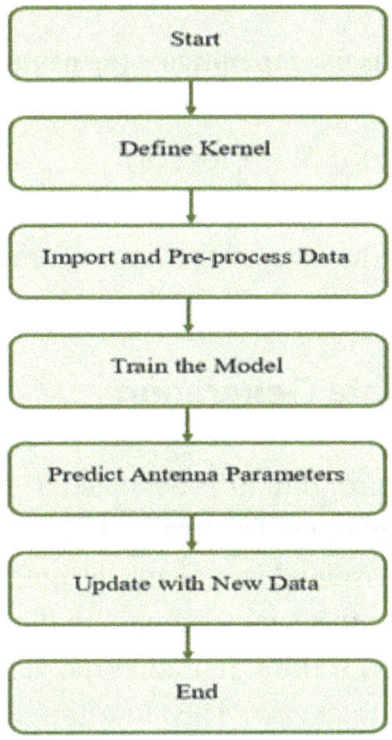

4.1 Gaussian Process Regression Approach

GPR is blended with optimization to predict Different boundaries of implantable antennas. GPR requires training data set of n input and output and output pairs. $\{(p_i, q_i) \mid i=1,2,3...r\}$, p_i is a input vector. The input vectors are collection of the lengths, widths, and the substrate height of the IMAs. Q_i

Is the output vector. A test data set is obtained as r^* given as $\{(p_i^*, q_i^*) | i = 1,2,3,..r^*\}$.The r is mapped to random variable as $[f_1, f_2, \cdots f_r]^T = [f(p_1), f(p_2) \cdots f(p_r)]^T$. The r^* is mapped into random variable as $[f_1^* f_2^* .. f_r^*]^T = [f_{(p1)}^*, f_{(p2)}^* .. f_{(pr)}^*]^T$. It follows Gaussian distribution. First consider mean as 0 as waiting or anticipating over r^*.

$$\begin{pmatrix} f \\ f^* \end{pmatrix} \sim B\left(0, \begin{matrix} S(P,P)S(P,P^*) \\ S(P^*, P)S(P^*,P^*) \end{matrix}\right) \tag{2}$$

Where S(a,b) is a gaussian distribution, a is the mean and b is the covariance matrix. P and P^* are regarded for the purpose of training. The q gives the posterior distribution. The mean k vector and the covariance matrix Σ could be found by,

$$k = S(P^*, P) S(P,P)^{-1} q \tag{3}$$

$$\Sigma = S(P^*,\ P^*) - S(P^*,P)S(P,P)^{-1}S(P,P^*) \tag{4}$$

The Gaussian likelihood is used to mitigate the negative log marginal likelihood.

$$\log k(q|P) = \frac{-1}{2}q^T S^{-1} q - \frac{1}{2}\log|S| - \frac{r}{2}\log(2\pi) \tag{5}$$

Where S is a matrix of the size of the hyper parameter known as Automatic Relevance Property (ARP).

4.2 Parameters and Data Generation

We discuss a ML based program. It is designed to model and predict three key antenna parameters. This includes frequency, length, and width using synthetic data. These parameters are essential in antenna design, particularly when considering factors like the dielectric material surrounding the antenna. The program uses Gaussian Process Regression (GPR) to predict the values. It depends on the data sample index and dielectric thickness. The algorithm models the relationship. It is by under varying conditions. The program provides a detailed simulation of how these parameters change across different data samples, influenced by the dielectric material's thickness.

4.2.1 Overview of Antenna Design and Importance of Parameters

Before beginning the details of the program, the significance of the parameters to be modelled is understood. The antenna performance is understood using various parameters. It includes size, shape and material properties for the given frequency. The three parameters generated in the program—frequency, length, and width—are integral to the design and functionality of an antenna. The parameters length and width of the given figure is considered. The figure 3 represents the dimensions of the antenna to be optimized. The parameters l_1, l_2, l_3, l_4, w_1 represent the lengths and with of the antenna.

1. **Frequency**: Wavelength of the incident wave determines the operating frequency of the antenna. The radiation pattern, efficiency and all parameters depend upon it. The key influencing factor of the frequency is the dielectric material which surrounds the antenna. Hence it is crucial in the choice of material.
2. **Length**: The radiation characteristic and resonant frequency is based on the length of the antenna. The length is adjusted to achieve the desired band. It also plays an important role in calculating the impedance of the antenna. The

lower frequency requires larger antenna and small antenna are suited for high frequency applications.

3. **Width**: The width of the antenna's structure affects the impedance and the bandwidth of the antenna. In certain designs, the width may be optimized to achieve a particular radiation pattern or to match the antenna's impedance with the transmission line. For many antenna designs, the width varies with the frequency and the dielectric material's properties. Each of these parameters must be carefully designed to meet the requirements of the system in which the antenna will be used, such as biotelemetry, communication, or sensor networks.

Figure 3. Structure of the Antenna to be optimized

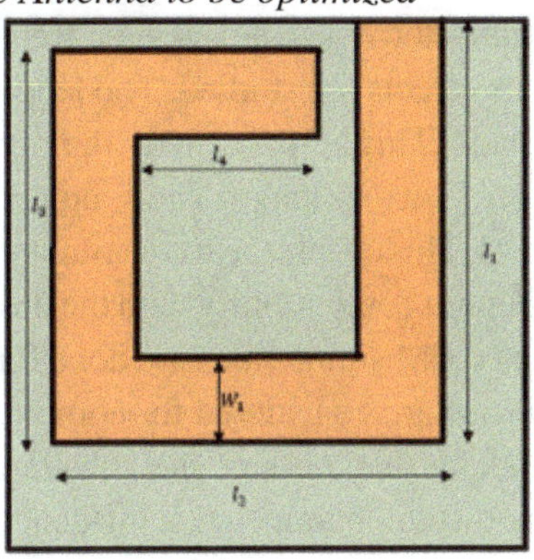

4.2.2 Data Generation Process

The fundamental of the study is data generation process. It helps to accurately model the antenna parameters. the parameters include length, width and frequency. These are obtained through simulation results. By using EM tools with relevant mathematical equations, antenna behaviour is modelled. The program calculates the parameters. It develops them as functions of sample index and substrate thickness.

A. Frequency Data Simulation

The important parameter of antenna function is the frequency. It is influenced by substrate thickness around the antenna. With increase in thickness, the resonant frequency decreases. It is because of EM waves travel slow through the material. This condition is documented in the antenna theory and is given as permittivity of the material. It impacts the effective wavelength of the waves. The simulation models the frequency by iteratively solving Maxwell's equations for various sample indices and dielectric thicknesses. With material properties such as relative permittivity and loss tangent the resonant frequency can be calculated. The simulated data shows a general trend of decreasing frequency with increasing dielectric thickness. However, slight variations are observed for specific sample indices, reflecting design-specific factors such as antenna geometry and coupling effects. With numerical techniques such as finite element or finite difference time domain (FDTD) methods, noise-free results are obtained. The results confirm that the frequency decreases consistently as the dielectric thickness increases, aligning with theoretical predictions and experimental observations in antenna research.

B. Length Data Simulation

The length of the antenna determines the resonant frequency. In this section, the length of the antenna and its effects are studied. The length is varied in small steps along different sample indices. This helps to study the impact on the performance. Each index is the simulation of various length The simulation gives different lengths to the antenna for each indices. This ensures real constraints within the design. Length values are calculated to maintain compatibility with the desired operating frequency and application requirements. EM simulations are done for every length. Boundary conditions, material properties are considered for computation. It accounts for the resonance and impedance. The thickness of the substrate is also initialized. The thickness is taken as the modifying parameter. It influences the electrical length of the antenna. Increase in thickness increases the antenna effective length. It alters the resonance behaviour. Systematic analysis of these simulation is required. As a modifying parameter, influencing the effective electrical length of the antenna. As the dielectric thickness increases, the antenna's effective length increases slightly, altering its resonant behavior. Studying the simulations, there is a direct relationship between length and resonant frequency. Long antenna favours small frequencies. The length values are constrained to practical ranges, such as between 10 mm and 120 mm, to reflect standard design considerations in communication and biotelemetry applications.

C. Width Data Simulation

The antenna width affects the impedance and bandwidth. In this study, width is modelled as a periodic function of index. The design choice suits the resonant structures and exhibits geometrical features. Simulations are conducted for antennas with varying widths, incorporating sinusoidal variations as a function of the sample index. To optimize the performance metrics the periodic variations are calculated. The substrate thickness is a secondary factor. It modifies the effective width of the antenna. Thick substrate results in increased width for better performance. These EM structures are modeled using advanced computational techniques. The EM behavior are precise. The computational tools generate data that resembles real world antenna performance. The simulation results highlight the sinusoidal variation in width, with clear modulation caused by dielectric thickness. These findings are consistent with the theoretical and practical behavior of antennas designed with periodic geometrical variations.

Under different substrate thickness and data indices, many simulations with varying parameter is found. This gives a comprehensive dataset that reflects a real antenna's behaviour. The simulations capture the intricate relationships between these parameters and demonstrate the critical role of dielectric material in influencing antenna performance. These results lay a robust foundation for further analysis, including advanced modeling techniques such as Gaussian Process Regression, to predict and optimize antenna parameters for specific application

4.3 What is Gaussian Process Regression (GPR)?

A Gaussian Process is a collection of random variables, any finite subset of which has a joint Gaussian distribution. GPR provides a probabilistic framework for regression, which means that it cannot only make predictions but also quantify the uncertainty associated with those predictions. This makes GPR particularly useful when the underlying data is noisy or when there are multiple sources of uncertainty.

In a typical regression problem, the goal is to find a function that maps the input variables (in this case, data sample index and dielectric thickness) to the output variables (frequency, length, and width). GPR does this by defining a distribution over possible functions that could explain the observed data.

The key advantage of GPR over traditional regression methods is its flexibility. Instead of assuming a specific functional form for the data (e.g., linear or polynomial), GPR learns the underlying relationships directly from the data. It also provides confidence intervals for its predictions, which allows us to understand the uncertainty in the predictions.

4.3.1 How GPR is Applied in the Program

Random variables of any finite subset of which has a joint Gaussian distribution is called as Gaussian Process. GPR provides a probabilistic framework for regression, which means that it cannot only make predictions but also quantify the uncertainty associated with those predictions. Hence when a noisy data with multiple sources of uncertainty is presented, GPR helps. In a typical regression problem, the goal is to find a function that maps the input variables (in this case, data sample index and dielectric thickness) to the output variables (frequency, length, and width). This is by defining the distribution over a function. This function could explain the observed data. The key advantage of GPR over traditional regression methods is its flexibility. Instead of assuming a specific functional form for the data (e.g., linear or polynomial), GPR learns the underlying relationships directly from the data. The confidence intervals for the predictions are given. This allows us to understand the predictions.

In this program, GPR is used to model the relationship between the data sample index, dielectric thickness, and each of the three antenna parameters—frequency, length, and width. The program applies GPR separately for each parameter. Training a separate model for frequency, length, and width is done. Figure 4 represents the proposed GPR model.

Frequency Prediction: The first GPR model is trained to predict the frequency based on the data sample index and dielectric thickness. It learns the frequency shift based in those two factors.

Length Prediction: A second GPR model is trained to predict the length. The length is based on the same inputs. The model shows the relationship between the sample index, dielectric thickness, and antenna length.

Width Prediction: A third GPR model predicts the width. It based on the data index and the substrate thickness. The sinusoidal variation in width is modelled. with the dielectric thickness modulating the amplitude or phase of the sinusoidal pattern. Each GPR models is trained with the data generated previously. The models learn the relationship between the input and the output variables. The input variables are sample index and dielectric thickness and the output variables are frequency, length, and width, allowing them to make predictions for these parameters across a range of data samples.

Figure 4. Proposed GPR Model

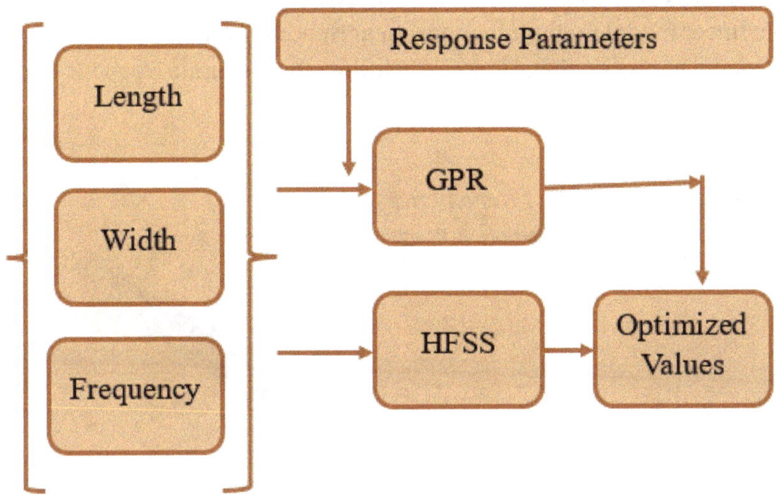

4.3.2 Prediction and Visualization

When the GPR models are trained, the predictions for the three parameters length, width and frequency can be obtained. These predictions are then compared to the actual synthetic data through scatter plots.

A. Data Sample vs Frequency

The Figure 5 shows the relationship between the data sample index and frequency. The frequency predicted is plotted as scatter points. The predicted value of frequency is based on the GPR model is represented as a straight line. The scatter points are colored according to the substrate thickness. It provides a visual representation of how the frequency changes with both the data sample index and dielectric thickness. The straight lines are the frequencies by the model. These lines should match the actual scatter points. As the substrate thickness increases, the frequency is expected to decrease. This trend should be visible in the scatter plot and the predicted lines.

Inference:

The predicted frequency follows a smooth and linearly increasing line. it suggests that the frequency response is influenced by the antenna's geometric properties. A linear increase indicates that. As the antenna dimensions change, the resonant frequency increases predictably.

Insights:

The relationship between the data sample index and the frequency might reveal how the design evolution (or fabrication process) affects the antenna's resonance. A simple linear model might miss non-linear trends or noise patterns. The Gaussian Process model identifies that. Hence providing more accurate frequency predictions for antenna designs.

Figure 5. Data Samples Vs. Frequencies

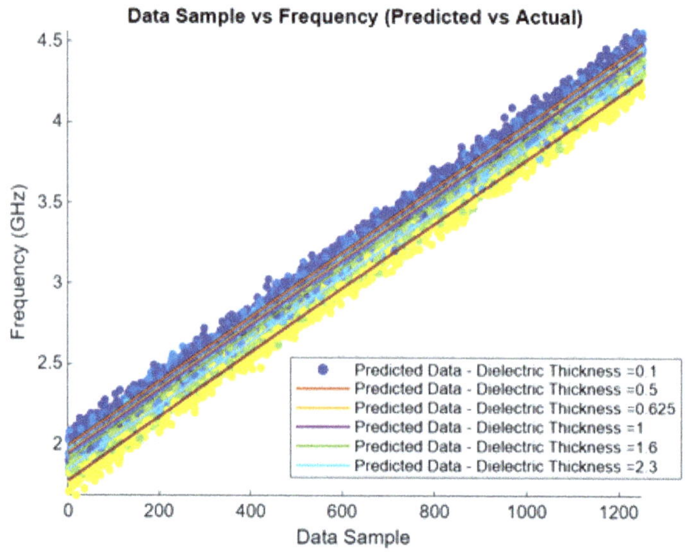

Figure 6. Data Samples Vs. lengths

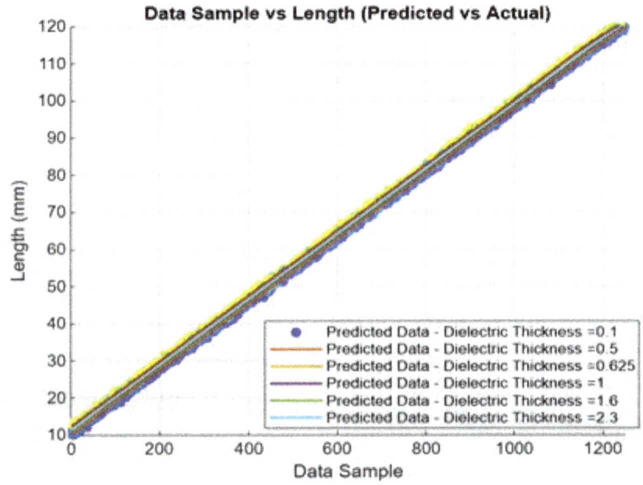

B. Data Sample vs Length

The Figure 6 shows the relationship between the data index and length. The dots represent the lengths. The straight line represents the predicted values. The scatter points are color-coded by substrate thickness. It is showing how the length varies with both the data index and the dielectric thickness. The trend of the line increases upwards. This shows increase in length with the samples. The predicted lines represent the GPR model output. It provides a smooth approximation of the length values across dataset. These lines also account for variations in substrate thickness. It gives an adjusted trend for different thickness level. The length is expected to increase linearly with the data index. As the substrate thickness increases, the length is adjusted upward. It should reflect in the predicted lines. This plot illustrates how well the GPR model captures the linear trend in the length data and its dependence on dielectric thickness.

Inference:

If the predicted length follows a non-linear pattern (e.g., quadratic or sinusoidal), it may indicate how design constraints or the optimization process evolve to meet specific resonance

rcquirements. The length is between 10 mm and 120 mm. A great portion of the data points should fall within this range. The model may predict an optimal length for a desired frequency.

Insights:

If there is a clear pattern, such as the length increasing over time with some noise, this could indicate that longer antennas are required for specific frequency bands, or certain lengths work better at certain frequencies. The model show that the length variation is controlling the antenna's resonant frequency, depending on the behavior of the predicted values. Noise patterns in the length data could point to the variability in production or design processes that might affect antenna size and performance.

C. Data Sample vs Width

The Figure 7 shows the relationship between the data index and width. Unlike the frequency plot and length plot, width demonstrates a sine like pattern. The scatter points show the actual width values for each data sample. The points are colored based on the substrate thickness. It helps to show how the sine pattern is influenced by the material. The predicted lines by the GPR model show the periodic nature of the width values. These smooth curves closely follow the sinusoidal variation of the scatter points, with slight adjustments to reflect the influence of dielectric thickness. The width should vary like a sine wave with the data index. The substrate thickness modifies the amplitude or phase of the sine pattern. The prediction lines should align with this pattern. It must provide a smooth representation of the data.

Inference:

The predicted width will follow a sinusoidal or oscillatory pattern. This may suggest that the width of the antenna has a cyclic or periodic effect on the antenna's performance (e.g., bandwidth tuning, impedance matching). If there is a clear correlation between width and frequency, it might indicate a strong design interaction where width plays a crucial role in tuning the antenna's performance for a given frequency range. Hence GPR model could handle non-linear relations like the sine variation.

Insights:

The periodic variation in width shows the design optimization process. Width is changed to achieve better impedance matching, minimize signal loss, or enhance bandwidth. The noise and small oscillations might indicate that while the width is being controlled, there are other factors that affect how the width influences the antenna's characteristics.

Figure 7. Data Samples Vs. Widths

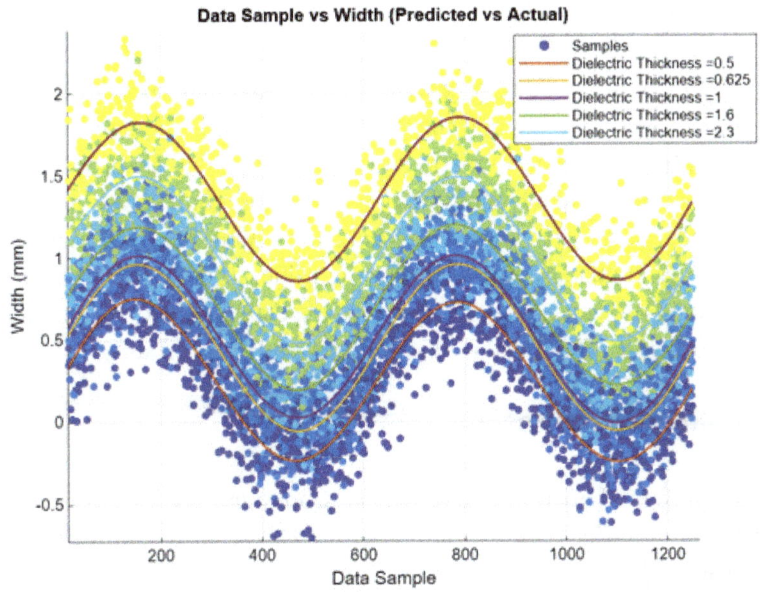

D. Insights into the trade-off between width and frequency:

Wider antennas are needed for higher frequencies. The width increases with frequency, a change in width compensates for other design factors like material properties or bandwidth requirements.

The program provides a comprehensive framework for modelling and predicting antenna parameters using Gaussian Process Regression. Data generation and training separate GPR models for every parameter demonstrates how these parameters vary with the data index and substrate thickness. The predictions generated by the GPR models are visualized through scatter plots that compare the actual data with the predicted values. The plots gives the accuracy of the GPR model. Insights to the trend is provided. Whether it's the slight decrease in frequency with increasing dielectric thickness, the linear increase in length with the sample index, or the sinusoidal variation in width, the GPR models effectively capture these behaviors and produce smooth, interpretable predictions. The designers can optimize antenna performance. They could make decisions in applications ranging from communication systems to biotelemetry devices.

E General Trends in Data

By examining the confidence intervals of the predictions, assess where the model is most confident or where predictions are more uncertain (e.g., at the extremes of data). If the residuals (differences between actual and predicted values) show a pattern, this could indicate that the model has not fully captured the underlying relationships. If the error between predicted and actual data are small and random, the model is doing a good job of predicting. This error is called residuals. This could refine the model or suggest that features need to be considered. High uncertainty in prediction at certain data samples of extreme values of data suggest regions where the model might not generalize. Also, where the data exhibits greater variance that is difficult to model effectively is also showed by the uncertainty.

4.4 Modelling Insights

By adjusting the hyperparameters of the Gaussian Process model (e.g., kernel function), you may find that a different kernel or variance settings better suit the data, improving predictions and generalizability. A data is known as overfit data if the model performs good on training data and poorly on unknown data. It is underfit if it performs poorly on training data. This analysis will inform about the problem complexity. It tells whether more features or data are needed. By adjusting the hyperparameters of the Gaussian Process model (e.g., kernel function), you may find that a different kernel or variance settings better suit the data, improving predictions and generalizability.

4.5 Data Quality and Manufacturing Insights

Fabrication defects and measurement inaccuracies denote that more detailed data is required to study the pattern. With noise data, these issues can be caused. The predictions could help identify ideal design parameters for frequency, length, and width.

4.6 Potential Applications

By understanding how the design parameters (length, width, and frequency) interact, one can optimize the antenna design for specific use cases (e.g., communication systems, wearable devices).

Design Iteration: These insights can inform how the antenna should evolve over the course of design iterations. Dimensions impact frequency in a predictable way and it reduce trial-and-error iterations.

If the results show that certain dimensions correlate strongly with better performance, these can be cross-referenced with material properties (e.g., dielectric constants) to select the most appropriate materials for different frequency bands.

5. ANTENNA DESIGN

A meander line antenna (MLA) consists of a radiating element with a like a folded path on one side of the dielectric substrate. Ground plane located on the opposite side. This configuration is useful for its compactness. It allows for a longer effective electrical length within a smaller physical space. Similar to traditional microstrip antennas, the performance of the meandered antenna depends on the choice of patch shape, dielectric substrate (with a relative permittivity $\varepsilon_r \leq 10.2$) and design frequency. To calculate the effective length of the meander line patch at the design frequency f_r, the formula is adjusted to account for the meandered structure:

$$L = \frac{c}{2f_d}\sqrt{2\Delta L} \tag{6}$$

Where, c is the speed of light, fd is the design frequency of the antenna, and ΔL represents the electrical length extension of the patch due to the fringing fields.

The width of the meandered patch is defined as:

$$w = \frac{c}{2f_d}\sqrt{\frac{1}{\varepsilon_r + 1}} \tag{7}$$

When the ratio W/h is greater than 1, the effective dielectric constant ϵef is given by:

$$\varepsilon_{ef} = \left(\frac{\varepsilon_r + 1}{2}\right) + \left(\frac{\varepsilon_r - 1}{2}\right)\sqrt{1 + \frac{12h}{w}} \tag{8}$$

where h is the height of the substrate.

The electrically extended length ΔL due to fringing fields is defined as:

$$\Delta L = 0.412h\frac{\left(\varepsilon_{ef} + 0.3\right)\left(\frac{w}{h} + 0.264\right)}{\left(\varepsilon_{ef} - 0.2.58\right)\left(\frac{w}{h} + 0.08\right)} \tag{9}$$

Generalized expression for resonant frequencies,

$$f_r = \frac{c}{2\sqrt{\varepsilon_{ef}}}\sqrt{\left(\frac{m}{L + 2\Delta L}\right)^2 + \left(\frac{n}{L + 2\Delta W}\right)^2} \quad (10)$$

where m and n are resonant modes of antenna.

5.1 Synthesis and Analysis

To synthesize and analyze the models, GPR technique was employed. The GPR is effective in handling small datasets. It makes it highly suitable for this investigation. Its performance depends significantly on the choice of mean and covariance functions. Covariance functions capture the relationships between the predictor variables and the response, and their properties are governed by a set of hyperparameters. In this study, four covariance functions—squared exponential, Matern 5/2, exponential, and rational quadratic—were selected for evaluation. Each function was tuned to identify its optimal performance. To further enhance the model's accuracy, all four covariance functions were optimized using a Bayesian optimizer. The Bayesian process was conducted in both isotropic and non-isotropic kernel functions. This ensured that the hyper parameters of the model provide accurate predictions. The best GPR model was then selected to predict key parameters of the meandered implantable antenna (IMA).

5.1.1. Prediction of Frequency

The frequency is the key parameter of the antenna. It is determined by the physical parameters such as length (L), width (W). In the proposed GPR model, the resonant frequency (fr) is the response parameter. The L, W, h, and ϵr was chosen as the predictor parameters. The choice of these parameters is such that the model can capture the complex relationships that exist in meandered antenna designs. This model highlights the input and output relationships of training the GPR model. This model provides an efficient method for predicting the frequency of meandered IMAs with a limited dataset.

5.1.2. Prediction of Patch Dimensions

The meandered IMAs reduces the overall size of the antenna maintaining its performance. These structure increases the path of surface current. In turn, it lowers the resonant frequency. To ensure compactness and optimized performance, prediction of the patch dimensions, patch length (L) and patch width (W), is essential. In this chapter, the GPR model was also applied to predict the patch dimensions. It is based on specific design parameters. These parameters included the dielectric constant of the substrate material (ϵr), the resonant frequency (fr), and the dielectric thickness (h). Using these variables as predictors the model provide accurate estimates of the patch dimensions. These are critical for achieving the desired resonant frequency. The relationships between the predictor parameters and the response variables are detailed. The combination of advanced GPR modelling and Bayesian optimization ensures that the proposed models provide robust predictions for both resonant frequency and patch dimensions. These models are particularly valuable for the design and optimization of meandered IMAs. It requires precise parameter tuning to meet the stringent requirements of implantable applications. The results demonstrate the efficacy of the GPR approach, even with a small dataset, and highlight its versatility in addressing complex design challenges. Sum of individual segments of the patch adds to half wavelength. The width of the patch is about 0.5 mm.

6. COMPUTED RESULTS AND VALIDATION

The modelling of resonant frequency, meandered patch length, and patch dimensions of the implantable meandered antenna (IMA) was conducted using a dataset generated through extensive simulations. The design vector for the IMA was defined as, where represents the patch length, is the dielectric thickness, is the slot length, and is the relative permittivity of the substrate. The design space for these parameters was established as follows: ranged from 13 mm to 48 mm with a step size of 0.5 mm, varied from 0.2 mm to 4 mm with a step size of 0.25 mm, ranged from 0.5 mm to 21 mm with a step size of 0.5 mm, and three different substrate materials with relative permittivity values of 3.5, 4.4, 10.2 were considered. This extensive range of design parameters allowed for an exploration of the IMA's behaviour under varying conditions.

A total of 1250 input vectors were generated. This is to ensure a uniform and unbiased representation of the design space. The simulations were performed on a computational setup consisting of an Intel(R) Core (TM) i5-6400 CPU, 16 GB of RAM, and a Windows 11 operating system. The simulated dataset was then used to evaluate the IMA's resonant frequency, patch length, and patch dimensions. This dataset provided valuable insights into the behaviour of the IMA, with the resonant frequency ranging from 0.4858 GHz to 7.8476 GHz, the length ranging from 0.5 mm to 21 mm, and the patch length ranging from 13 mm to 48 mm. These results demonstrated the wide operating range and design flexibility of the IMA.

To model and predict the resonant frequency, slot length, and patch dimensions of the IMA, supervised learning techniques were employed. A total of 20 regression models were investigated, including Linear Regression, Tree Regression, Ensemble Regression, Support Vector Machine Regression (SVMR), Gaussian Process

Regression (GPR), and Artificial Neural Networks (ANN). The dataset was used for training and testing sets. With 1250 samples used for training and 202 samples reserved for testing the models. Each regression model was evaluated based on its ability to accurately predict the key parameters of the IMA. Performance metrics such as root mean squared error (RMSE), coefficient of determination (), and mean absolute error (MAE) were used to assess the models' effectiveness. Additionally, the minus log-likelihood value was considered to evaluate the models' ability to handle nonlinear behavior, particularly for the resonant frequency.

Among the other models, the Gaussian Process Regression (GPR) models demonstrated superior performance. To improve the accuracy of the GPR models, Bayesian optimization was employed to optimize. The aim is to optimize hyperparameters and minimize the mean squared error (MSE). The optimized GPR models for predicting the resonant frequency, length, and width exhibited the configurations as given by the figure 8 is:

1. The GPR model for resonant frequency prediction utilized a nonisotropic rational quadratic kernel function. It is considered with a zero-basis function. The value of sigma was optimized to 11.5016, and the model converged after 30 iterations. This resulted in a highly accurate model with minimal MSE. It is capable of handling the nonlinear behavior of the frequency.
2. For length prediction, the GPR model has an isotropic exponential kernel function. It is with a constant basis function. The sigma value was optimized to 6.9112, and the kernel scale parameter 0.27482. The model achieved convergence after 50 iterations, providing predictions for the length.
3. The GPR model for patch length prediction used an isotropic exponential kernel function with a constant basis function. The sigma value was optimized to 0.13653. The kernel scale parameter was 7.8696. The model had 50 iterations to converge. Patch length had with minimal MSE.

The optimized GPR models were evaluated using the reserved test dataset of 202 samples. The performance of the models was validated by comparing the predicted values against the actual values. It is done for the resonant frequency, width, and patch length. The results indicated that the GPR models provided highly accurate predictions, with most data points symmetrically distributed around the diagonal line in the actual versus predicted plots. These plots demonstrated the reliability and precision of the GPR. It predicts the key parameters of the IMA.

Figure 8. Minimum MSE vs number of iterations for different iteration of (a) frequency, (b) length and (c) width

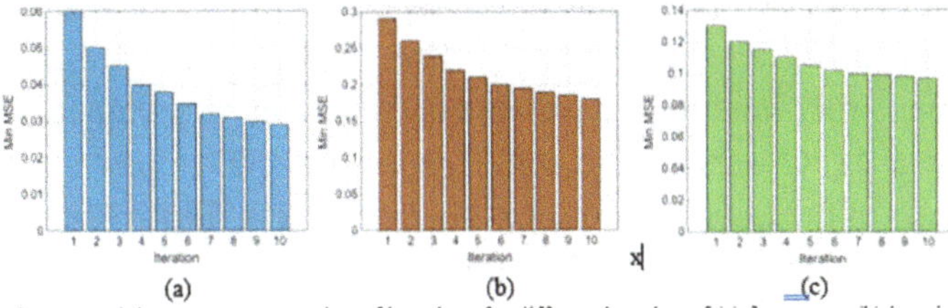

To visualize the optimization process, convergence plots were generated for each parameter. The plots depicted the minimum MSE responses at each iteration. The convergence plots clearly illustrated the efficiency of the Bayesian optimization process in minimizing the MSE for the GPR models.

The performance of the GPR models was compared with other regression models. This includes SVMR, Tree Regression, and Ensemble Regression. The comparison revealed that the GPR models consistently outperformed the other models in terms of RMSE, and MAE. The optimized GPR models achieved the lowest RMSE and the highest values, confirming their suitability for accurately predicting the resonant frequency, slot length, and patch length of the IMA. The results demonstrated that the GPR models were particularly effective in handling the nonlinear behavior of the resonant frequency, which was a critical aspect of the IMA's design.

The findings have significance for the design and optimization of antennas. The accurate prediction of key parameters enables the efficient design of IMAs for biotelemetry. The GPR models provide a powerful tool for antenna designers. It allows to explore the design space and achieve optimal performance with minimal computational effort. The Bayesian optimization enhances the efficiency of the modeling process. It makes sure the models are tuned to minimum MSE.

In conclusion, the modeling and prediction of resonant frequency, slot length, and patch length for the meandered implantable antenna were successfully achieved using optimized GPR models. The results confirmed the high accuracy and reliability of the GPR models. It makes them an invaluable resource for the design and optimization. The study highlights the importance of leveraging advanced regression techniques and optimization methods to address the challenges of antenna design, paving the way for the development of innovative solutions for biotelemetry and related applications.

The explanation for the Minimum MSE (Mean Squared Error) graphs by the MATLAB program is given. The optimization process and the performance of the GPR models used to predict parameters. These graphs illustrate the iterative process of hyperparameter tuning to minimize the error and improve model accuracy.

The graph demonstrates that the best sigma value is 0.75653. With a constant basic function and an isotropic exponential kernel function is shown. The kernel scale parameter is set to 7.8696. It reflects the model's capability to minimize prediction error for patch length effectively. The graphs underscore the importance the appropriate kernel functions, basis functions, and hyperparameters to optimize model performance. Observing the convergence of MSE values across iterations, the effectiveness of the Bayesian optimizer in fine-tuning the GPR models becomes evident. These visualizations not only validate the predictive accuracy of the models but also highlight the robustness of the optimization process for various antenna parameters in the simulated dataset.

7. ANTENNA VALIDATION IN IMPLANTABLE ENVIRONMENT

Designing implantable antennas for the 2.45 GHz ISM band addresses unique challenges due to the higher frequency. The compact size and performance requirements are particular because higher frequencies have smaller wavelengths. It results in reduced skin depth in human tissues and increased sensitivity to impedance mismatches. One of the design considerations is miniaturization. At 2.45 GHz, the wavelength is 12.24 cm in free space. It reduces in tissues due to their high dielectric constants. To manage this, techniques like meandered geometries, spiral structures are used. Also, the use of high-permittivity substrates is employed to reduce the antenna's physical size of 5x5 mm^2, without compromising the performance.

The interaction with biological tissues is a critical factor at 2.45 GHz. The tissues exhibit higher losses due to dielectric and conductive properties at this frequency. This causes attenuation of EM waves. It can impact the antenna's radiation efficiency and overall performance. Designing proper impedance matching is crucial. This is done to maximize power transfer and minimize signal reflections. The antennas are typically tuned using simulations in multilayer tissue. The phantoms replicate the dielectric properties of skin, fat, and muscle. Accurate modeling makes sure that the antenna performs under real-world implantation conditions.

Material choice is another important consideration. The conductive and dielectric materials used in the antenna must be biocompatible. It must be resistant to the corrosive environment of the human body. Metals like gold and titanium are commonly used for conductors. Biocompatible dielectrics such as PDMS, parylene-C, or liquid crystal polymer (LCP) are often used for encapsulation. At 2.45 GHz, the encapsulation material also insulates the antenna from body fluids, which can cause to degrade its performance by altering its impedance and radiation properties.

Figure 9. 3 layered model of human tissues with Antenna Packaging

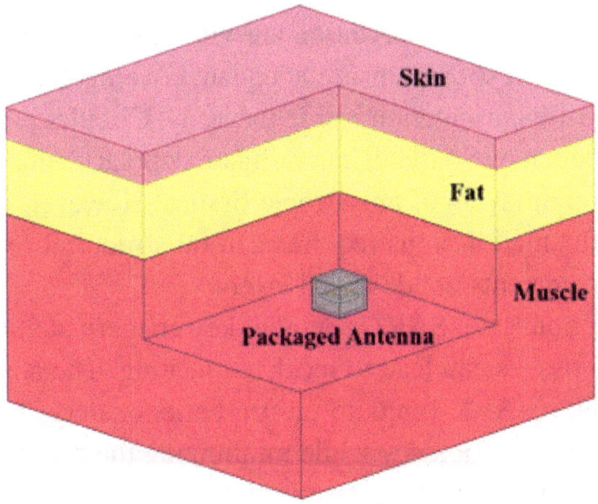

Table 1. Comparison of Designed Antenna with Standard Implantable Antenna Parameters

Parameter	Designed Antenna	Typical Benchmark
Resonant Frequency	2.45 GHz	2.4–2.48 GHz (ISM Band)
Bandwidth (−10 dB)	200 MHz	≥100 MHz for ISM applications
Return Loss (S11)	−27.6 dB	< −10 dB acceptable; < −20 dB preferred
Maximum Gain	−20 dB	−30 dB to −10 dB typical for implantables
SAR @ 1 W	244 W/kg	Limit: 1.6 W/kg (1g tissue) / 2 W/kg (10g tissue)
SAR-Compliant Input Power	6.56mW (FCC), 8.2mW (ICNIRP)	≤10 mW for SAR-safe operation
Antenna Size	5x5 mm^2	<30 mm typical for implants
Radiation Pattern	Distorted Omni-directional	Direction-dependent, affected by tissue
Simulation Medium	Skin–Fat–Muscle	3-layer or multi-layer models used
Efficiency	Moderate (qualitatively inferred)	Low to moderate (~−15 dB to −10 dB typical)

Bandwidth and efficiency are critical metrics for 2.45 GHz implantable antennas, especially since the ISM band demands robust communication in noisy and lossy environments. Achieving a sufficiently wide bandwidth (e.g., around 100–300 MHz) is essential to accommodate variations caused by tissue heterogeneity and antenna misalignment during implantation. Techniques like slots, varying the antenna geometry, or using metamaterials help enhance bandwidth, maintaining a compact form factor.

The radiation pattern and polarization are vital. For reliable biotelemetry, the antenna radiation pattern should ensure adequate coverage to external devices. It must be regardless of its orientation inside the body. Circular polarization is often preferred at this frequency. So that it can mitigate depolarization effects caused by signal passage through layers of anisotropic tissues. However, achieving circular polarization in such small designs requires careful structural modifications, such as adding parasitic elements or additional layers.

Specific Absorption Rate (SAR) is another key concern at 2.45 GHz, as the energy absorption by tissues can lead to localized heating. The antenna design must comply with regulatory SAR limits to ensure patient safety. This often involves optimizing the antenna's efficiency while minimizing the power radiated into surrounding tissues. Additionally, long-term reliability and durability are critical, as the antenna must function consistently over time, minimizing the need for surgical replacements. By addressing these considerations, implantable antennas operating at 2.45 GHz can deliver high-performance wireless communication while ensuring patient safety and comfort.

7.1 Heterogenous medium

For implantable antenna, the implant environment is often modeled. It is done by using a three-layer structure to mimic human tissue as shown in the figure 9. This simplified representation for the fundamental interaction of EM waves with biological tissues The layers include:

Skin: This is the outermost layer, typically modeled as the first interface for electromagnetic waves. Skin has a low water content. This leads to lower permittivity and conductivity. Its thickness usually ranges from 2–4 mm.

Fat: Beneath the skin, fat serves as the second layer. It has lower permittivity and conductivity than muscle. It is due to its minimal water content. This makes it a barrier for EM wave. Fat thickness varies between individuals and body regions.

Muscle: The innermost layer is muscle. It has high water content. Due to this, it has higher permittivity and conductivity. It dominates the implant environment's energy absorption. It influences the antenna performance due to its high attenuation.

This three-layer model is used in simulations to estimate antenna performance. This includes the return loss, gain, efficiency, and specific absorption rate (SAR) of implantable antennas. It provides a balance between computational efficiency and anatomical accuracy. Despite its simplicity, the model effectively predicts the antenna's behavior in typical implantation scenarios, guiding design optimization. For more precise studies, additional layers (e.g., bone, organs) or more complex heterogeneous models can be used. This approach ensures the design is both practical and safe for real-world applications.

Figure 10. S parameter plot of the designed antenna

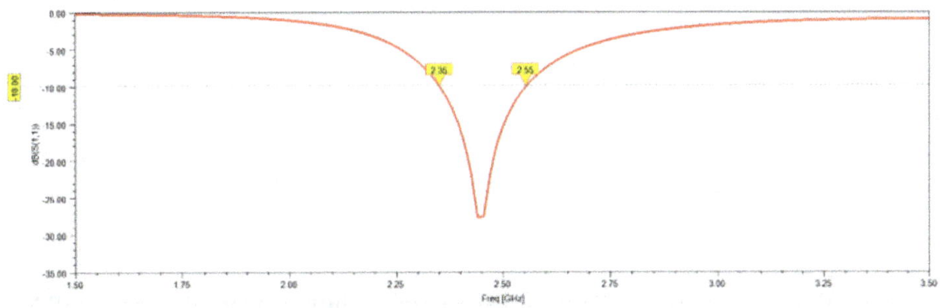

The implantable antenna which is designed demonstrates impressive performance in terms of S-parameters and bandwidth. Hence, making it suitable for biotelemetry applications. The return loss (S11) is -27.6 dB at 2.45 GHz represented by the figure 10. It is indicating good impedance matching and low signal reflection. Such a low return loss makes sure of the efficient signal transmission. It is critical for reliable wireless communication between implantable medical devices and external systems. The antenna also achieves a wide bandwidth, spanning from 2.35 GHz to 2.55 GHz, covering a range of frequencies within the Industrial, Scientific, and Medical (ISM) band. The wide bandwidth provides a great ease in signal transmission. It ensures stable performance even in dynamic biological environments. The dielectric properties of tissues tend to vary.

The combination of a deep return loss and broad bandwidth highlights the robustness of the antenna's design, particularly in overcoming challenges such as miniaturization and tissue-induced detuning effects. This performance can be attributed to careful material selection. Optimized geometrical design, and advanced impedance-matching techniques help to achieve this. The antenna has the capability to maintain high performance under rough tissue conditions. It makes it a promising candidate for applications like health monitoring, drug delivery. Furthermore, its efficiency at 2.45 GHz aligns well with the ISM band requirements. It is enabling the use of widely available communication protocols. The antenna characteristics not only meet but exceed the essential criteria for modern implantable devices. It ensures reliable and effective wireless communication in real-world biomedical scenarios. Table 1 represents the comparison of the antenna parameters with standards.

7.2 Current distribution and Radiation mechanism:

The meandered implantable antenna operates at 2.45 GHz. It is with a quarter-wavelength structure. It utilizes efficient radiation and impedance matching mechanisms for biotelemetry applications. The primary radiation mechanism stems from the meandered structure of the antenna. It concentrates the surface current distribution along its length. At the resonant frequency, the antenna exhibits a maximum current density of 307 A/m as shown in figure 11. It is concentrated near the central segments of the meandered paths. This behavior ensures that the majority of the antenna length participates in radiating EM waves efficiently. The design reduces the antenna size without trading-off its performance. It is a crucial requirement for implantable applications.

The quarter wave-length aids in impedance matching, as the reactive components of the antenna (inductive and capacitive) are minimized at resonance, ensuring efficient power transfer between the antenna and the feeding network. The meandering increases the effective electrical length while maintaining a compact physical footprint, enabling resonance at 2.45 GHz. This design choice not only supports miniaturization but also ensures compatibility with tissue-loading effects. The biological tissues, with their high dielectric properties, influence the antenna's performance by increasing the effective permittivity around the antenna, which shifts the resonance to a lower frequency. To overcome this, the design is optimized inside the human tissue. It ensures minimal detuning.

Radiation is achieved through the alternating current distribution along the meandered segments. It creates constructive interference of the radiated waves. Additionally, the compact size and geometry of the antenna enhance its adaptability to body-implant scenarios while maintaining sufficient radiation efficiency. The deep return loss of -27.6 dB at 2.45 GHz reflects excellent impedance matching, minimizing signal reflection and maximizing power delivery to radiate effectively. Thus, the antenna achieves a balance between compactness, effective radiation, and robust impedance matching.

Figure 11. The current distribution of the proposed antenna

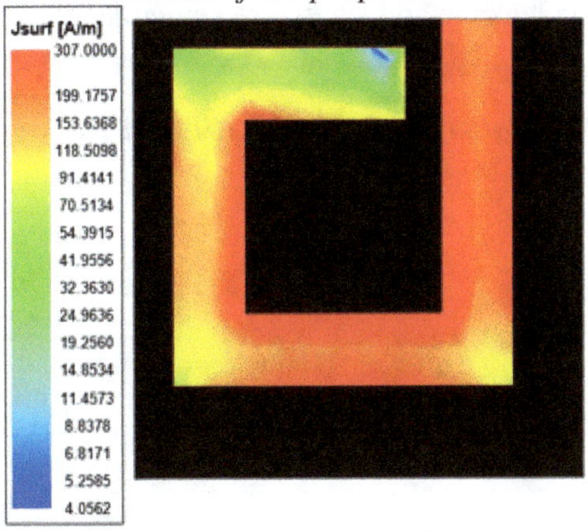

The radiation pattern of an antenna is studied at ϕ = 0° and ϕ = 90°. The maximum gain is -20 dB. This provides view into its performance in specific planes. The ϕ = 0° plane is the azimuthal plane aligned with the x-axis. The ϕ = 90° plane is plane at the y-axis. The radiation pattern is the variation of gain as a function of the elevation angle (θ). It measured from the z-axis. A maximum gain of -20 dB indicates that the antenna radiates very little energy represented by figure 12 (b). It is less compared to an isotropic radiator. It is common for implantable antennas. The antennas designed for low-power applications like biotelemetry. This low gain is often a result of the antenna's small size, lossy materials, or the energy absorption caused by its placement in human tissues.

Table 2. State of art comparison

Work	Algorithm	Size	Frequency	Gain
(Essa et al., 2024)	-	50x50 mm^2	2.45 GHz	-14.9 dBi
(Yahya et al., 2024)	GPR	40x42 mm^2	0.915 GHz	-12.5 dBi
(Sharma & Pandey, 2021)	GPR	40 x 40 mm^2	0.48-7.84 GHz	5.9 dB
This work	GPR	5x5 mm^2	2.45 GHz	-20 dBi

The radiation pattern at ϕ = 0°, ϕ = 90° displays how the antenna radiates in the vertical plane and the orthogonal plane as in figure 12 (a). For implantable antennas, these patterns are distorted due to the surrounding environment. This results in asymmetrical shapes. It causes reduced performance in certain directions. The patterns at these angles can provide information about the antenna directional characteristics. It also explains the efficiency of the antenna. By analysing these cross-sectional patterns, designers can evaluate the antenna's suitability for applications where radiation leakage and safety are key concerns.

Figure 12. (a) Radiation Pattern at 2.45 GHz (b) gain Vs Frequency

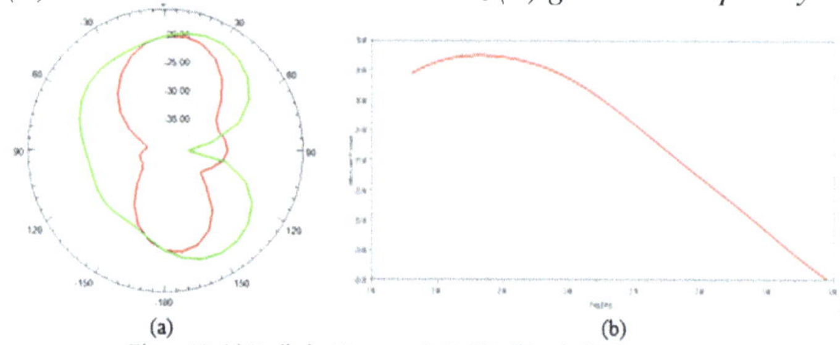

7.3 SAR Analysis

The Specific Absorption Rate (SAR) is the rate at which energy is absorbed by human tissue when exposed to EM fields. The designed antenna has an average SAR of 244 W/kg at an 1W input power. The input power must be optimized to meet safety limits. In the United States, the Federal Communications Commission (FCC) sets the SAR limit at 1.6 W/kg averaged over 1 gram of tissue. The International Commission on Non-Ionizing Radiation Protection (ICNIRP) limits as 2 W/kg averaged over 10 grams of tissue. SAR is directly proportional to input power. The acceptable input power can be calculated with the ratio of the regulatory SAR limit to the measured SAR. For FCC the input power must be reduced to approximately 6.56 mW. For ICNIRP, it should be lowered to around 8.2 mW. These reductions ensure the device meets the safety standards. Hence minimizing the risk of thermal effects on tissues. This calculation shows the importance of optimizing antenna design and system performance to achieve efficient operation within permissible SAR levels. For applications involving wearable or implantable devices where user safety is paramount. This is represented by figure 13. Table 2 discuss the state of art of the antenna design.

Figure 13. SAR of the antenna placed in human body

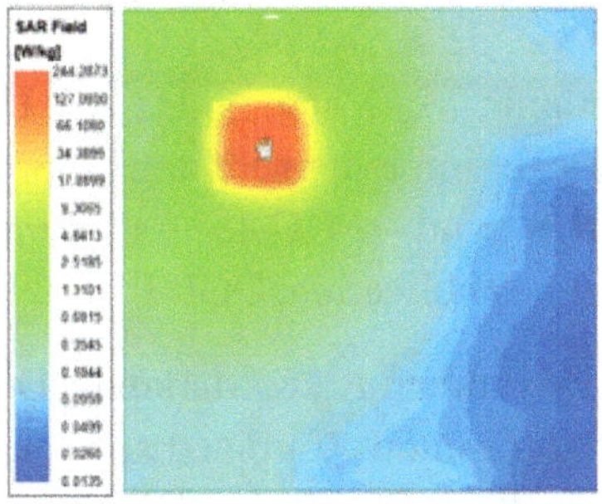

8. CONCLUSION

This study demonstrates the effectiveness of Gaussian Process Regression (GPR). GPR is used as a surrogate modelling technique for characterizing and optimizing compact microstrip antennas. Its applications extend to implantable antenna design. By investigating the dependencies on various parameters, the GPR model achieved low MSE values of 0.06, 0.3, 0.14 for predicting resonant frequency, slot length, and patch length, respectively. These results underscore the accuracy and reliability of the GPR approach in capturing complex relationships between antenna parameters. For implantable antennas, such as the meandered monopole with a bandwidth of 250–300 MHz under tissue, the GPR model can be instrumental in optimizing gain, bandwidth, and SAR levels. This reduces the need for exhaustive simulations. GPR enables rapid design iterations. It ensures compliance with SAR limits 6.56 mW for FCC and 8.2 mW for ICNIRP. It also maintains a desirable performance metrics. The integration of advanced regression techniques into antenna design reduces computational costs. It also provides valuable insights into parameter dependencies. Hence it paves the way for innovative solutions in biotelemetry applications. Future work will focus on extending the GPR to incorporate additional antenna performance metrics, such as circular polarization and multiband operation, to further enhance its utility in biomedical applications.

REFERENCES

Abbas, M., Khan, S. U., Aamir, M., Khan, N. U., Ali, U., Ullah, S., Basir, A., & Björninen, T. (2024). Compact implantable antenna for enhanced Bio-Telemetry communication in the MICS band. *IEEE Access : Practical Innovations, Open Solutions*, *1*, 1. Advance online publication. DOI: 10.1109/ACCESS.2024.3512950

Anwar, U., Ajijola, O. A., Shivkumar, K., & Markovic, D. (2022). Towards a Leadless Wirelessly Controlled Intravenous Cardiac Pacemaker. *IEEE Transactions on Biomedical Engineering*, *69*(10), 3074–3086. DOI: 10.1109/TBME.2022.3161415 PMID: 35320081

Bose, M., & Karuppiah, V. (2024). Metamaterial inspired superstrate loaded miniaturized quad port MIMO antenna for 5G C-Band applications. *Optics Communications*, *131123*. Advance online publication. DOI: 10.1016/j.optcom.2024.131123

Cao, Y., Brubaker, M. A., Fleet, D. J., & Hertzmann, A. (2015). Efficient Optimization for Sparse Gaussian Process Regression. *IEEE Transactions on Pattern Analysis and Machine Intelligence*, *37*(12), 2415–2427. DOI: 10.1109/TPAMI.2015.2424873 PMID: 26539847

Essa, A., Almajali, E., Mahmoud, S., Amaya, R. E., Alja'Afreh, S. S., & Ikram, M. (2024, June). "Wireless Power Transfer for Implantable Medical Devices: Impact of Implantable Antennas on Energy Harvesting,". *IEEE Open Journal of Antennas and Propagation*, *5*(3), 739–758. DOI: 10.1109/OJAP.2024.3392160

Fei, P., Shen, Z., Wen, X., & Nian, F. (2015). A Single-Layer circular polarizer based on hybrid meander line and loop configuration. *IEEE Transactions on Antennas and Propagation*, *63*(10), 4609–4614. DOI: 10.1109/TAP.2015.2462128

Frontoni, E., Romeo, L., Bernardini, M., Moccia, S., Migliorelli, L., Paolanti, M., Ferri, A., Misericordia, P., Mancini, A., & Zingaretti, P. (2020). A decision support system for diabetes chronic care models based on general practitioner engagement and EHR data sharing. *IEEE Journal of Translational Engineering in Health and Medicine*, *8*, 1–12. DOI: 10.1109/JTEHM.2020.3031107 PMID: 33150095

Heo, J., Jang, D., Choo, H., Na, D.-Y., & Bae Park, Y. (2025). Real-Time Duct Prediction and Propagation Analysis Using Genetic Algorithm-Based RFC and Parabolic Equation Method. *IEEE Access : Practical Innovations, Open Solutions*, *13*, 471–479. DOI: 10.1109/ACCESS.2024.3522376

Iqbal, A., Al-Hasan, M., Mabrouk, I. B., & Denidni, T. A. (2024). A compact Self-Triplexing antenna for head implants. *IEEE Transactions on Antennas and Propagation*, *72*(11), 8207–8214. Advance online publication. DOI: 10.1109/TAP.2024.3450332

Jacob, S., Bose, M., & Karuppiah, V. (2024). Artificial Magnetic Conductor-Backed Microstrip-Fed Dual-Band antenna for 5G wearable sensor nodes and C Band application in WBAN network. In *Advances in information security, privacy, and ethics book series* (pp. 221–241). DOI: 10.4018/979-8-3693-2786-9.ch009

Kerkhoff, A. J., & Ling, H. (2007). Design of a Band-Notched Planar Monopole antenna using genetic algorithm optimization. *IEEE Transactions on Antennas and Propagation*, *55*(3), 604–610. DOI: 10.1109/TAP.2007.891563

Khattak, R. Y., Ahmed, Q. Z., Shoaib, S., Lazaridis, P. I., & Alade, T. (2024). Recent advances in antennas for biotelemetry and healthcare applications. *IEEE Open Journal of Antennas and Propagation*, *1*(6), 1499–1522. Advance online publication. DOI: 10.1109/OJAP.2024.3462289

Liu, Y., Chen, Y., Lin, H., & Juwono, F. H. (2016). A novel differentially fed compact Dual-Band implantable antenna for biotelemetry applications. *IEEE Antennas and Wireless Propagation Letters*, *15*, 1791–1794. DOI: 10.1109/LAWP.2016.2536735

Malik, J., Kim, S., Seo, J. M., Cho, Y. M., & Bien, F. (2022). Minimally invasive implant type electromagnetic biosensor for continuous glucose monitoring System: In vivo evaluation. *IEEE Transactions on Biomedical Engineering*, *70*(3), 1000–1011. DOI: 10.1109/TBME.2022.3207240 PMID: 36112548

Meenambal, B., Jacob, S., & Vasudevan, K. (2024). Metamaterial-Inspired concatenated dual ring antenna design for 5G NR C-Band applications. In *Advances in information security, privacy, and ethics book series* (pp. 200–220). DOI: 10.4018/979-8-3693-2786-9.ch008

Mohammed, N., Cluff, K., Griffith, J., & Loflin, B. (2019). "A Non-invasive Wearable Readout System for Bio-fluid Phenomena Detection," *2019 IEEE 16th International Conference on Wearable and Implantable Body Sensor Networks (BSN)*, Chicago, IL, USA, pp. 1-4, DOI: 10.1109/BSN.2019.8771042

Peng, N. J., Li, N. K., & Irwin, G. (2008). A new Jacobian matrix for optimal learning of Single-Layer neural networks. *IEEE Transactions on Neural Networks*, *19*(1), 119–129. DOI: 10.1109/TNN.2007.903150 PMID: 18269943

Pucci, L., Bacchielli, T., & Giorgetti, A. (2025). "Cooperative Maximum Likelihood Target Position Estimation for MIMO-ISAC Networks," in *IEEE Wireless Communications Letters*, DOI: 10.1109/LWC.2025.3548446

Sarker, N., Podder, P., Mondal, M. R. H., Shafin, S. S., & Kamruzzaman, J. (2023). Applications of Machine Learning and Deep Learning in antenna Design, Optimization, and Selection: A review. *IEEE Access : Practical Innovations, Open Solutions, 11*, 103890–103915. DOI: 10.1109/ACCESS.2023.3317371

Sharma, K., & Pandey, G.P. (2021). "Efficient modelling of compact microstrip antenna using machine learning", *International Journal of Electronics and Communications* DOI: 10.1016/j.aeue.2021.153739

Sherene, J., Vasudevan, K., Maheshwari, G. U., & Meenambal, B. (2024). Design and analysis of AMC-Based anti-symmetric Dual L-Shaped antenna for WBAN Application. In *Lecture notes in networks and systems* (pp. 179–194). DOI: 10.1007/978-981-97-1943-3_16

Son, J., & Cha, H. (2020). An implantable neural stimulator IC with anodic current pulse modulation based active charge balancing. *IEEE Access : Practical Innovations, Open Solutions, 8*, 136449–136458. DOI: 10.1109/ACCESS.2020.3012028

Wang, Q., Long, B., Wang, W., Lv, K., Tang, Y., & Xie, M. (2024). "An Enhanced Hard-Decision-Based ML Receiver for UsIM-SS Intra-Body Systems," *2024 IEEE 10th International Symposium on Microwave, Antenna, Propagation and EMC Technologies for Wireless Communications (MAPE)*, Guangzhou, China,, pp. 1-3, DOI: 10.1109/MAPE62875.2024.10813890

Yahya, M. S., Soeung, S., Rahim, S. K. A., Musa, U., Ba Hashwan, S. S., & Haque, M. A. (2024). Machine Learning-Optimized Compact Frequency Reconfigurable Antenna With RSSI Enhancement for Long-Range Applications. *IEEE Access : Practical Innovations, Open Solutions, 12*, 10970–10987. DOI: 10.1109/ACCESS.2024.3355145

Zhang, J., Xu, J., Chen, Q., & Li, H. (2023, August). Machine-Learning-Assisted Antenna Optimization With Data Augmentation. *IEEE Antennas and Wireless Propagation Letters, 22*(8), 1932–1936. DOI: 10.1109/LAWP.2023.3269811

Chapter 6
Adapting 5G Network Slicing Approaches for Wi-Fi:
Strategic Insights for Advancing Multi-Band Resource Management in 6G Networks

S. Thangam
http://orcid.org/0000-0003-2251-3651
Amrita School of Computing, Amrita Vishwa Vidyapeetham, Bengaluru, India

Amireddy Shashank Reddy
http://orcid.org/0009-0006-3260-6535
Amrita School of Computing, Amrita Vishwa Vidyapeetham, Bengaluru, India

Madithilveedu Sheethal Mol
http://orcid.org/0009-0002-2204-2036
Amrita School of Computing, Amrita Vishwa Vidyapeetham, Bengaluru, India

Dishan Narsimhachari Achar
http://orcid.org/0009-0004-6184-1038
Amrita School of Computing, Amrita Vishwa Vidyapeetham, Bengaluru, India

Jesy Janet Kumari
http://orcid.org/0000-0002-3318-1632
Amrita School of Computing, Amrita Vishwa Vidyapeetham, Bengaluru, India

DOI: 10.4018/979-8-3373-0852-4.ch006

ABSTRACT

This research explores the applicability of employing the concepts of multi-band slicing proposed in the context of 5G networks to Wi-Fi networks, thus, opening the way to adopting enhanced resource management strategies that will be essential for 6G networks. Multi-band slicing enables the dynamic supply of resources at multiple communication bands, which is a crucial feature when it comes to meeting the complex and often very high 6G requirements such as ultra-reliable low-latency communication (URLLC), enhanced mobile broadband (eMBB), and massive machine-type communication (mMTC). These techniques accommodate the increase in next-generation wireless communication while efficiently managing bandwidth and offering adaptable control over traffic flow. This research is a comprehensive and primary suggestion for applying slicing mechanisms in 6G architectures, as it provides guidance for constructing efficient and scalable networks for the next generation of communications based on the needs and requirements posed by new technologies and services.

I INTRODUCTION

The advancement in 5G technology has transformed the telecommunication industry with the advancement in speed, latency, and density. The most significant feature of 5G is called network slicing, where the actual configuration of the network is partitioned to provide multiple networks over the same physical equipment. This capability has transformed various use cases across the board, such as streaming multimedia, low latency services, and industrial IoT (Kaloxylos, 2018; Kumari, Thangam, & Raja, 2023; Zhang, 2019). However, conventional WLANs cannot support slicing and therefore not very effective in responding to the several needs of today's wireless services (Richart et al., 2020).

This research proposes integrating 5G-inspired network slicing into Wi-Fi networks. The system can efficiently allocate resources to optimize throughput, latency, and spectrum efficiency by enabling Wi-Fi access points to support multiple logical slices with distinct characteristics. Such slicing empowers Wi-Fi to handle diverse traffic requirements and enhances its utility in increasingly complex environments. This approach not only improves Wi-Fi's ability to address modern demands but also bridges the gap between 5G and WLANs (Carmo et al., 2018).

Looking ahead to 6G, the proposed framework lays the foundation for next-generation wireless networks. Considering that 6G is likely to offer terabit speeds, ultra-low latency, and intelligent connectivity, the application of additional techniques like real-time optimization with the neural network will improve Wi-Fi slicing. These developments establish Wi-Fi as an essential facilitator of future wireless systems as the system complexity and requirements for 6G and beyond increase. The rest of the sections in the paper are organized as follows. The related work is discussed in Section 2, Section 3 contains the proposed methodology, process description, and System Design, and Section 4 contains results. Lastly, Section 5 contains the conclusion and future work.

II RELATED WORK

Olimid and Nencioni (Olimid & Nencioni, 2020) give a comprehensive security analysis of the 5G network slicing with special emphasis on the security threats and approaches to the real issues. They talked about the services lifecycle for securing tenant domains and outlined the possible threat that can be mitigated by adopting slice isolation as the possible mitigation measure.

Zhang et al. cooperated in proposals for an overall logical architecture of the network slicing (Zhang et al., 2017) and joint power and sub-channel allocation for spectrum-sharing in two-tier systems. On self-organized networking, concerns handover from one access network to the next and Interference control. Zhang et al. also mentioned some of the issues such as the reconstruction of network architecture and other interfacing with other 5G technologies; they stated that they deserve further study in the future.

Chahbar et al. (Chahbar et al., 2021) presented a comprehensive literature evaluation of the (E2E) 5G NS models, and the standardization scope, opportunities, and challenges for various SDOs. They assessed how slicing models could be added to RAN, CN, and TN and proposed an architecture. However, their work aggregates various contributions from a particular domain into a single E2E network slicing framework.

Thangam et al. (Thangam & Kirubhakaran, 2012) developed a QoS-aware service selection architecture in pervasive context, achieving solutions to instability. Incorporating cross layer mechanisms and using cooperative nodes that have high server connectivity, their QSMP algorithm offered the best choice of service according to the different QoS demands. According to simulation, existence of delay, packet drop rates increased and delivery ratios have proven it suitable for healthcare in rural areas. This work demonstrates the possibility of applying the presented solutions at the Internet of Things and smart city scales.

Dangi et al. (Dangi et al., 2022) described the state of the art of machine learning (ML) in security for 5G network slicing. The authors discussed how ML and deep learning are achievable for different stages of the network slicing life cycle in planning, designing, deployment, and monitoring. This paper deals with the identification of the possible security threats associated with 5G network slicing and to address them with the help of ML algorithms. It offers a roadmap about the effects of implementing ML-based methods by giving a taxonomy of network slice security throughout different application areas.

Su et al. (Su et al., 2019) have analyzed the task of resource management by employing 5G network slicing focusing on the scalability, elasticity, and isolation of resources. The authors present models of resource distribution based on the application of game theory and linear programming. The general scope of the proposal is resource management through SDN and NFV for small RAN and CN. Su et al. also raised questions related to the end-to-end coordination and isolation in 5G networks.

Josip Lorincz et al. (Lorincz et al., 2024) proposed a comprehensive overview of network slicing to improve the energy efficiency of 5G networks. They discussed specific slicing cases, approaches to resource sharing, and the state of AI development in managing network slices. In this regard, the authors argue that if network slicing is adopted as a strategic business model, it will reduce operating cost and CO2 emissions that are in line with global energy standards.

Amika Pal Sundan et al. in (Sundan et al., 2020) presented a probabilistic-based spectrum slicing for energy and spectral efficiency analysis in 5G networks. They have also employed certain user density approaches in different geographical areas and the Hidden Markov Model for spectrum allocation. Their approach was an attempt to finding a way of informing about the real-time adjustments in the delivery of services of the network and the resource.

Sunday Oladejo et al. (Oladejo & Falowo, 2019), introduced a greedy resource allocation algorithm for total power saving of the slice networks in 5G using stochastic geometry. For this, they have utilized the mixed-integer linear fractional optimization approach to maximize the QoS for many slices and considered the sustainability and efficiency of the networks.

Alessandro Lieto et al. (Lieto et al., 2020) tackled the dynamic aspect of network slicing in radio access networks when it comes to the use of game theory for resource allocation among multiple players. This market-oriented approach allowed tenants to adapt quickly to changes in the usage of resources to minimize costs as well as to make profits without compromising on the quality of services being provided.

Bho Matthiesen et al. (Matthiesen et al., 2018) explained the way of enhancing the throughputs and power consumption in the network slicing in case of having Heterogenous QoS demands. They presented the dynamic bandwidth and power control algorithms, suggesting that they are better than the static ones in 5G systems.

K. Jaya Sampath et al. in (Abhivarun et al., 2023) have proposed a new High-Level Synthesis based channel coding paradigm for the 5G network to improve the throughput and latency. The study looked at LDPC codes that were employing new approaches like the double diagonal structures by parity check matrices and the layered decoding algorithm. They successfully maximized the use of areas and resources provided to them and provided solutions adhering to the standards of 5G technology.

M. V. Sri Shiva Shourya Sharma et al. (Gokul & Sankaran, 2021) proposed a deep learning-based detection mechanism for mitigating Resource Depletion Attacks in 5G networks. With LSTM and GRU models, they obtained detection accuracies of 97.62% & 99.07%. The approach involved experimentalism with UERANSIM and Open5gs with an emphasis on the security of the 5G architecture where deep learning was deemed as important.

Gokul NB et al. (Sharma et al., 2024) put forward an algorithmic load balancing defense mechanism for RDAs specifically for base stations in the 5G network. Their model employed NS-3 simulations to assess the effects of botnets and negative connections. They effectively used least connection methods for distributing loads and showed robustness in minimizing the effects of attacks without additional hardware.

Abhivarun S N et al. in (Sampath et al., 2021) proposed a MIMO antenna system for 5G mobiles in the N77 band. They employed wide-band radiation isolation and distinct active components such as PIN and varactor diodes with an isolation level above 17 dB and a gain factor higher than 8 dB. Hence, their design focuses on the practical aspect of efficient isolation as well as compactness that are relevant for the 5G smartphones.

While significant research exists on network slicing in 5G networks, its adaptation to Wi-Fi remains underexplored. Most studies focus on resource allocation or QoS guarantees individually, lacking comprehensive frameworks for diverse traffic types in Wi-Fi. Additionally, there is limited investigation into energy efficiency for IoT deployments and real-time adjustments of bandwidth and power, essential for handling mixed traffic. These gaps highlight the need for tailored, dynamic slicing algorithms for Wi-Fi.

Methodology

The methodology introduces how resources across three slices, enhanced Mobile Broadband (eMBB), massive Machine-Type Communication (mMTC), and Ultra-Reliable Low-Latency Communication (URLLC) adjusts parameters such as channel width, transmission power, and modulation schemes in real- time to highlight the performance improvements in throughput, latency, and reliability. The traditional single band Wi-Fi setup is compared against rule-based dynamic slicing algorithm inspired by 5G to optimize Wi-Fi networks. Both approaches are evaluated in the same simulation environment using the **ns-3 simulator** integrated with **ns3-gym**, ensuring a fair comparison. The design and implementation of the dynamic slicing algorithm target the optimization of network resources. The single-band and dynamic slicing Wi-Fi techniques are explained below, with a focus on their implementation and key features.

Figure 1. Architecture of the network slicing framework in the dynamic Wi-Fi optimization approach

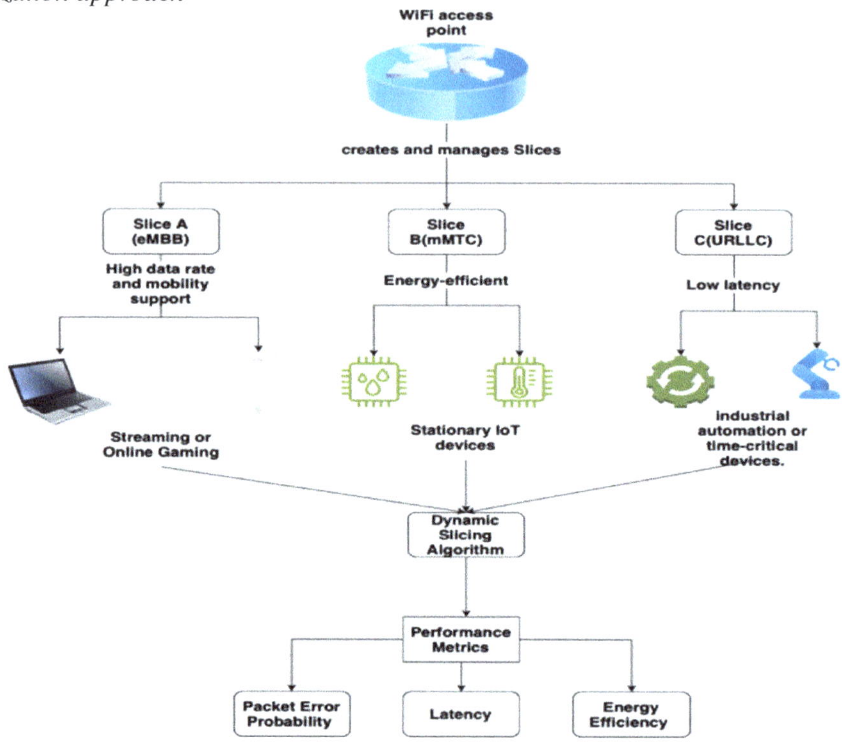

a. Single Band Wi-Fi Simulation

The single-band Wi-Fi simulation is used as a reference for comparing the efficiency of the proposed dynamic slicing algorithm. In this configuration, all the devices operate in a single channel that has a defined bandwidth of 20 MHz only. The transmission power is to be fifteen dBm, and the Modulation and Coding Scheme (MCS) needs to be specified as five. This setup is reminiscent of the problem faced by Wi-Fi, wherein resources are shared without considering certain types of traffic.

Due to the similarity in the distribution of resources, there is always high interference and deficient performance, especially under conditions where there is both voice and data traffic. The throughput, latency, and packet loss are metrics summed up over all devices and used to conclude the drawbacks of this approach.

b. Dynamic Slicing Algorithm

The **Dynamic Slicing Algorithm for Wi-Fi Networks** allows the modification of parameters on the go to cater to different types of traffic for the various slices. The algorithm changes between the mMTC, eMBB, and URLLC states according to the current evaluations of throughput and latency, as shown in Algorithm 1.

b. 1 Slice Characteristics

Table 1. Slice Characteristics

Slice	Description	Throughput Range	Latency Range
mMTC	Designed for IoT devices with low data rates and energy efficiency.	30–50 kbps	100–300 ms
eMBB	Supports high-bandwidth applications like video streaming.	80–100 Mbps	50–100 ms
URLLC	Ensures ultra-low latency and high reliability for critical applications.	20–40 Mbps	< 5 ms

Every slice is designed to facilitate certain traffic types with specific performance characteristics. The specifications of these slices in terms of their throughput and latency are provided in Table 1

b. 2 Slice Specific Details

Table 2. Slice Transition according to Algorithm.

Traffic Type	State Transition Condition	Action	Optimization Focus
mMTC	Latency > 100 ms or Throughput < 10 Mbps	Increase channel width (max 40 MHz), adjust power (max 20 dBm)	Energy efficiency for IoT devices
eMBB	Latency > 50 ms or Throughput < 50 Mbps	Increase channel width (max 80 MHz), adjust power (up to 20 dBm)	High throughput for video streaming
URLLC	Latency > 5 ms or Packet Loss > 0.001%	Maximize channel width (80 MHz), set power to 20 dBm	Low latency and critical reliability

The dynamic slicing algorithm's behavior is described in detail in Table 2, including the state transition rules, actions, and optimization objectives for each traffic type. The table shows the flexibility of the algorithm with the number of resources for mMTC, eMBB, and URLLC slices depending on their real-time requirements. For example, in mMTC if delay goes up to one hundred ms or if the data rate drops down to 10 Mbps, the algorithm increases the channel width (up to 40 MHz) and optimizes the transmit power for better operation. In eMBB, either high latency or low throughput triggers the algorithm to assign up to 80 MHz bandwidth and greater transmission power for high throughput data applications. For URLLC the algorithm assigns greater weight for ultra-reliable low-latency communications by tuning bandwidth and transmission power if latency is above 5ms or packet loss is above 0.001%. This also serves to prioritize URLLC traffic so that it avails itself at the required performance levels. This detailed breakdown shows how the algorithm effectively changes based on different traffic demands and provides the platform enough throughput for each traffic type.

b. 3 Parameter Adjustment Logic

The algorithm dynamically adjusts three critical parameters—channel width, transmission power, and modulation and coding scheme (MCS) in a way that each slice requires. Channel width, transmission power, controlling signal reliability and range, is carefully optimized for each slice. Modulation and coding schemes are adjusted based on real-time channel conditions to optimize the speed and reliability of data transmission. Higher MCS values can deliver higher data rates but do require signal strength the algorithm controls to meet the required level.

This dynamic adjustment mechanism allows the algorithm to effectively balance resource allocation and performance optimization, ensuring that each slice achieves its intended goals. The mMTC slice conserves energy for IoT devices, the eMBB slice delivers consistent high throughput for bandwidth-intensive applications, and the URLLC slice guarantees low latency and high reliability for critical operations, outperforming traditional single-band Wi-Fi configurations.

c. Performance Metrics

Table 3. Performance Metrics

Metric	Description
Throughput (Mbps)	Measures the average data transfer rate. High throughput indicates better performance.
Latency (ms)	Represents the delay experienced by packets. Lower latency is critical for URLLC applications.
Packet Loss (%)	It indicates the percentage of lost packets. Lower values reflect greater reliability.

Table 3 lists the key performance metrics used to evaluate dynamic Wi-Fi slicing. Performance is evaluated using five key metrics, each of which is critical to understanding the benefits of dynamic slicing.

The metrics are logged and analysed for each traffic slice under both single-band and dynamic slicing configurations. This approach provides a comprehensive understanding of the algorithm's impact on Wi-Fi performance.

III IMPLEMENTATION AND RESULTS

The research is implemented using the ns-3 network simulator, integrated with the ns3-gym extension to enable real-time Python-based control. extension that allows Python to control the events in real-time. Such a setup is useful for arriving at accurate simulations and assessments of the dynamic slicing algorithm. The environmental conditions are set to mimic real life working or home space, thus making it easier to compare single-band conventional Wi-Fi with the use of slicing techniques.

The specific layout of the simulation area is 20 × 10-meter environment, where Wi-Fi access points are installations at a height of approximately 3 meters from the floor. Devices are randomly placed throughout the area while some devices are moved around to simulate real world usage. The mobility of the user is chosen as a random walk model and the speeds of pedestrian's range in between 1m/sec to 2m/s.

Table 4. Results for Dynamic slicing

Slice	Iteration	Step	Throughput	Latency	Packet loss	txPower	Channel Width
eMBB	1	1	95	56	0	15	20
eMBB	1	2	95	72	865	15	40
mMTC	1	1	40.7	111	0	14	20
mMTC	1	2	40.7	171	2.9	13	20
URLLC	1	1	34.5	6	0	15	20
URLLC	1	2	34.5	2	252	20	40

The research focuses on three distinct traffic categories as shown in Table 4. enhanced Mobile Broadband (eMBB), massive Machine-Type Communication (mMTC), and Ultra-Reliable Low-Latency Communication (URLLC). Each category is assigned certain traffic and mobility characteristics to mimic different application scenarios. Characteristic features of all simulation experiments include the duration of the simulation being 20 seconds, and steps during the experiment being completed each second. To increase the reliability of results, all experiments are repeated 10 times.

Table 5. Results for Single-band environment

Iteration	Step	Throughput	Latency	Packet loss	txPower	Channel Width
1	1	55.44	0	0	15	20
1	2	55.44	289	575	15	20
1	3	55.44	577	884	15	20
1	4	55.44	866	951	15	20
1	5	55.44	1155	1028	15	20

As shown in Table 5, different performance parameters were evaluated under the dynamic Wi-Fi slicing and single-band Wi-Fi, including the throughput, latency, packet loss, transmission power, and channel width. The generated plots illustrate how dynamic slicing adheres to the demand for certain slices and then optimizes based on use cases such as eMBB, mMTC and URLLC.

The findings shown in Table 4. and Table 5. indicate that such performance advantages are achievable due to dynamic slicing when contrasted with single-band Wi-Fi. Such performance benefits relative to single-band Wi-Fi are made possible by dynamic slicing, as shown by these results.

a. Latency Distribution by Slice and Network Type

Figure 2. Latency Distribution by Slice and Network Type

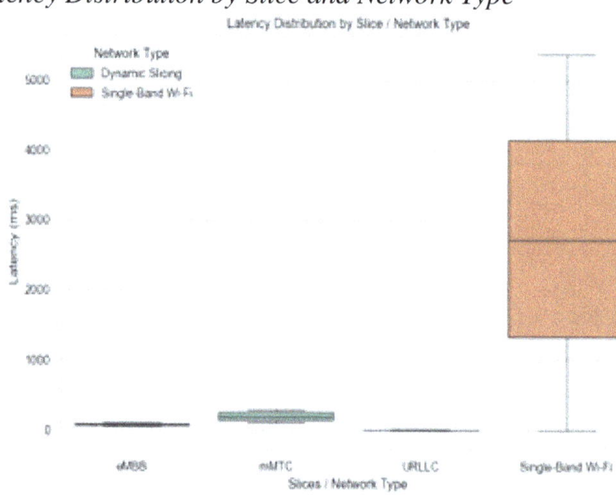

As per the latency distribution graph, dynamic slicing wins by a clear margin for meeting slice-specific requirements. Latency is kept low (min 1–10 ms range) for URLLC and can provide real-time communication. With mMTC devices, as expected, the latency is medium in the one hundred–three hundred ms range due to energy efficiency being paramount here and thus low latency is less of a focus for the algorithm. On the other hand, eMBB scores 50–one hundred ms latency attuned for applications with high-throughput requirements such as video streaming. On the other hand, single-band Wi-Fi shows much higher latency numbers over five thousand ms in many cases because there are no optimizations on a specific slice. This demonstrates that single-band Wi-Fi is inefficient at coping with traffic diversity demands illustrated in Fig. 3

b. Latency Comparison Over Simulation Steps

Figure 3. Latency Comparison (Dynamic Slicing vs Single-Band)

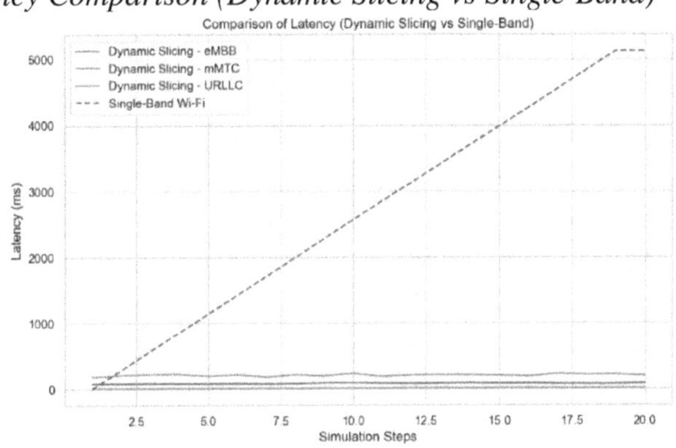

The temporal analysis is shown in Fig. 4. The advantages of dynamic slicing become even stronger and further reinforce the benefits of dynamic slicing. Latency stays constant for all the slices in which time frame, portraying the stability of the simulated algorithm. Hence, it retains its low latency for URLLC and preserves the upper limits of mMTC and eMBB remaining within their respective bounds In contrast, the latency of single-band Wi-Fi continuously increases, showing a linear growth tendency with simulation steps. The resultant increase is one symptom of congestion and resource contention, validating how limited single-band Wi-Fi becomes in a dynamic environment.

c. Overall Channel Width Comparison

Figure 4. Overall Channel Width Comparison

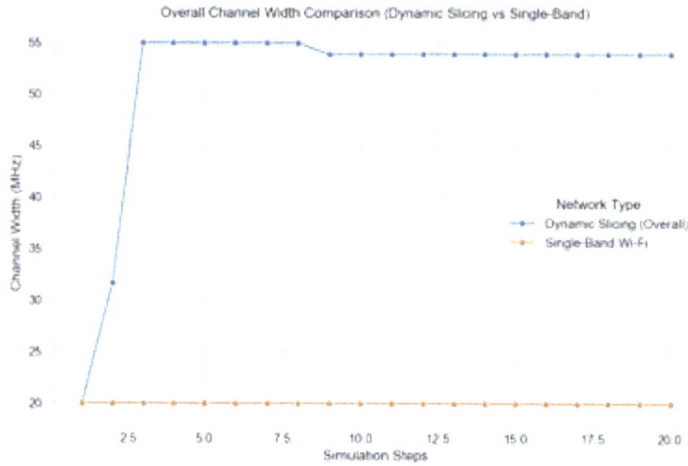

From the channel width graph, we can observe how dynamic slicing is able to arrange resources in an increasing manner. The overall channel width rises to as much as 55 MHz in total (for all slices), though this is done to satisfy the demands of such slices as eMBB, proven by the versatility of the algorithm under discussion. On the other hand, single-band Wi-Fi continues to operate a channel breadth of 20 MHz, which limits its potential to support different traffic patterns. This inability to vary bandwidth depending on the current network traffic is one factor that leads to its poor performance based on the results captured in Fig. 5

d. Overall, Packet Loss Comparison

Figure 5. Overall Packet Loss Comparison

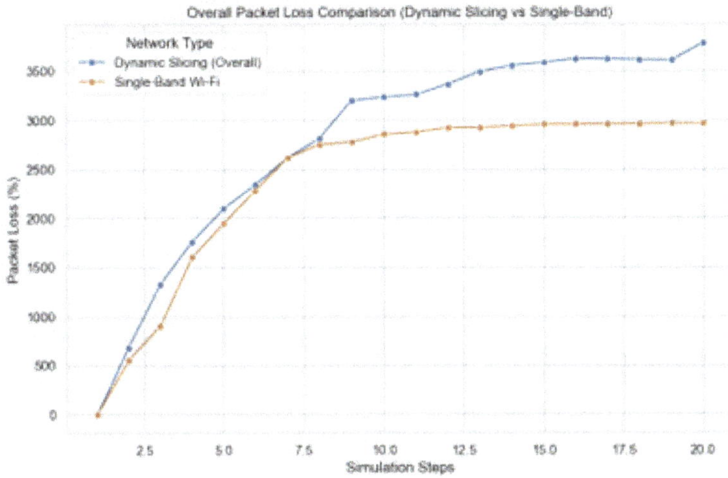

Relative packet loss shows the effects of dynamic slicing for network dependability. In the first phase of simulation, the breakdown of system availability into slices shows it to have a much lower packet loss than what was recorded in the single-resource approach in dynamic slicing that can assign resources with slice priority. However, as the simulation progresses and traffic demands increase, packet loss stabilizes at a slightly higher level due to the prioritization of critical slices like URLLC and eMBB. Single-band Wi-Fi keeps a constant level of packet loss and challenges in not being able to adjust to the difference in traffic load or support important applications as illustrated in Fig. 6

e. Overall Throughput Comparison

Figure 6. Overall Throughput Comparison

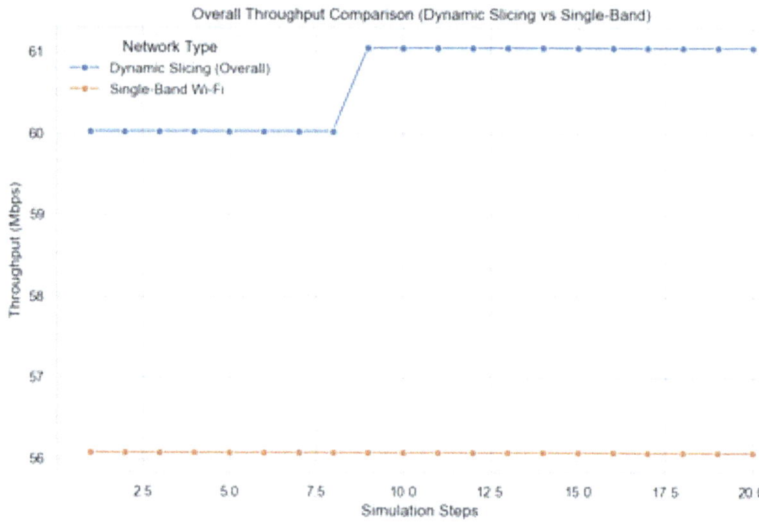

The throughput graph in Fig. 7 also illustrates how dynamic slicing yields higher cumulative throughput. Dynamic slicing fulfils a sustained throughput of around 60-61 Mbps, and eMBB optimizations contribute significantly. Therefore single-band Wi-Fi rates reach 56 Mbps, which marks a pathetic performance since it does not have a mechanism of providing resources to high-priority applications dynamically. In fact, thanks to slice-specific bandwidth allocation and optimized transmission power, dynamic slicing has significantly improved throughout.

f. Throughput Comparison by Slice

Figure 7. Throughput Comparison by slice

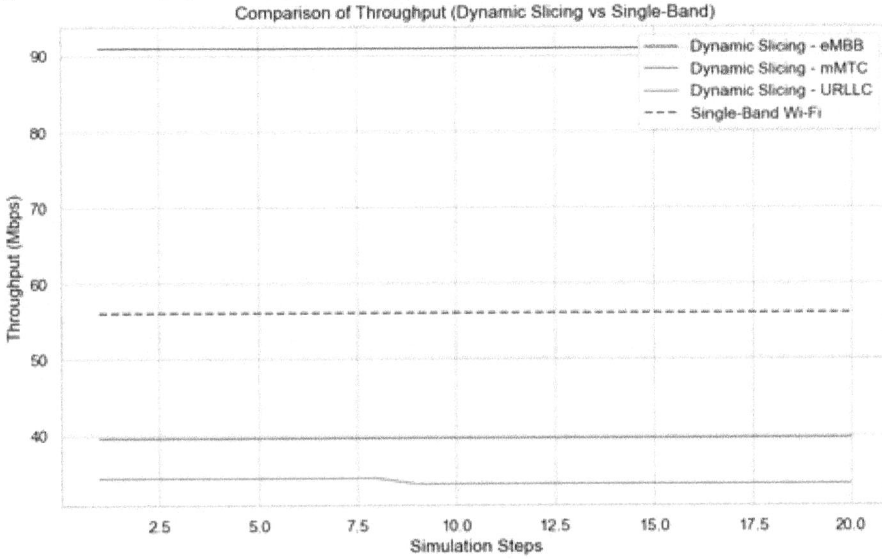

Additional comparisons plotted in terms of throughput by slice in the Fig. 8 reveal further gain on the dynamic slicing technique offers eMBB slice, in terms of highest throughput greater than 90 Mbps since the algorithm targets to meet the demands of the throughput–oriented traffic. While mMTC and URLLC slices maintain stable throughputs of almost 40 Mbps and 30 Mbps respectively, to meet their respective traffic types. On the other hand, single band Wi-Fi distributes throughput evenly across all types of traffics without differentiating between high bandwidth or latency sensitivity applications. This uniformity especially reveals the inefficiency of single-band Wi-Fi in mixed-traffic operation mode.

g. Throughput Distribution by Slice and Network Type

Figure 8. Throughput Distribution by slice and Network Type

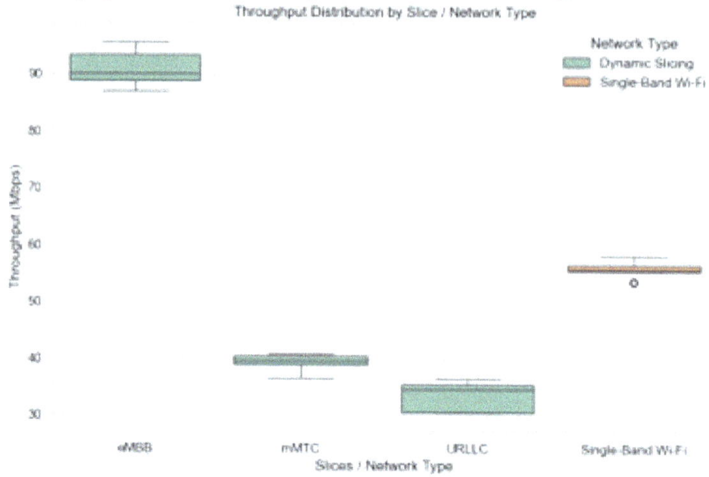

To drive the point home about how reliable dynamic slicing is, the throughput distribution graph in Fig.9 is presented. Two examples demonstrate how narrow variability of throughput is in each slice means that the performance is equitably split across simulation steps. On the other hand, single-band Wi-Fi shows greater fluctuation in throughput, which portrays the problems that exist in dealing with different traffic types. The tight distributions in dynamic slicing also confirm that this approach provides stable throughput, even when the network conditions are different.

Overall Comparison

The results clearly show that dynamic Wi-Fi slicing significantly outperforms single-band Wi-Fi across key metrics, demonstrating its adaptability to varying traffic demands. Throughput improves by 8.14%, with dynamic slicing achieving an average of 60.64 Mbps compared to 56.08 Mbps for single-band Wi-Fi. Latency sees a dramatic reduction of 95.59%, dropping from 2685.18 ms in single-band to just 118.41 ms, essential for time-sensitive applications like URLLC. Packet loss improves by 16.56%, while transmission power decreases by 17.69%, emphasizing energy efficiency without performance loss. Dynamic slicing also increases channel width by 157.16%, enabling better resource allocation for high-throughput and low-latency demands.

In contrast, single-band Wi-Fi suffers from high latency, limited throughput, and an inability to adapt to diverse traffic needs, making it inefficient for mixed-traffic environments. These results position dynamic Wi-Fi slicing as a superior solution for improving wireless network performance in future systems.

IV CONCLUSION

This research shows the effectiveness of dynamic Wi-Fi slicing, grounded on 5G network slicing principles. Dynamic slicing consistently provides lower latency, and higher throughput, and is more resource-efficient than traditional single-band Wi-Fi by tailoring the resource allocation according to the nature of traffic. The results show that dynamic slicing has great promise in satisfying various types of wireless communication scenarios, especially for mixed-traffic environments.

With the proposed approach, slices such as eMBB, mMTC and URLLC can receive specialized resource optimizations like dynamic channel width and transmission power adjustments. These adaptations enhance throughput and decrease latency for time-sensitive applications, validating the concept in contemporary wireless architectures. However, the study also identifies areas for improvement, such as packet loss, which remains a limitation of the current dynamic slicing framework. Addressing this issue can further enhance the robustness and reliability of dynamic slicing systems.

Looking ahead, the insights from this study pave the way for extending these concepts to the realm of 6G networks. With the advent of 6G, which promises even greater speeds, ultra-low latency, and massive connectivity, the principles of dynamic slicing can be expanded and refined. Future research can explore the integration of advanced techniques, such as neural network-based models, to further enhance the system's ability to dynamically adapt to traffic demands. By leveraging the predictive and adaptive capabilities of machine learning, future implementations can achieve faster decision-making and more precise resource allocation, leading to an overall improvement in network performance.

A further key future research path is correcting the larger packet loss in implemented dynamic slicing seen herein. To prevent packet loss and to guarantee reliable communication (especially in the context of URLLC), strategies like improved error correction, retransmission methods or predictive models can be investigated.

This work not only demonstrates the short-term benefits of 5G principles to realize dynamic Wi-Fi slicing but serves as a basis for expanding support in 6G. By leveraging improved resource allocation approaches and advanced machine learning techniques, future implementations can address scalability issues and unleash full potential of dynamic slicing in future wireless systems.

REFERENCES

Abhivarun, S. N., Sankar, A., Harshan, K., Priharini, S., & Natarajamani, S. (2023, November). Design and Performance Evaluation of MIMO Antenna for 5G Smartphones. In 2023 IEEE 3rd International Conference on Applied Electromagnetics, Signal Processing, & Communication (AESPC) (pp. 1-4). IEEE.

Carmo, M., Jardim, S., Neto, A., Aguiar, R., Corujo, D., & Rodrigues, J. J. P. C. (2018). Slicing WiFi WLAN-Sharing Access Infrastructures to Enhance Ultra-Dense 5G Networking. In: *IEEE International Conference on Communications (ICC)* DOI: 10.1109/ICC.2018.8422769

Chahbar, M., Diaz, G., Dandoush, A., C'erin, C., & Ghoumid, K. (2021). A Comprehensive Survey on the E2E 5G Network Slicing Model. *IEEE Transactions on Network and Service Management, 18*(1), 49–62. DOI: 10.1109/TNSM.2020.3044626

Dangi, R., Jadhav, A., Choudhary, G., Dragoni, N., Mishra, M. K., & Lalwani, P. (2022). ML-Based 5G Network Slicing Security: A Comprehensive Survey. *Future Internet, 14*(4), 116. DOI: 10.3390/fi14040116

Gokul, N., & Sankaran, S. (2021, December). Modeling and defending against resource depletion attacks in 5g networks. In 2021 IEEE 18th India Council International Conference (INDICON) (pp. 1-7). IEEE.

Kaloxylos, A. (2018). A Survey and an Analysis of Network Slicing in 5G Networks. *IEEE Communications Standards Magazine, 2*(1), 60–65. DOI: 10.1109/MCOMSTD.2018.1700072

Kumari, J. J. J., Thangam, S., & Raja, A. S. (2023, October). Efficacious Routing Approaches in Vehicular Ad hoc Network: An Empirical Study. In *2023 First International Conference on Advances in Electrical, Electronics and Computational Intelligence (ICAEECI)* (pp. 1-7). IEEE.

Kumari, J. J. J., Thangam, S., & Raja, A. S. (2023, December). An Adaptive on Demand Modified Ant Colony Optimization Routing for VANETs. In International Conference on Intelligent Systems Design and Applications (pp. 480-492). Cham: Springer Nature Switzerland.

Kumari, J. J. J., Thangam, S., & Raja, A. S. (2024, March). QoS CBSC: An Enhanced Metaheuristic Strategy on QoS-Cloud-based Service in Cloud. In 2024 5th International Conference on Intelligent Communication Technologies and Virtual Mobile Networks (ICICV) (pp. 685-691). IEEE.

Lieto, I., Malanchini, I., Mandelli, S., Moro, E., & Capone, A. (2020). Malanchini, S. Mandelli, E. Moro, A. Capone: Strategic Network Slicing Management in Radio Access Networks. *IEEE Transactions on Mobile Computing*, *21*(4), 1434–1448. DOI: 10.1109/TMC.2020.3025027

Lorincz, J., Kukuruzovi'c, A., & Blažević, Z. (2024). Z. Blaˇzevi'c: A Comprehensive Overview of Network Slicing for Improving the Energy Efficiency of Fifth-Generation Networks. *Sensors (Basel)*, *24*(10), 3242. DOI: 10.3390/s24103242 PMID: 38794095

Matthiesen, B., Aydin, O., & Jorswieck, E. A. (2018). Throughput and Energy-Efficient Network Slicing. In: WSA 2018; 22nd International ITG Workshop on Smart Antennas, pp. 1–6

Oladejo, S., & Falowo, O. (2019). An Energy-Efficient Resource Allocation Scheme for 5G Slice Networks. In: *Proc. Southern Afr. Telecommun. Netw. Appl. Conf.*, pp. 1–4

Olimid, R. F., & Nencioni, G. (2020). 5G Network Slicing: A Security Overview. *IEEE Access : Practical Innovations, Open Solutions*, *8*, 99999–100009. DOI: 10.1109/ACCESS.2020.2997702

Richart, M., Baliosian, J., Serrat, J., Gorricho, J.-L., & Ag¨uero, R. (2020). Slicing With Guaranteed Quality of Service in WiFi Networks. *IEEE Transactions on Network and Service Management*, *17*(3), 1822–1837. DOI: 10.1109/TNSM.2020.3005594

Sampath, K. J., Kumar, N. K., Yeswanth, K., Snehith, K., & Anooja, B. (2021, June). An efficient channel coding architecture for 5G wireless using high-level synthesis. In 2021 5th International Conference on Trends in Electronics and Informatics (ICOEI) (pp. 674-680). IEEE.

Sharma, M. S. S. S., Sai, M. T. V., & Sankaran, S. (2024, August). Deep Learning based approach for detection of Resource Depletion Attacks in 5G Networks. In 2024 IEEE 5th India Council International Subsections Conference (INDISCON) (pp. 1-6). IEEE.

Su, R., Zhang, D., Venkatesan, R., Gong, Z., Li, C., Ding, F., Jiang, F., & Zhu, Z. (2019). Resource Allocation for Network Slicing in 5G Telecommunication Networks: A Survey of Principles and Models. *IEEE Network*, *33*(6), 172–179. DOI: 10.1109/MNET.2019.1900024

Sundan, A. P., Jha, R. K., & Gupta, A. (2020). Energy and Spectral Efficiency Optimization Using Probabilistic Based Spectrum Slicing (PBSS) in Different Zones of 5G Wireless Communication Network. *Telecommunication Systems*, *73*(1), 59–73. DOI: 10.1007/s11235-019-00598-0

Thangam, S., & Kirubhakaran, E. (2012). QoS-aware Service Selection architecture for pervasive environment. *European Journal of Scientific Research*, *80*(2), 159–168.

Zhang, H., Liu, N., Chu, X., Long, K., Aghvami, A.-H., & Leung, V. C. M. (2017). Network Slicing Based 5G and Future Mobile Networks: Mobility, Resource Management, and Challenges. *IEEE Communications Magazine*, *55*(8), 138–145. DOI: 10.1109/MCOM.2017.1600940

Zhang, S. (2019). An Overview of Network Slicing for 5G. *IEEE Wireless Communications*, *26*(3), 111–117. DOI: 10.1109/MWC.2019.1800234

Chapter 7
Analysis of Intermodulation Distortion in Integrated RF Coexistence

Ayushi Sanjay Jaiswal
http://orcid.org/0009-0003-8480-9332
Visvesvaraya National Institute of Technology, Nagpur, India

Ankit A. Bhurane
http://orcid.org/0000-0001-8181-7685
Visvesvaraya National Institute of Technology, Nagpur, India

ABSTRACT

The integration of multiple wireless technologies within the same system on a chip has significantly enhanced connectivity and device interoperability. However, the close proximity of antennas and the non-linear behavior of RF front-end components introduce signal coupling, leading to the generation of intermodulation and harmonic distortions. These distortion not only degrade system performance but also cause spectral leakage. As modern wireless systems increasingly rely on multi-radio coexistence, understanding the underlying mechanisms of these distortions becomes crucial. This chapter provides a comprehensive analysis of intermodulation distortions, examining their spectral characteristics, contributing factors, and impact on system efficiency. Additionally, it explores key design parameters that influence distortion severity and discusses mitigation strategies aimed at minimizing interference. By addressing these challenges, the chapter offers valuable insights into optimizing wireless communication systems for improved reliability and regulatory compliance.

DOI: 10.4018/979-8-3373-0852-4.ch007

INTRODUCTION

Multiple wireless technologies like Wi-Fi, Bluetooth and Zigbee are becoming more prevalent as the applications in wireless communication expand. These wireless technologies are essential and cater to different use cases based on range, power consumption, and data transfer rates. Wi-Fi, which is governed by IEEE 802.11 standards, provide high speed connectivity for internet access and local networking, making it suitable for homes, offices, and public spaces. (Pahlavan & Krishnamurthy, 2021) Bluetooth technology, which is based on IEEE 802.15.1 standard, is optimized for short range and low power communication, and is commonly used for connecting personal devices such as smartphones, headphones, and wearables. (Bluetooth SIG, Inc., n.d.) Another steadily growing wireless technology is Zigbee, which follows the IEEE 802.15.4 standard and is designed for low power, low data rate applications, often employed in smart home automation and industrial IoT systems. (Ramya, Shanmugaraj, & Prabakaran, 2011) Together, these technologies enhance user experience and operational efficiency across various domains.

As modern devices increasingly require seamless connectivity across multiple protocols, the integration of multiple radios within a single system has largely increased. Implementation of multiple radios on the same system on chip (SoC) improves the range and device efficiency. However, this integration leads to generation of several distortions. The major causes of these unwanted distortion signals are the closely positioned antennas on the same chip, supporting multiple radios, and the overlapping frequency bands in which these wireless radios operate. (Garroppo et al., 2011)

When two antennas are in close proximity on a SoC, some part of the transmit signal from one antenna may leak into the RF transmit chain of another. The nonlinear behavior of the components and devices present immediately before the transmitting antennas in the RF transmit chain leads to coupling of the two signals due to less isolation in between the collocated antennas. This results in the generation of undesired signals, like harmonics or the intermodulation products, which occur at frequencies near the intended transmission channels. Additionally, other undesired signals may be present at the output. However, this chapter primarily focuses on distortions caused by harmonics and intermodulation products.

Another reason for concern is that WLAN and Narrowband technologies (Bluetooth/ Zigbee) operate in the same 2.4GHz unlicensed Industrial, Scientific and Medical (ISM) frequency band. Since the two transmitted signals are in this same frequency range the likelihood of coupling of the two signals is significantly high, which consequently leads to distortions. (Challoo et al., 2012) The frequency bands adjacent to the ISM spectrum are licensed by regulatory bodies and designated for specific applications. However, distortions originating from signals within the

ISM band can extend into these neighboring bands, causing interference with other systems. Such interference not only disrupts proper functionality but also breaches regulatory compliance, making it a critical issue.

This chapter examines distortion signals generated due to the coupling of signals transmitted through co-located antennas operating within the same frequency range. The primary focus of the chapter is on intermodulation distortions while providing a brief overview of harmonic distortions. It begins by examining the nonlinearities in the components and devices that lead to signal coupling, followed by detailed analysis of the two types of distortion signals. Subsequently, a concise discussion highlights the importance of mitigating these distortions. The chapter then delves into the coexistence scenarios, addressing system applications, related parameters, and specifications. In addition, it explains the key parameters that significantly impact the distortions. Finally, the chapter concludes by presenting two mitigation techniques for intermodulation distortions, supported by relevant examples. Overall, this chapter aims to provide a comprehensive understanding of distortions caused in wireless networks and equip readers with practical approaches to minimize their impact in wireless communication systems.

TX-TX INTERFERENCE: NONLINEARITY AND DISTORTION EFFECTS

This section elucidates the concept of system non-linearity, employing numerical equations to provide a clear explanation. Subsequently, it describes two distinct distortion signals that arise as a consequence of non-linear system behavior. Illustrative figures are incorporated in the section to facilitate a more comprehensive understanding of the concepts.

Analysis of Nonlinear Systems

A system (or component) is linear if its output can be expressed as linear combination (superposition) of responses to individual inputs. For example, consider two inputs to a linear system $x_1(t)$ and $x_2(t)$, then the relation to the output components can be respectively shown as

$$x_1(t) \rightarrow y_1(t)\,,\, x_2(t) \rightarrow y_2(t)$$

where the arrows indicate the operation of the system and y(t) represents the output, then the following equation holds true

$$a x_1(t) + b x_2(t) = a y_1(t) + b y_2(t)$$

for any constant values of a and b. Any system that does not adhere to this condition can be defined as nonlinear. Nonlinear systems, therefore, cannot be accurately represented by a simple proportional relationship between input and output. A common approximation for the output of a nonlinear system uses a polynomial expansion and can be approximated as

$$y(t) \approx \alpha_1 x(t) + \alpha_2 x^2(t) + \alpha_3 x^3(t)$$

where the output is clearly not directly proportional to the input. Here α_1, α_2 *and* α_3 are the linear, second order and third order coefficients respectively. These coefficients determine the strength of the corresponding nonlinear terms and how much they contribute to the overall output of the system. While many RF circuits are simplified with linear models for small signal analysis, these simplifications break down when dealing with large and complex signals. In such cases, the inherent nonlinearities of the components become crucial. For the sake of simplicity, we often focus on memoryless (independent of past inputs) and time varying nonlinear systems. The following analysis will delve into the causes and consequences of nonlinear behavior in components and devices, particularly within the context of wireless communication.

Characterization of Nonlinear Behavior in RF Transmission

To enable the transmission for wireless technologies in the RF bands, the RF transmission chain typically comprises of key components like the digital-to-analog converters (DACs), mixers, filters, power amplifiers and antennas. All these components play a critical role in processing and transmitting the desired signal. A simple RF transmit chain for one transmitting antenna is shown in figure 1. Ideally the analog signal that is given as an input to the transmit chain should get converted and transmitted through the antenna. But the nonlinear behavior of devices and components often results in undesired signals, in addition to the intended output for a given input signal. In this RF transmit chain it is the power amplifier (PA in figure 1) that majorly contributes to the nonlinear behavior. Therefore, in practical scenarios for a power amplifier, the output corresponding to a specific input signal, comprises the amplified version of the input along with the additional distortion signals. These distortions are either the harmonics which occur at the integer multiples of the center frequency or the intermodulation distortions that arise from the coupling of two signals. Both these distortion signals are discussed briefly in the following sub sections.

Figure 1. RF Transmission Chain with One Antenna

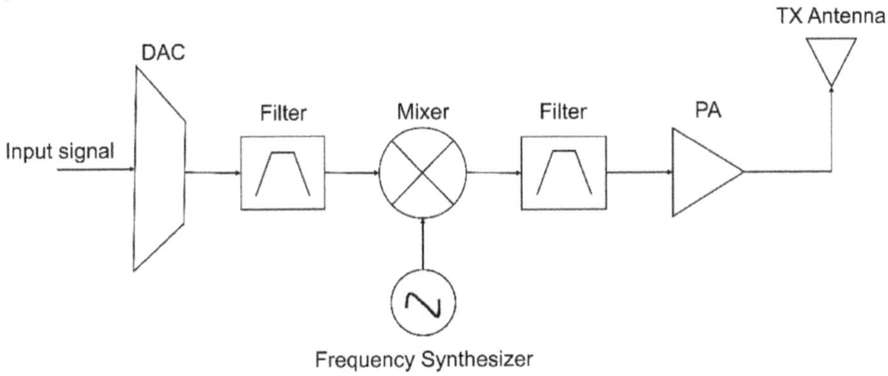

Harmonic Distortions

Harmonics of a signal refer to integer multiples of a fundamental frequency present in a periodic signal. The fundamental frequency is the first harmonic denoted by f1, and the higher harmonics are the subsequent integer multiples of this value (Attia, El-Metwally, & Fahmy, 2010). These harmonics play a crucial role in shaping the signal's overall frequency spectrum, influencing its waveform characteristics and spectral content. Mathematically, the harmonics of a given signal can be expressed as nf1, where n is a positive integer (i.e. n = 2,3,4, etc.). For instance, if a signal has a fundamental frequency of f1, its second harmonic will be 2f1, the third harmonic will be 3f1, and so on.

In practical applications, harmonics are significant in various domains such as signal processing, wireless communication, and audio engineering. While harmonics are essential in generating desired waveforms (such as square or sawtooth waves), excessive harmonic content in any system can lead to power quality issues, interference, and electromagnetic compatibility challenges. These harmonics contribute to the overall signal shape.

If a sinusoid is applied to a nonlinear system, the output does exhibit the harmonics. In equation representing the output of the non-linear system, if x(t) = $A\cos\omega t$, where ω is the sinusoidal frequency of x(t), then

$$y(t) = \alpha_1 A\cos\omega t + \alpha_2 A^2 \cos^2 \omega t + \alpha_3 A^3 \cos^3 \omega t$$

$$y(t) = \alpha_1 A\cos\omega t + \frac{\alpha_2 A^2}{2}\left(1 + \cos 2\omega t\right) + \frac{\alpha_3 A^3}{4}\left(3\cos\omega t + \cos 3\omega t\right)$$

$$y(t) = \frac{\alpha_2 A^2}{2} + \left(\alpha_1 A + \frac{3\alpha_3 A^3}{4}\right)\cos\omega t + \frac{\alpha_2 A^2}{2}\cos 2\omega t + \frac{\alpha_3 A^3}{4}\cos 3\omega t$$

The term with the input frequency in the above equation is the fundamental frequency and the higher order terms are the harmonics. From the above expansion, (Razavi, 2012) we can make two observations. First, even order harmonics result from α_j with even j and vanish if the system has odd symmetry, i.e., if it is fully differential. In reality, however, mismatches corrupt the symmetry, yielding finite even order harmonics. Second, it is evident that the amplitude of the nth harmonic comprises a term proportional to A^n alongside other terms proportional to higher powers of A. When A is small, these higher order terms can be neglected, allowing us to approximate the nth harmonic's growth as being directly proportional to A^n. In simple terms, as the order of harmonic signals increases, their amplitudes are reduced.

The occurrence of these harmonics are at increasingly high separations from the desired channel frequencies. Therefore, interference due to harmonics in the RF signal transmissions can be effectively managed. The lower order harmonics can be eliminated using filters while the higher order harmonics can be disregarded due to their low signal strength.

Intermodulation Distortions

The other type of distortion signals that arises in the output of a nonlinear system are the intermodulation distortions (IMD). These interfering signals occur only when there are two or more signals entering the nonlinear system. For example, let's consider two signals with fundamental frequencies f_1 and f_2 respectively, that are applied as the input. The output will then consist of amplified signals, harmonics of both the frequencies and the IMD products. The intermodulation distortions are the results of the sum and difference of the two fundamental frequencies and with their respective integer harmonics. For example, intermodulation of f_1 with $2f_1$, $3f_1$ or f_2, $2f_2$, $3f_2$ will result in products like $f_1 \pm 2f_1$, $f_1 \pm 3f_1$, $f_1 \pm f_2$, $f_1 \pm 2f_2$, $f_1 \pm 3f_2$ respectively. Similarly, the second signal with fundamental frequency f_2 will also be coupled with f_1 and harmonics of both the fundamental frequencies and form additional intermodulation products.

To understand this with respect to sinusoidal signals consider the input to the nonlinear system as

$$x(t) = A_1 \cos\omega_1 t + A_2 \cos\omega_2 t$$

Thus,

$$y(t) = \alpha_1 (A_1 \cos\omega_1 t + A_2 \cos\omega_2 t)$$

$$+ \alpha_2 (A_1 \cos\omega_1 t + A_2 \cos\omega_2 t)^2 + \alpha_3 (A_1 \cos\omega_1 t + A_2 \cos\omega_2 t)^3$$

On expanding the terms on the right side of the equation and discarding the DC components, following intermodulation products are formed

$$\omega = \omega_1 \pm \omega_2 : \alpha_2 A_1 A_2 \cos(\omega_1 + \omega_2)t + \alpha_2 A_1 A_2 \cos(\omega_1 - \omega_2)t$$

$$\omega = 2\omega_1 \pm \omega_2 : \frac{3\alpha_3 A_1^2 A_2}{4} \cos(2\omega_1 + \omega_2)t + \frac{3\alpha_3 A_1^2 A_2}{4} \cos(2\omega_1 - \omega_2)t$$

$$\omega = 2\omega_2 \pm \omega_1 : \frac{3\alpha_3 A_2^2 A_1}{4} \cos(2\omega_2 + \omega_1)t + \frac{3\alpha_3 A_2^2 A_1}{4} \cos(2\omega_2 - \omega_1)t$$

And the fundamental frequency here can be represented as:

$$\omega = \omega_1, \omega_2 : \left(\alpha_1 A_1 + \tfrac{3}{4}\alpha_3 A_1^3 + \tfrac{3}{2}\alpha_3 A_1 A_2^2\right)\cos\omega_1 t + \left(\alpha_1 A_2 + \tfrac{3}{4}\alpha_3 A_2^3 + \tfrac{3}{2}\alpha_3 A_2 A_1^2\right)\cos\omega_2 t$$

The above equations are referred to from the book by Razavi (2012), it provides explanation of the two distortion signals, the harmonics and the intermodulation products along with a brief of nonlinearity.

The harmonics and the intermodulation distortion products are identified by their order, which is calculated as the sum of coefficients of the two adding frequencies. The following table 1 lists some of the examples of IMD products and their respective orders.

Table 1. Order of different Intermodulation Distortion Products and Harmonic Distortions

Sr. No.	Distortion signal (harmonics/ IMD products)	Order of the product
1.	f_1, f_2	1st order
2.	$2f_1$, $2f_2$, $f_1 + f_2$, $f_1 - f_2$	2nd order
3.	$3f_1$, $3f_2$, $2f_1 + f_2$, $2f_1 - f_2$, $f_1 + 2f_2$, $f_1 - 2f_2$	3rd order

Similarly, higher order intermodulation products are evaluated in the same manner, and the number of possible combinations increases as the order rises. This results in a larger set of intermodulation products, which can further contribute to interference and signal degradation.

Figure 2. IMD Products formed due to fundamental frequencies f_1 and f_2

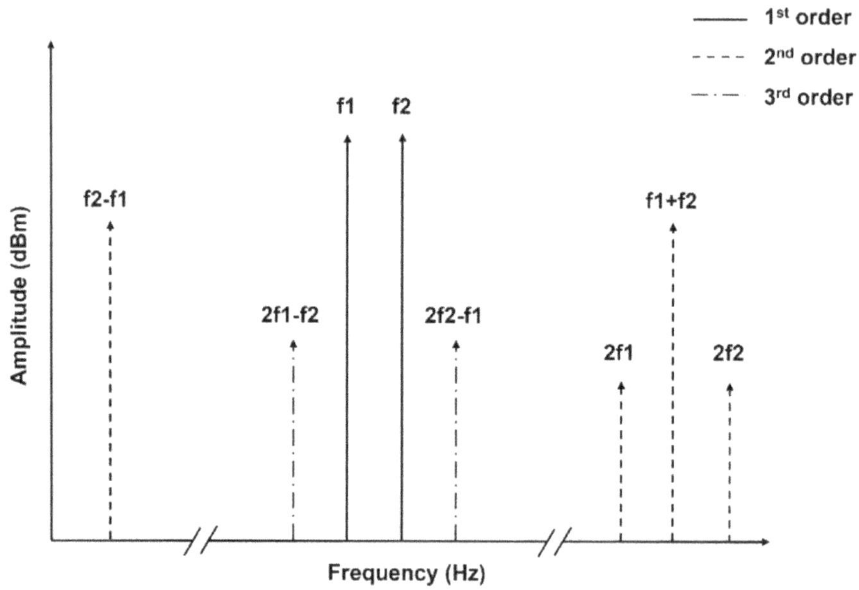

Figure 2 shows the intermodulation products formed due to fundamental frequencies f_1 and f_2 till third order which lie close to the desired frequencies. Other products of the same order arise at distant frequencies, hence, are of less concern and are not shown.

The Effects of Distortion: Why it Matters?

This section briefly explains the crucial need to consider distortion signals. The primary reason for concern is that these distortion signals affect other applications operating either in the same ISM band or in neighboring licensed bands. All the frequencies in the ISM and neighboring bands are reserved for specific applications. Any device that generates distortions that hamper the functionality of other operating devices is not considered suitable and violates the regulatory measures. Figure 3 shows a simple representation of how intermodulation distortions caused due to applications in the 2.4 GHz ISM band occur outside the frequency range. Here, a Bluetooth narrowband signal is transmitted at a channel with center frequency 2459 MHz and a WLAN signal is transmitted at 2472 MHz channel frequency. Clearly the third order distortions occurring at $2f_1$ - f_2 (2446 MHz) and $2f_2 - f_1$ (2485 MHz) can interfere with the applications that are already functional on these respective frequencies.

Figure 3. Impact of IMD Products on other applications in the same and neighboring frequency band

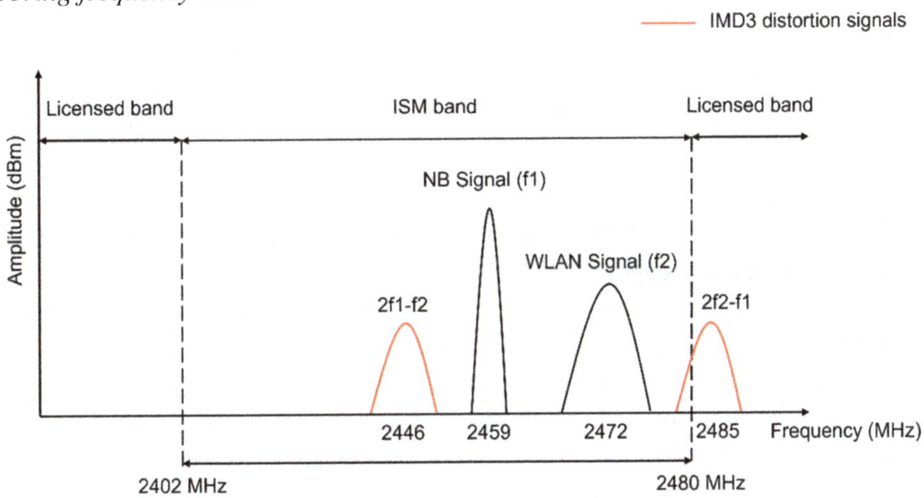

Note that the distortions due to the other two third order IMD products viz. $2f_1 + f_2$ (7390 MHz) and $2f_2 + f_1$ (7403 MHz) are not shown in the figure because these frequencies are very far from the desired frequencies and may not cause any issues due to interference.

Therefore, the devices should be designed such that they neither create distortions, nor get affected by any external interferences, that is all the devices should comply with the interference and compatibility standards. There are various international bodies that define specifications for these parameters, for example 3rd Generation Partnership Project (3GPP), International Telecommunication Union (ITU) and Federal Communications Commission (FCC) in the United States etc. This chapter adheres to the standards defined by the FCC. The threshold for the intermodulation distortions given by FCC is followed in the entire chapter as the regulatory measure.

MANAGING INTERFERENCE: THE COEXISTENCE CHALLENGE

Coexistence of multiple wireless technologies is the need of growing communication applications. But the distortions due to coupling of signals transmitted through collocated antennas are difficult to eliminate. (Zhu et al., 2007) Note that intermodulation distortions occur only when there are two or more radios operating in close vicinity and in the same frequency range. Out of all the possible distortions due to harmonics and IMD products, the third order intermodulation distortion signals (IMD3) are of great concern. (Males-Ilic, Milovanovic, & Budimir, 2001) These products are formed at frequencies which are very close to the desired transmit channel frequencies and cannot be easily removed using filters.

This chapter majorly focuses on coexistence of 2.4 GHz Wireless LAN (WLAN) radio and 2.4 GHz Narrowband (Bluetooth/ IEEE 802.15.4) technology along with their respective IMD3 products. Within the 2.4 GHz range of ISM bands, WLAN utilizes thirteen channels each with 20 MHz bandwidth for transmission (number of channels may change with different signal bandwidth). On the other hand, narrowband technology operates with seventy-nine channels, each with 1 MHz bandwidth in the same frequency range.

WLAN supports different modulations depending on the utilized frequency spectrum. This technology has evolved through several generations, each improving upon the previous one. Early standards like IEEE 802.11b used simpler modulations like Direct-Sequence Spread Spectrum and phase-shift keying, offering lower data rates. Later, IEEE 802.11g introduced OFDM, boosting speeds significantly. IEEE 802.11n brought in MIMO and wider channels, further increasing throughput. Subsequent standards like IEEE 802.11ac, IEEE 802.11ax (Wi-Fi 6), and IEEE 802.11be (Wi-Fi 7) have continued this trend, employing higher-order modulation schemes (like QAM), wider channels, and more advanced techniques like OFDMA and MU-MIMO to achieve gigabit and multi-gigabit speeds, improved spectral efficiency, and better performance in congested environments. (Hiertz et al., 2010)

In the case of narrow band radios, Bluetooth supports different rates, power consumption by devices and offers different ranges for transmission. Bluetooth is broadly categorized into two types: Bluetooth Classic, which supports point-to-point communication with data rates of 1 Mbps (basic data rate/ BR) or 2–3 Mbps (enhanced data rate/ EDR), and Bluetooth Low Energy (BLE), which supports point-to-point, broadcast, and mesh topologies with transmission rates ranging from 125 Kbps to 2 Mbps. (Bluetooth SIG, Inc., n.d.) Devices using BLE consumes less power as compared to those using BT classic. However, BT classic provides a higher transmission range than BLE, making it more suitable for devices which require higher coverage. (Shorey & Miller, 2000) (Bisdikian, 2002)

There are multiple scenarios that might arise when WLAN and Narrow band radios operate simultaneously, that can lead to generation of distortion signals. The possibilities depends on which radio signal is desired, and which is the interferer or which one amongst the two is at a higher priority based on the application they are supporting. The wireless signal or technology which dominates in the shared frequency band or interferes with the transmit signal of another radio signal is called an aggressor. Whereas the signal or technology that suffers from interference caused due to aggressor is known as victim. These terminologies are only defined for the purpose of explanation in this chapter and are not standard. Victim's performance degrades due to dominant presence of the undesired signal (aggressor) in the shared frequency band. In cases when two collocated antennas supporting two wireless technologies are present on the SoC, both the radios (antennas) act as victim and aggressor, because the transmission occurs simultaneously. It then depends on the powers that the two radios are individually transmitting that defines the strength of the IMD3 products formed.

Figure 4 shows two antennas on a single chip and the intermodulation distortion products formed due to coupling of transmit signals from them, respectively. In the coexistence of two wireless technologies, both act as victims and aggressors simultaneously. In the figure, the intermodulation distortions are shown, assuming that the transmission is ON at both the antennas. (Ophir, Bitran, & Sherman, 2004)

Figure 4. Demonstration of IMD3 generation due to simultaneous transmissions

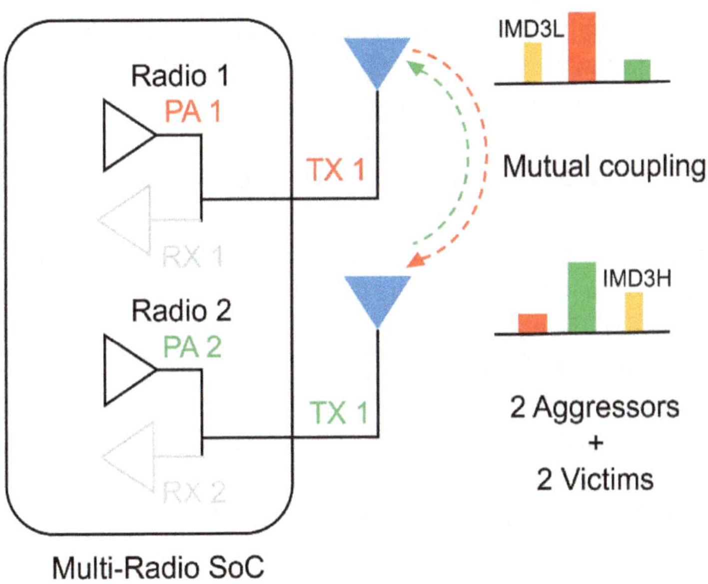

In figure 4 signals transmitted through both the antennas are shown. Both these signals are transmitted at the same Tx power. Since both the radios act as victim and aggressor due to simultaneous transmission, IMD3 spurious signal is formed at both the antennas. At antenna 1 the strength of the aggressor signal coming from antenna 2 is less than what was transmitted from the second power amplifier and vice versa. This difference is caused because of the isolation that is maintained between the two collocated antennas. If the isolation between the two antennas are increased the amount of aggressor power reaching the victim power amplifier will reduce, eventually resulting in lower IMD3 distortion signal amplitudes. Here, IMD3L represents the IMD3 product formed at the lower frequency component and IMD3H represents the higher frequency component. Therefore, it is the received power of the aggressor signal at victim power amplifier that determines the strength of the IMD3 spurious signal.

FACTORS INFLUENCING IMD3 PRODUCT LEVELS

This section briefly talks about some of the attributes of the wireless technologies which impact the occurrence of IMD3 products. The transmitting signal power, WLAN modulation, Bluetooth data rate, frequency separation in between the two transmitting channels and the isolation between the two collocated antennas are some of the most important parameters that affect the IMD3 signal strength significantly. The following are the descriptions of each of these parameters and how their variations affect the IMD3 spurious emissions.

Transmit Power by the Wireless Technologies

The primary constraint of transmitting antennas that impact the IMD3 spurious signal amplitude are the signal powers they transmit. It highly impacts how much power does victim antenna transmit and how much aggressor power reaches the victim power amplifier. If we consider a constant isolation between the two collocated antennas, increasing the aggressor transmitting power will naturally increase the amount of it reaching the victim RF chain. And at a constant aggressor received power at the victim power amplifier, the amplitude of the IMD3 distortion signal increases with the increasing victim transmit powers. Therefore, not all combination of powers from the aggressor and victim can be transmitted simultaneously.

WLAN Standards and Modulation

Secondly the amplitude of the IMD3 spurious signals varies depending on the specific WLAN modulation scheme, such as IEEE 802.11b, 802.11a, 802.11g,

802.11n, 802.11ac, and 802.11ax. This variation is due to differences in their channel bandwidth, modulation complexity, and transmission power levels.

When Bluetooth operates alongside IEEE 802.11b, IMD3 is generated primarily due to frequency overlap. However, the interference is relatively less severe because 802.11b employs a simpler DSSS modulation and has a narrower 22 MHz channel bandwidth, which results in lower spectral density and reduced nonlinear interactions.

In contrast, IEEE 802.11g and 802.11a utilize Orthogonal Frequency Division Multiplexing (OFDM), which increases spectral efficiency but also makes them more susceptible to intermodulation distortions when coexisting with Bluetooth. IEEE 802.11g operates in the 2.4 GHz band, where Bluetooth is also active, leading to more significant IMD3 effects due to overlapping frequency components. IEEE 802.11a, operating in the 5 GHz band, generally avoids direct Bluetooth interference but can still exhibit nonlinear distortion effects in multi-radio environments.

IEEE 802.11n, which supports both 20 MHz and 40 MHz channel bandwidths, experiences more pronounced IMD3 when operating in the 40 MHz mode due to the increased spectral footprint. The use of Multiple-Input Multiple-Output (MIMO) technology in 802.11n can also introduce additional nonlinearities, as multiple concurrent signal paths increase the likelihood of intermodulation.

With IEEE 802.11ac, which extends channel bandwidths up to 160 MHz and operates exclusively in the 5 GHz band, IMD3 effects can become even more significant in multi-radio environments where Bluetooth or other nearby transmissions introduce strong nonlinear interactions. The increased power levels and high-order modulation schemes (such as 256-QAM) in 802.11ac lead to stronger IMD3 products that can degrade both Wi-Fi and Bluetooth performance.

The latest IEEE 802.11ax standard, also known as Wi-Fi 6, employs advanced techniques such as Orthogonal Frequency Division Multiple Access (OFDMA) and Target Wake Time (TWT) to improve spectral efficiency and reduce interference. However, due to its wider bandwidths (up to 160 MHz) and high order modulation (1024-QAM), the susceptibility to IMD3 is worsened, especially when coexisting with Bluetooth in dense network environments.

Therefore, the severity of IMD3 depends on the Wi-Fi modulation scheme, with older IEEE standards like 802.11b experiencing less severe interference due to DSSS and narrower bandwidths, whereas newer IEEE standards such as 802.11n, 802.11ac, and 802.11ax are more affected due to their use of OFDM, wider channel bandwidths, higher transmission power, and complex modulation schemes.

Bluetooth Data Rate

Another impact of intermodulation between WLAN and Bluetooth differs due to the difference between Bluetooth basic rate and Bluetooth enhanced data rate because of variations in their modulation schemes used and data rates. Bluetooth BR uses simpler Gaussian Frequency Shift Keying (GFSK) modulation and operates at a lower data rate of 1 Mbps, resulting in less severe interference with WLAN, especially IEEE 802.11b standard. In contrast, Bluetooth EDR employs more complex π/4-DQPSK or 8-DPSK modulation and functions at higher data rates (2-3 Mbps), which produces stronger interference with WLAN, particularly the IEEE 802.11n standard. The higher data rate and broader signal bandwidth of Bluetooth EDR increase the likelihood of generating IMD3, causing more significant interference compared to Bluetooth BR.

Frequency Separation Between the Transmitting Channels

Frequency separation between the two transmitting channels also impact the amplitudes of IMD3 products formed due to coupling of the signals transmitted over them. It plays a crucial role in deciding the severity of IMD3 products. When two signals, from Bluetooth and WLAN, are transmitted on close frequencies, the likelihood of producing strong IMD3 products increases because the coupling occurs more readily when the frequencies interact. If the frequencies are well separated, the chance of significant interaction decreases, leading to weaker IMD3 products. Higher frequency separation reduces the overlap between the spectral components of the transmit signals, which can minimize the distortions. Thus, ensuring high frequency separation between the transmitting channels helps in mitigating the impact of IMD3 distortions.

Isolation Between the Collocated Antennas

The final critical parameter that directly influences the generation of IMD3 distortion signals is the isolation in between the two transmitting antennas. Isolation basically defines the separation between the two antennas, required to reduce the coupling or interference in between the signals they transmit. Lower isolations between the collocated antennas leads to higher signal leakage from one antenna to another, which eventually results in intermodulation distortions. On the other hand, maintaining higher isolation between the antennas reduce the chances of leakage or the signal strength and thus generation of IMD3 products. This helps in ensuring optimal coexistence of technologies like WLAN and Bluetooth.

While the parameters influencing IMD3 distortions appear straightforward to implement, the requirements of specific applications may not always be met if these parameters are strictly optimized to prevent IMD3 products. For a device to support desired functionality, operational range, and efficiency, it is important to transmit radio signals with specific characteristics. For instance, IEEE 802.11n WLAN modulation provides better performance than IEEE 802.11b, but the IMD3 generated due to it is much more severe than IEEE 802.11b scenario. Similarly, while IMD3 products generated by Bluetooth EDR are typically more intense than those produced by Bluetooth BR, EDR offers significantly higher data rates and better performance, making it suitable for data intensive applications.

In certain cases, maintaining high frequency separation between transmitting channels is not feasible due to application specific requirements. For example, some applications necessitate transmission over closely spaced frequency channels, which increases the risk of IMD3 interference. Additionally, there are use cases where antenna isolation must be as minimum as possible, further increasing the interference due to IMD3. These constraints are not solely dictated by application needs but also by the design limitations inherent to SoC architectures.

As a result, eliminating IMD3 spurious emissions is not entirely possible. Instead, optimization techniques are employed to reduce their amplitude and ensure the emissions remain within the regulatory limits, maintaining a balance between application performance and compliance.

KEY TECHNIQUES FOR REDUCING IMD3 DISTORTION

WLAN technology and Bluetooth supports transmit powers over a varied range. Wi-Fi transmit power isn't a fixed value but varies based on several factors. These include the specific Wi-Fi standard (e.g., 802.11ax vs. 802.11b), the frequency band used (2.4 GHz, 5 GHz, or 6 GHz), regional regulatory limits, the type of device (access point vs. client), and even the chosen channel. While maximum transmit power can reach around 20-30 dBm, the optimal setting is often lower (around 10-17 dBm) to minimize interference and maximize network efficiency.

On the other hand, Bluetooth is designed for short-range, low-power communication, unlike Wi-Fi's longer-range focus and operates only in the 2.4 GHz frequency band. Bluetooth devices are categorized into power classes, with Class 2 being the most common, operating around 10 dBm (10 milliwatts). This is significantly lower than the typical Wi-Fi transmit power.

In cases when two wireless technologies coexist in the same frequency band, the power that each antenna transmits defines the strength of the IMD3 signal that is generated. The interference due to IMD3 products is negligible when the two radios transmit low powers, but it becomes prominent as the transmit powers increase.

Another aspect of the coexisting radios is the priority in which they can transmit. Priority of a wireless technology in a coexistence scenario is determined by the specific requirements of the application they support. For example, consider a device that supports communication through both Wi-Fi and Bluetooth technologies.

The device is connected to the network via Wi-Fi for browsing and simultaneously needs to transfer files over the Bluetooth connection. Naturally transfer of files must take higher priority, since latency in file transfer is critical and could negatively impact user experience. Thus, transmission from the Bluetooth antenna will be given a higher priority and WLAN transmission is assigned to be a lower priority. It is essential to understand that such priority decisions are made by the device's resource management system, which evaluates the importance of each application's performance needs. Therefore, the priority of the radio entirely depends on the application that the two radios are supporting.

In a coexistence scenario where both the radios act simultaneously as victim and aggressor, they can also have either high or low priority depending on the applications and functionality. This means that being in low or high priority is not related to whether the radio is a victim or aggressor. The following are two significant techniques that can be used to either completely eliminate the IMD3 spurious emissions or considerably reduce the interference caused by them.

Tx Arbitration: A Method for IMD3 Suppression

The basic technique used to eliminate the intermodulation products is to completely arbitrate one of the radio signals. The wireless technology that is at a lower priority is completely aborted in this case. Since only one radio then transmits, the chances of generation of IMD3 products is entirely removed. The signal that was requested first by the system and has a higher priority depending on the supported application, continues to be transmitted. The arbitration decision is made based on the amplitude of the IMD3 product that is generated by intermodulation of two signals that appear at the input of the power amplifier. If the IMD3 amplitude value is less than the regulatory specification defined by FCC on the logarithmic scale, then the distortions are within the maximum acceptable limits and the simultaneous transmission can be allowed. Conversely if the measured IMD3 product amplitude is higher than the defined FCC margin, the combination of transmit powers requested by the two simultaneous radios will lead to excessive distortions and the concurrent transmission is not given a grant. FCC has defined a standard value of -41.25 dBm as the maximum allowed amplitude of IMD3 products formed due to intermodulation and coupling in coexistence of multiple wireless radios.

When the amplitude of the IMD3 spurious signal is measured on the dBm scale then values are typically negative. If the IMD3 amplitude is above the FCC specification, it implies that it is less negative compared to the regulatory margin and can be visualized as being above the threshold. For example, if the amplitude of the IMD3 signal is -40 dBm, -39.5 dBm, -38 dBm or any other more positive value it will violate the FCC regulatory specification. On the other hand, if the IMD3 amplitudes are more negative than -41.25 dBm that is -42.5 dBm, -43 dBm or higher it remains within the allowable range, permitting simultaneous transmission of the radios. Figure 5 shows an example of simultaneous transmission of WLAN and Narrow band signal and the respective IMD3 product which is formed. WLAN here is the aggressor and transmits 20 dBm of power at the transmitting antenna. Narrow band radio is the victim which is transmitting 13 dBm of power at its antenna. The figure shows the IMD3 measured at the victim power amplifier. The narrow band signal is transmitted at 2.430 GHz channel frequency. The WLAN signal is transmitted at 2.472 GHz channel frequency and its component that reaches the victim PA is shown in the figure. The reduction in the strength of the aggressor signal at the victim power amplifier is due to 15 dB isolation that is maintained between the collocated antennas. Since the IMD3 is measured at the narrow band power amplifier, the product that will be more severe out of all the possible third order products will occur at 2.388 GHz frequency which is closer to narrow band transmitting channel and occurs at the difference of twice of narrow band channel frequency, and WLAN channel frequency. Clearly the amplitude of the IMD3 product generated is beyond the FCC regulatory specification of -41.25 dBm. Therefore, this scenario of simultaneous transmission of 13 dBm of narrow band power and 20 dBm of WLAN power cannot be allowed, and the lower priority radio can be arbitrated to meet the regulatory requirements. There are some other higher order IMD products shown in the figure. Their amplitudes, which is -50 dBm, are below the regulatory threshold and thus can be disregarded.

Figure 5. Simultaneous transmission of WLAN and Narrow band signal and the respective IMD3 Product

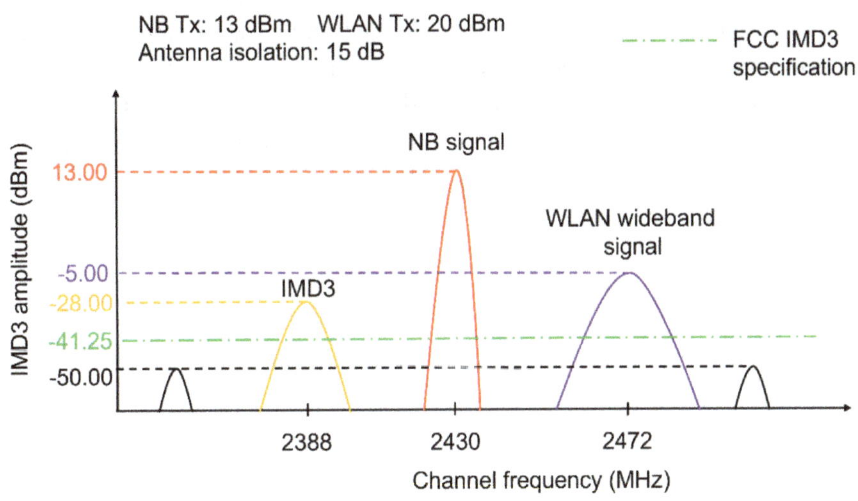

Different IMD3 products are formed for different combinations of the transmitted powers by the two coexisting wireless technologies. The strength of the distortion signals can be measured in a two-dimensional sweep over the entire range of power that the radios support, on a spectrum analyzer. For each combination of transmit powers by the two radios, an arbitration decision can be made, whether they are simultaneously allowed or not.

Although this technique appears easy to implement, it is not the most efficient way of reducing IMD3 distortion signals. Removing a radio from the transmitting frequency range will not generate IMD3 distortions but the concept of coexistence itself will be violated. The efficiency and scope of application of the device will greatly decline. Also, the channel for the lower priority radio that will be arbitrated will thus remain unused, affecting the overall throughput. Therefore, making an arbitration decision can be used as a deciding basis for some other technique but cannot be ideally implemented.

Cancellation Technique for IMD3 Reduction

The next technique used to minimize the IMD3 signals and interference caused due to them is known as the cancellation technique. This technique proposes a method where an inverse replica of the aggressor transmit signal is generated at the victim transmitting output, which subsequently reduces the undesired IMD3 product. (Tam et al., 2024) Consider a scenario where Wi-Fi and narrow band coexist on the same SoC, where Wi-Fi is the victim and Bluetooth is the aggressor. IMD3 products are formed when a signal from the aggressor antenna enters the victim antenna and couples with the victim's desired signal. In this method a similar signal, to the interfering signal from aggressor, with a 180 degrees phase shift is generated at the victim. Since the two signals are out of phase and have the same magnitudes, they tend to cancel out each other. By performing this cancellation, the aggressor's interfering signal is eliminated thus reducing the amplitude and strength of the IMD3 spurious signal formed. This technique is illustrated in figure 6 (Tam et al., 2024). For the example shown both the wireless technologies transmit maximum possible powers to assess the worst case scenarios.

Figure 6. Demonstration of Cancellation technique to mitigate IMD3 spurious emissions

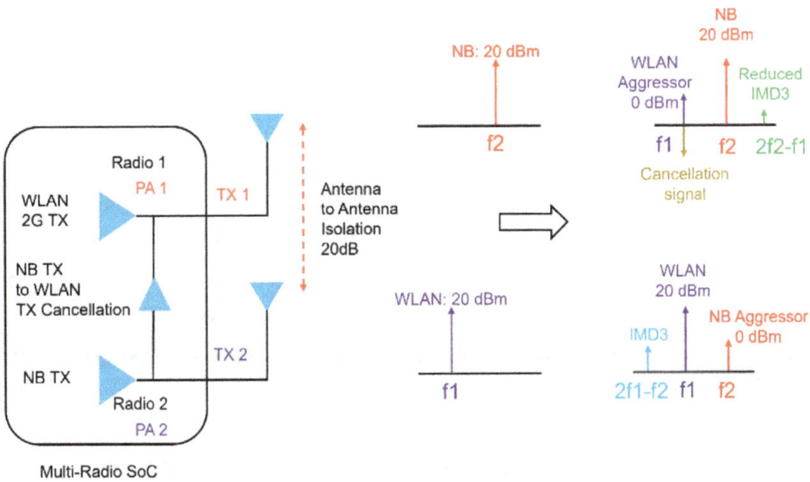

To implement this optimization technique on the hardware, some additional components like attenuator, phase interpolator, an extra power amplifier for cancellation, voltage detector and most essentially a feed forward routing path from the aggressor to the victim PA (Tx to Tx) are required. This technique of IMD3 cancellation is usually applied to applications that require low isolation in between the collocated transmitting antennas.

Cancellation technique has some drawbacks like the increased complexity of the hardware implementation. Additionally, the power levels that this technique can support for creating inverted signals and cancellation of IMD3 are also limited to a maximum value. On the other hand, the arbitration technique is easy to understand and simple to implement. Therefore, each technique comes with its own set of advantages and disadvantages. The choice of an appropriate technique depends critically on the specific requirements of the application, including factors such as performance, complexity, resource availability, and compliance with the regulatory standards.

SUMMARY

This chapter explores the growing challenge of signal distortion in modern wireless systems, driven by the increasing integration of multiple radio technologies like Wi-Fi, Bluetooth, and Zigbee onto single devices. As these technologies, each serving distinct use cases, coexist and often share frequency bands, particularly the 2.4 GHz ISM band, the close proximity of antennas on system on chips creates signal leakage and coupling. (He et al., 2021) This coupling, worsened by nonlinearities in RF components, leads to the generation of unwanted distortion products, primarily the harmonics and intermodulation products. The chapter focused on these distortions, analyzing their origins in component nonlinearities and examining the specific characteristics. It emphasized the importance of mitigating these distortions, especially given their potential to interfere with adjacent licensed bands and violate regulatory compliance. The chapter then described the coexistence scenarios, discussing relevant system applications, parameters, and specifications. It then concludes by presenting two specific mitigation techniques for intermodulation distortion, providing practical insights for minimizing their impact on wireless communication systems. Ultimately, this chapter provides a comprehensive understanding of the causes and consequences of signal distortion in co-located multi-radio systems, equipping readers with the knowledge necessary to design and implement robust and compliant wireless communication devices.

REFERENCES

Attia, H. A., El-Metwally, M., & Fahmy, O. M. (2010). Harmonic distortion effects and mitigation in distribution systems. *The Journal of American Science*, *6*(10), 173–183.

Bisdikian, C. (2002). An overview of the Bluetooth wireless technology. *IEEE Communications Magazine*, *39*(12), 86–94. DOI: 10.1109/35.968817

Bluetooth, S. I. G. Inc. (n.d.). *Tech overview*. Bluetooth Technology Website. https://www.bluetooth.com/learn-about-bluetooth/tech-overview/

Challoo, R., Oladeinde, A., Yilmazer, N., Ozcelik, S., & Challoo, L. (2012). An overview and assessment of wireless technologies and co-existence of ZigBee, Bluetooth and Wi-Fi devices. *Procedia Computer Science*, *12*, 386–391. DOI: 10.1016/j.procs.2012.09.091

Garroppo, R. G., Gazzarrini, L., Giordano, S., & Tavanti, L. (2011, June). Experimental assessment of the coexistence of Wi-Fi, ZigBee, and Bluetooth devices. In *2011 IEEE International Symposium on a World of Wireless, Mobile and Multimedia Networks* (pp. 1-9). IEEE. DOI: 10.1109/WoWMoM.2011.5986182

He, S., Zhang, B., Luo, J., & Qu, M. (2021, May). Research on coexistence and mutual interference of Wi-Fi and Bluetooth. In *2021 International Conference on Microwave and Millimeter Wave Technology (ICMMT)* (pp. 1-3). IEEE. DOI: 10.1109/ICMMT52847.2021.9617988

Hiertz, G. R., Denteneer, D., Stibor, L., Zang, Y., Costa, X. P., & Walke, B. (2010). The IEEE 802.11 universe. *IEEE Communications Magazine*, *48*(1), 62–70. DOI: 10.1109/MCOM.2010.5394032

Males-Ilic, N., Milovanovic, B., & Budimir, D. (2001, December). Low intermodulation amplifiers for RF and microwave wireless systems. In *APMC 2001. 2001 Asia-Pacific Microwave Conference (Cat. No. 01TH8577)* (Vol. 3, pp. 984-987). IEEE. DOI: 10.1109/APMC.2001.985282

Ophir, L., Bitran, Y., & Sherman, I. (2004, September). Wi-Fi (IEEE 802.11) and Bluetooth coexistence: Issues and solutions. In *2004 IEEE 15th International Symposium on Personal, Indoor and Mobile Radio Communications (IEEE Cat. No. 04TH8754)* (Vol. 2, pp. 847-852). IEEE.

Pahlavan, K., & Krishnamurthy, P. (2021). Evolution and impact of Wi-Fi technology and applications: A historical perspective. *International Journal of Wireless Information Networks*, *28*(1), 3–19. DOI: 10.1007/s10776-020-00501-8

Ramya, C. M., Shanmugaraj, M., & Prabakaran, R. (2011, April). Study on ZigBee technology. In *2011 3rd international conference on electronics computer technology* (Vol. 6, pp. 297-301). IEEE. DOI: 10.1109/ICECTECH.2011.5942102

Razavi, B., & Behzad, R. (2012). *RF microelectronics* (Vol. 2). Prentice hall.

Shorey, R., & Miller, B. A. (2000, December). The Bluetooth technology: merits and limitations. In *2000 IEEE international conference on personal wireless communications. Conference proceedings (Cat. No. 00TH8488)* (pp. 80-84). IEEE. DOI: 10.1109/ICPWC.2000.905777

Tam, S. W., Razzaghi, A., Wong, A., Narravula, S., Xu, W., Loo, T., Kambale, A., Lowrance, A., Carnu, O., Lin, Y., & Tsang, R. (2024). A 28 nm CMOS Dual-Band Concurrent WLAN and Narrow Band Transmitter With On-Chip Feedforward TX-to-TX Interference Cancellation Path for Low Antenna-to-Antenna Isolation in IoT Devices. *IEEE Journal of Solid-State Circuits*, *59*(5), 1301–1311. DOI: 10.1109/JSSC.2024.3352532

Zhu, J., Waltho, A., Yang, X., & Guo, X. (2007, August). Multi-radio coexistence: Challenges and opportunities. In *2007 16th International Conference on Computer Communications and Networks* (pp. 358-364). IEEE.

KEY TERMS AND DEFINITIONS

Collocated Antennas: Collocated antennas are multiple antennas placed in close proximity to each other, usually on the same device or platform. This arrangement is common in multi-radio systems to support various wireless technologies, but it leads to challenges like signal interference and coupling between the antennas.

Concurrent Transmission: Concurrent transmission refers to the simultaneous transmission of data by two or more devices or systems, often over the same medium or frequency band. This significantly increases data throughput and network capacity, but it also introduces challenges in managing interference and ensuring reliable communication.

Coupling: Signal coupling refers to transfer of energy from one signal or circuit to another. This occurs due to various mechanisms, such as electromagnetic induction, capacitive coupling, or shared impedance, leading to unwanted interference or signal corruption.

IEEE 802 Standard: The IEEE 802 standard is a family of standards developed by the Institute of Electrical and Electronics Engineers (IEEE) that define protocols for various networking technologies, primarily focusing on local and metropolitan area networks (LAN/MAN). It covers a wide range of technologies, including Ethernet (802.3), Wi-Fi (802.11), Bluetooth (802.15), and others, specifying how devices communicate over wired and wireless media.

Memoryless System: It is a system whose output at any given time depends only on the input at that same time. It has no memory of past input, and its current output is entirely determined by the present input.

Multi-radio Coexistence: Multi-radio coexistence refers to the ability of multiple radio technologies (like Wi-Fi, Bluetooth, Zigbee, cellular, etc.) to operate simultaneously within the same device or in close proximity without causing significant interference to each other. It involves managing the shared radio frequency spectrum and mitigating interference to ensure that each radio can perform its intended function effectively.

Narrow Band Signals: The signals that have small bandwidth and occupy less portion of the total frequency spectrum are called narrow band signals. These signals can transmit information efficiently over longer distances with less power but have limitations over the amount of data that can be sent. Examples include Bluetooth and Zigbee technology.

Power Amplifier: A power amplifier is an electronic circuit that increases the power of a signal. It is typically used as the final stage in a transmission system to drive a load, such as a speaker or an antenna. Its primary function is to deliver sufficient power to the load while minimizing distortion and maximizing efficiency.

RF Transmit Chain: It is the path that a signal takes from baseband to radio frequency. It consists of stages like mixing, filtering, amplification, and finally, radiation by the antenna. It is used to efficiently and linearly convert the information signal into a powerful RF signal suitable for wireless transmission.

System on Chip: A System on a Chip (SoC) is an integrated circuit that combines all or most components of a complete electronic system onto a single chip.

Time Variant System: It is a system whose characteristics or parameters change over time. That is the relationship between the input and output varies depending on when the input is applied to the system. Mathematically, the system's transfer function or impulse response is a function of time.

Chapter 8
Advancements in 5G and 6G Technologies:
The Next Generation of Nanosheet Devices

Raj Kumar S.
http://orcid.org/0009-0002-0499-9619
Velammal College of Engineering and Technology, India

Kavitha K.
http://orcid.org/0000-0002-9159-9690
Velammal College of Engineering and Technology, India

ABSTRACT

The next generation of nanosheet-based semiconductor devices marks a pivotal evolution in modern electronics, as industry and research converge to address the scaling limitations of traditional FinFET architectures. Nanosheets, with their enhanced electrostatic control and high on-off current ratios, are emerging as the cornerstone of advanced semiconductor technologies. This chapter explores the advancements in nanosheet transistor design, materials innovation, with an emphasis on their scalability beyond 3nm technology nodes. Furthermore, the discussion extends to device reliability, energy efficiency, and potential applications in artificial intelligence (AI), high-performance computing, and edge devices. By delving into the synergistic collaboration between process technology and device architecture, this chapter underscores the transformative role of nanosheet technologies in shaping the future of semiconductor devices and redefining the boundaries of Moore's Law.

DOI: 10.4018/979-8-3373-0852-4.ch008

INTRODUCTION

The transition from 4G to 5G technology has marked a significant milestone in wireless communication, offering enhanced data transfer rates, reduced latency, and improved reliability. With 5G networks being deployed globally, researchers are already focusing on the next generation—6G—which aims to revolutionize connectivity with even greater speeds, ultra-low latency, and intelligent automation (Agiwal et al., 2016). The continuous advancements in semiconductor technology are a crucial factor in achieving these goals, and among them, nanosheet transistors have emerged as a promising innovation for meeting future demands (Osseiran et al., 2014).

Nanosheet devices represent the next stage in transistor evolution, succeeding FinFETs as the dominant architecture for high-performance and energy-efficient computing. These transistors utilize stacked nanosheets with a gate-all-around (GAA) structure, enabling better electrostatic control, higher drive currents, and improved scalability compared to traditional FinFETs (Akyildiz et al., 2020). As the industry moves towards sub-3 nm process nodes, nanosheet transistors provide a viable solution to continue Moore's Law, which predicts the doubling of transistor density approximately every two years (Rappaport et al., 2013).

The demand for higher data rates and lower power consumption in 5G and 6G networks requires innovative approaches in semiconductor fabrication. The integration of nanosheet transistors in next-generation processors and network devices enables higher processing power, improved efficiency, and better thermal management. These benefits are essential for supporting technologies such as terahertz (THz) communication, artificial intelligence-driven networks, and advanced edge computing applications (Rappaport et al., 2015).

Moreover, 6G technology aims to introduce breakthroughs in areas such as quantum communication, AI-assisted network optimization, and ultra-reliable low-latency communication (URLLC). The increased processing capabilities of nanosheet transistors play a vital role in enabling these advancements by providing the required computational power at reduced energy costs (Andrews et al., 2014). The ability to support high-frequency operations, especially in the THz spectrum, makes nanosheet devices ideal candidates for future wireless infrastructure (Saad et al., 2019).

Despite their numerous advantages, nanosheet transistors face challenges in large-scale manufacturing, including process complexity, material limitations, and variability in device performance (Bhandari et al., 2021). Advanced fabrication techniques such as Extreme Ultraviolet (EUV) Lithography and Atomic Layer Deposition (ALD) are being developed to address these issues and ensure the mass production of nanosheet-based chips. As research progresses, continuous improvements in nanosheettechnology will be essential to optimize performance and enhance reliability for future wireless applications (Shafi et al., 2017).

This paper aims to explore the role of nanosheet transistors in the evolution of 5G and 6G technologies. It will discuss their design principles, fabrication methods, performance benefits, and challenges while highlighting their impact on next-generation communication systems (Boccardi et al., 2014). By understanding the capabilities and limitations of nanosheet devices, researchers and engineers can develop innovative solutions to overcome technological barriers and pave the way for future advancements in wireless communication (Tsiftsis et al., 2020).

EVOLUTION FROM 5G TO 6G

The transition from 5G to 6G marks a transformative era in wireless communication, introducing advanced capabilities beyond those of its predecessor. While 5G has already revolutionized data speeds, reduced latency, and enabled massive connectivity, 6G aims to push these parameters even further by leveraging emerging technologies such as artificial intelligence, quantum communication, and terahertz (THz) frequencies.

Figure 1. Understanding 5G and the Glimpse of 6G: The Evolution of Connectivity

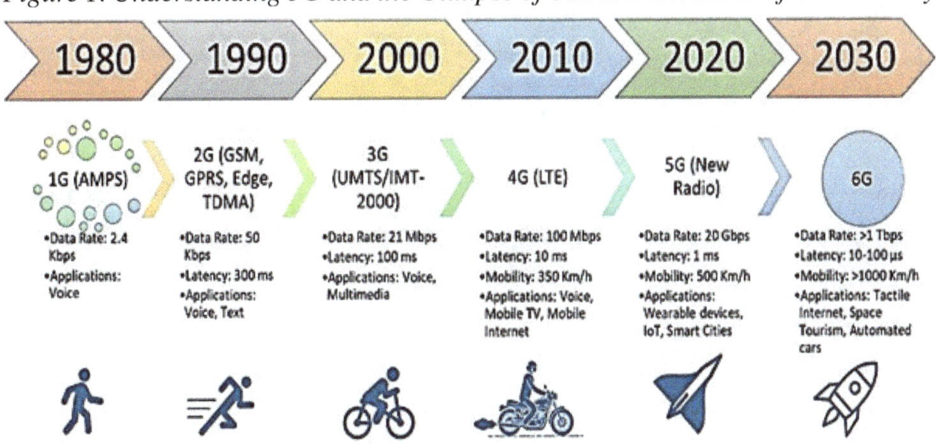

(Courtesy: pulse/understanding-5g-glimpse-6g-evolution-connectivity-dr-manpreet-puri/)

Limitations of 5G Technology

5G networks, though a significant upgrade from 4G, have some inherent limitations that necessitate the development of 6G:

- **Spectrum Congestion:** Despite the use of millimeter-wave (mmWave) frequencies, the growing number of connected devices is leading to spectrum saturation.
- **Latency Constraints:** 5G achieves ultra-low latency (~1ms), but future applications such as holographic communication and remote surgeries require even lower latency levels.
- **Energy Consumption:** The high power requirements of 5G infrastructure contribute to operational costs and carbon footprints, calling for more energy-efficient solutions.

Figure 2. Technological development of communication systems towards 6G

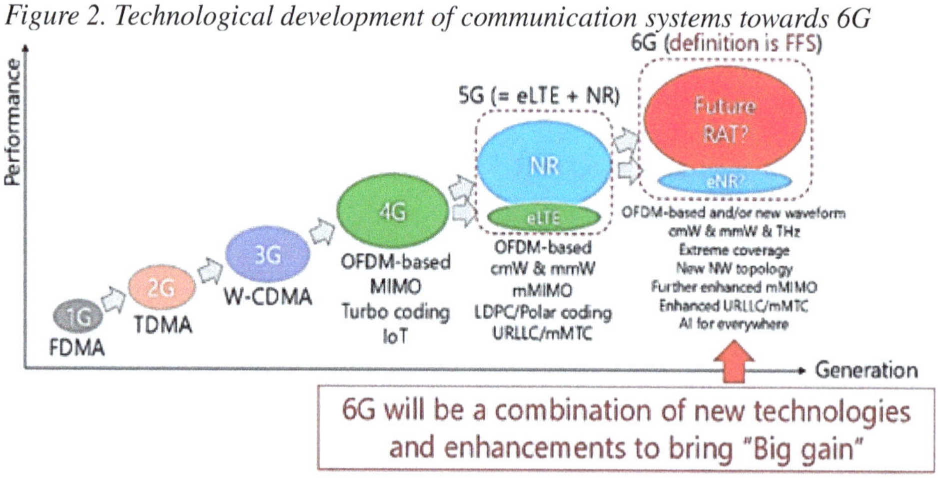

(Courtesy: Yoshihisa et al. [21])

Key Features of 6G

6G aims to overcome the limitations of 5G and introduce groundbreaking advancements, including:

- **Terahertz (THz) Communication:** Operating in the THz spectrum (0.1–10 THz) will allow ultra-fast data rates up to 1 Tbps, significantly enhancing wireless performance.
- **AI-Driven Networks:** Artificial intelligence and machine learning will be embedded into network architectures to optimize real-time traffic management and predictive maintenance.
- **Quantum Communication:** By integrating quantum cryptography, 6G will provide enhanced security and enable ultra-secure data transmission channels.
- **Holographic Connectivity:** 6G will support real-time holographic projection, enabling next-generation applications in telepresence and remote collaboration.
- **Ubiquitous Connectivity:** Unlike 5G, which is limited by infrastructure, 6G will leverage satellite and aerial networks to ensure seamless global coverage.

Nanosheet Transistors: Enabling 6G Performance

The evolution to 6G will require semiconductor innovations to keep pace with increasing computational demands. Nanosheet transistors provide the necessary advantages:

- **Higher Performance Scaling:** Compared to FinFETs, nanosheet transistors offer superior scaling potential for advanced semiconductor nodes beyond 3 nm.
- **Lower Power Consumption:** The improved electrostatic control of gate-all-around (GAA) nanosheet transistors reduces leakage currents, enhancing power efficiency.
- **Increased Switching Speed:** With reduced resistance-capacitance delays, nanosheet devices enable ultra-high-frequency operations essential for THz communication.

Figure 3. Key Features of 6G

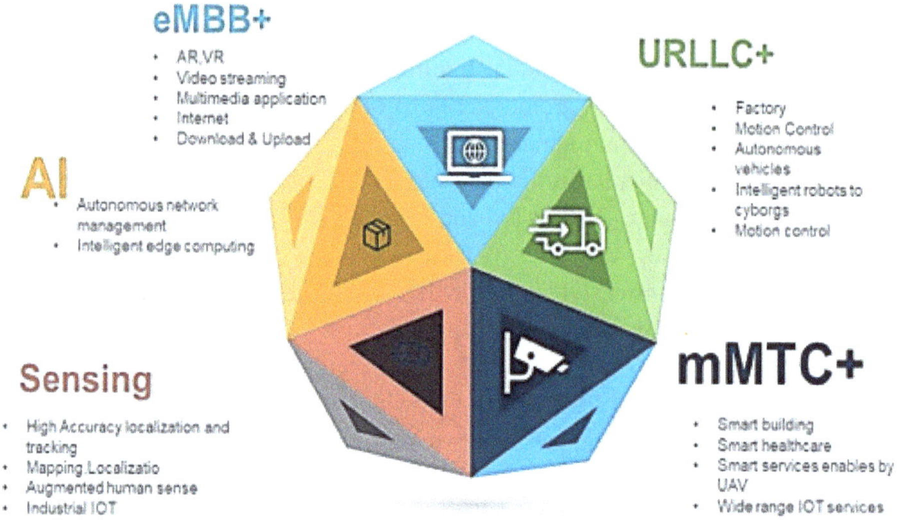

(Courtesy: pulse/overview-6g-use-cases-5g-learning/)

Future Prospects and Challenges

Despite the promising advantages of 6G, several challenges must be addressed before large-scale deployment:

- **Manufacturing Complexities:**Nanosheet transistor fabrication requires sophisticated techniques like EUV lithography, increasing production costs.
- **Infrastructure Development:** The integration of 6G networks with existing telecom infrastructure demands significant investments in new hardware and software.
- **Regulatory and Security Concerns:** Governments and regulatory bodies must establish global standards to ensure secure, fair, and efficient 6G implementation.

The transition from 5G to 6G represents a major technological leap, driven by semiconductor advancements such as nanosheet transistors. With ongoing research and development, these innovations will unlock unprecedented capabilities, shaping the future of wireless communication (Chen et al., 2014; Wang et al., 2014).

NANOSHEET TRANSISTORS: A BREAKTHROUGH IN SEMICONDUCTOR TECHNOLOGY

Introduction to Nanosheet Transistors

Nanosheet transistors are an innovative advancement in semiconductor technology, marking a significant departure from traditional FinFET designs. These devices feature a gate-all-around (GAA) architecture, where multiple stacked nanosheets serve as conducting channels, improving electrostatic control and enhancing performance at advanced process nodes. With increasing demands for higher efficiency and miniaturization in semiconductor devices, nanosheet transistors are expected to replace FinFETs as the industry standard beyond the 3 nm process node (Dang et al., 2020; Yang et al., 2022).

Advantages of Nanosheet Transistors

- **Superior Electrostatic Control:** The GAA structure offers better control over the channel, reducing leakage currents and improving energy efficiency.
- **Higher Drive Currents:** Multiple stacked nanosheets allow increased current flow, improving transistor performance for high-speed computing.
- **Scalability:** Unlike FinFETs, nanosheet transistors can scale efficiently beyond 3 nm, ensuring the continuation of Moore's Law.
- **Lower Power Consumption:** Reduced short-channel effects enable lower operating voltages, decreasing overall power consumption in semiconductor devices.

Manufacturing Process of Nanosheet Transistors

The fabrication of nanosheet transistors involves several advanced semiconductor processing techniques, including:

- **Extreme Ultraviolet (EUV) Lithography:** Enables precise patterning of nanosheets at sub-5 nm nodes.
- **Atomic Layer Deposition (ALD):** Allows precise thickness control for nanosheet layers, improving device consistency.
- **Selective Etching Techniques:** Used to define individual nanosheets and create multi-stack structures, ensuring uniformity and reliability.
- **Strained Silicon Engineering:** Enhances carrier mobility, improving device speed and performance.

Comparison with FinFETs

Table 1. Comparison of Nanosheet Transistors and FinFETs

Feature	FinFETs	Nanosheet Transistors
Scaling Potential	Limited	Higher
Power Efficiency	Moderate	High
Drive Current	Lower	Higher
Electrostatic Control	Good	Superior
Manufacturing Complexity	High	Higher

Applications in 5G and 6G Technologies

- **5G Base Stations:**Nanosheet transistors enhance power efficiency and performance in network processing units.
- **6G Terahertz Communication:** High-frequency operations enabled by nanosheet devices improve wireless communication speeds.
- **AI and Edge Computing:** Lower power consumption allows seamless integration in AI-driven network architectures.
- **Quantum and Secure Communication:** Improved transistor performance supports encryption and secure data transmission.

Challenges and Future Prospects

Despite their advantages, nanosheet transistors face several challenges:

- **Complex Fabrication Processes:** The transition from FinFETs to nanosheets requires significant investment in new manufacturing techniques.
- **Material Limitations:** Finding optimal materials for nanosheet channels and dielectrics is essential for improving reliability and efficiency.
- **Cost Implications:** Advanced fabrication techniques increase production costs, requiring economies of scale for commercialization.

Figure 4.Challenges and Future Prospects of 6G

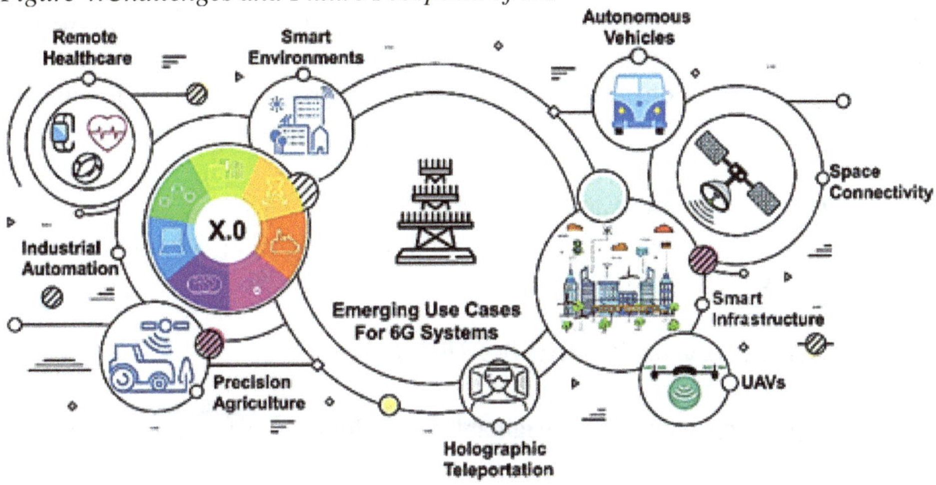

(Courtesy: Akyildiz et al. [2])

The future of nanosheet transistors remains promising as research efforts continue to optimize their performance and scalability. As 5G and 6G networks evolve, nanosheet devices will play a crucial role in enabling high-speed, energy-efficient communication systems.

IMPACT OF NANOSHEET DEVICES ON 5G AND 6G

Nanosheet transistors are expected to play a critical role in the evolution of 5G and the development of 6G communication networks. As wireless technology progresses, the demand for higher performance, lower power consumption, and improved scalability in semiconductor devices continues to rise. Nanosheet transistors, with their gate-all-around (GAA) architecture, provide superior electrostatic control and enhanced efficiency, making them ideal for next-generation network infrastructure. This section explores the impact of nanosheet devices on the capabilities, performance, and efficiency of 5G and 6G technologies (Fakhimi et al., 2024; You et al., 2021).

Enhancing Network Performance

Nanosheet transistors contribute significantly to the performance improvements in 5G and 6G networks through:

- **Higher Data Transmission Speeds:** The ability of nanosheet devices to operate at higher frequencies enables faster data rates, supporting speeds beyond 1 Tbps in 6G networks.
- **Lower Power Consumption:** Advanced electrostatic control minimizes leakage currents, reducing power consumption in network processing units and data centers.
- **Improved Switching Efficiency:** Faster transistor switching speeds lead to lower latency, essential for real-time applications such as autonomous vehicles and remote surgeries.

Energy Efficiency in 5G and 6G Networks

With the growing demand for sustainable technology, nanosheet devices provide significant energy efficiency benefits:

- **Reduced Heat Generation:** The lower power dissipation of nanosheet transistors decreases thermal output, improving the longevity of network hardware.
- **Optimized Power Scaling:**Nanosheet transistors support dynamic voltage scaling, allowing adaptive power consumption based on network demand.
- **Integration with AI-driven Network Optimization:** AI-based algorithms leverage nanosheet-powered processors for real-time energy-efficient network management.

Support for Advanced 6G Applications

Nanosheet devices enable key technological advancements required for future 6G applications:

- **Terahertz (THz) Communication:** The superior speed and efficiency of nanosheet transistors make them ideal for high-frequency operations in the THz spectrum (0.1–10 THz).
- **Holographic and Immersive Communication:** Enhanced processing capabilities support high-resolution, real-time holographic transmissions.
- **Quantum and Secure Communication:**Nanosheet transistors enable low-power cryptographic computations essential for quantum encryption.

Role in Edge Computing and AI-driven Networks

- **AI Processing at the Edge:** The high performance of nanosheet transistors allows for real-time AI inference, improving network automation and predictive maintenance.
- **Low-latency Processing:** Faster transistor switching times ensure ultra-reliable low-latency communication (URLLC), crucial for industrial automation and healthcare applications.
- **Scalability for IoT Devices:**Nanosheet-based chips enhance processing efficiency for massive IoT (mIoT) applications in smart cities and industrial deployments.

While nanosheet transistors offer substantial benefits, challenges remain:

- **Manufacturing Complexity:** The fabrication of nanosheet devices requires precision engineering and advanced semiconductor materials.
- **Cost Considerations:** The transition from FinFETs to nanosheets involves increased production costs and investment in new fabrication facilities.
- **Material Limitations:** Research is ongoing to optimize nanosheet materials for higher performance and longevity.

Nanosheet transistors are set to revolutionize the future of wireless communication by enabling higher speeds, greater energy efficiency, and seamless integration with AI-driven networks. Their impact on 5G and 6G technologies will be transformative, paving the way for next-generation applications that demand ultra-fast, secure, and sustainable connectivity (Kachorovskii et al., 2021; Zhang et al., 2019).

EMERGENCE OF 5G AND 6G TECHNOLOGIES

The rapid advancement of communication technologies has ushered in the development and deployment of 5G, with 6G already in the early stages of research and development. These technologies promise ultra-fast data rates, ultra-low latency, and enhanced connectivity across numerous devices and platforms. Central to these advancements are nanosheet transistors, which offer superior performance in managing data transmission and computational tasks. Despite the promise of transformative benefits, the implementation and operation of these systems are not without significant challenges. Two major categories of limitations are Implementation Complexity and Energy Consumption.

IMPLEMENTATION COMPLEXITY

5G Networks

The rollout of 5G technology introduces new infrastructure demands. Unlike previous generations, 5G operates on higher frequency bands, which have limited range and penetration. To maintain strong and reliable signals, dense deployments of small-cell networks are required. These small cells must be strategically placed on streetlights, buildings, and utility poles. Urban areas, though well-equipped, face logistical challenges in securing space and regulatory approval. Rural and remote regions often lack the necessary investment and existing infrastructure. This results in a digital divide, where certain areas benefit significantly earlier than others. Moreover, coordination among multiple stakeholders—governments, telecom operators, and equipment vendors is essential but complex. The integration with legacy systems also adds to the challenge, requiring careful planning and interoperability testing.

6G Networks

While 5G is still being rolled out, research into 6G has begun, targeting commercial deployment around 2030. 6G aims to leverage technologies such as terahertz (THz) frequencies, quantum communication, and AI-driven network management. The use of terahertz waves can support massive data rates but suffers from extremely short range and sensitivity to obstacles. This will necessitate an even denser and more intelligent infrastructure than 5G. The incorporation of AI for real-time optimization and quantum technologies for secure communication will increase system complexity. These enhancements require breakthroughs in materials science, device fabrication, and computational frameworks. Deployment timelines are likely to be extended due to the need for global standardization and technical maturity. Security protocols and data governance will also need to evolve rapidly to support the new capabilities.

ENERGY CONSUMPTION

High Power Demands of Advanced Devices

The high-speed and high-capacity features of 5G and 6G come at the cost of significantly increased energy consumption. Nanosheet transistors, while offering enhanced performance, are power-intensive due to their miniaturized architecture and processing load. Base stations for 5G already consume 2 to 3 times more energy than their 4G counterparts. With 6G expected to handle exponentially more data, energy demands will further escalate. Mobile devices will require more advanced battery technologies to keep up with faster processing and communication tasks. Data centers will face increased cooling and power supply challenges to support continuous, high-throughput operations.

This surge in energy requirements poses both environmental and economic concerns. Efforts are underway to design energy-efficient hardware and explore renewable energy solutions for network operations. Nevertheless, achieving sustainability while maintaining high performance remains a significant hurdle. Power optimization across devices, networks, and infrastructure will be a critical focus in the coming years. The convergence of 5G/6G technologies with nanosheet innovations marks a new era in global connectivity and digital transformation.

CHALLENGES AND FUTURE PROSPECTS

While nanosheet transistors offer remarkable advantages for 5G and 6G applications, they also present several challenges that must be addressed before widespread adoption. These challenges include manufacturing complexities, cost implications, material limitations, and performance consistency. Additionally, researchers continue to explore ways to further optimize nanosheet technology to meet the growing demands of future wireless communication networks (Knap et al., 2023).

Manufacturing Challenges

The production of nanosheet transistors involves several intricate steps that increase fabrication complexity:

- **Advanced Lithography Requirements:** The transition to sub-3 nm nodes necessitates the use of extreme ultraviolet (EUV) lithography, which is costly and requires precise process control.
- **Material Engineering Limitations:** Finding the right materials for the nanosheets and dielectric layers remains a challenge in achieving optimal transistor performance and longevity.
- **Process Variability:** Variations in nanosheet thickness, alignment, and gate formation can lead to inconsistent device behavior, affecting yield rates in mass production.

Cost and Economic Factors

- **High Production Costs:** The development and implementation of nanosheet transistor technology require significant investments in new fabrication plants and equipment.
- **Limited Early Adoption:** Due to high costs, only major semiconductor manufacturers with advanced fabrication capabilities can integrate nanosheet technology in the near term.
- **Scalability Concerns:** The transition from FinFET to nanosheet devices demands considerable industry-wide adaptation, requiring time and resources.

Performance Optimization

- **Power Efficiency Trade-offs:** Although nanosheet transistors provide superior electrostatic control, optimizing them for low-power applications without compromising performance remains a challenge.
- **Thermal Management Issues:** As transistor density increases, heat dissipation concerns arise, necessitating innovative cooling solutions for nanosheet-based processors.
- **Device Reliability and Stability:** Long-term reliability testing is essential to ensure that nanosheet transistors maintain performance over extended operational lifetimes.

Future Research Directions

- **Material Innovations:**Ongoing research explores alternative materials, such as 2D materials and new high-k dielectrics, to enhance nanosheet transistor performance.
- **Hybrid Architectures:** Combining nanosheets with other emerging semiconductor technologies, such as complementary FETs (CFETs), could improve efficiency and scalability.
- **AI-driven Design Optimization:** Machine learning and AI algorithms are being utilized to optimize nanosheet device designs for improved performance and reduced power consumption.

Roadmap for Commercialization

- **Standardization Efforts:** Industry collaboration is needed to develop common standards and manufacturing guidelines for nanosheet transistors.

- **Pilot Production and Early Deployment:** Initial applications will likely be in high-performance computing (HPC) and AI-driven data centers before integration into consumer electronics.
- **Integration into 6G Infrastructure:** As 6G research advances, nanosheet transistors will be critical in supporting the high-frequency, low-latency requirements of next-generation networks.

Nanosheet transistors hold immense potential to revolutionize semiconductor technology for 5G and 6G applications. However, their widespread adoption faces several technical, economic, and scalability challenges. Addressing these issues requires a concerted effort from industry leaders, researchers, and policymakers to drive innovation and ensure the successful commercialization of nanosheet technology. As advancements continue, nanosheet devices are poised to become a cornerstone of next-generation wireless communication (Li et al., 2022; Zong et al., 2019).

CONCLUSION

The evolution from 5G to 6G requires advanced semiconductor technologies to meet future performance demands. Nanosheet transistors provide an innovative solution with superior efficiency and scalability. Despite challenges in fabrication and integration, they hold great promise for next-generation wireless communication networks. Nanosheet transistors represent a transformative breakthrough in semiconductor technology, poised to drive the evolution of 5G and 6G networks. Their superior electrostatic control, higher drive current, and energy efficiency make them a key enabler of next-generation wireless communication. As the demand for faster, more reliable, and power-efficient networks grows, nanosheet devices will play a crucial role in achieving these objectives. The integration of nanosheet transistors in wireless infrastructure will facilitate advancements in high-frequency communication, AI-driven network optimization, and ultra-low latency applications. Furthermore, their potential in enabling terahertz communication and quantum-secure networks makes them essential for the future digital landscape. However, challenges such as manufacturing complexity, material limitations, and cost factors must be addressed to ensure successful commercialization. Continued research and collaboration between industry and academia are essential to refine nanosheet transistor technology. Efforts in standardization, pilot production, and optimization strategies will further accelerate their adoption in mainstream semiconductor manufacturing. As nanosheet transistors mature, they will not only enhance 5G capabilities but also lay the foundation for the seamless deployment of 6G networks. In conclusion, nanosheet devices are set to revolutionize the semiconductor industry, providing the necessary computational power and efficiency for advanced communication technologies. Their successful implementation will shape the future of wireless networks, enabling smarter, faster, and more sustainable connectivity for the digital era.

REFERENCES

Agiwal, M., Roy, A., & Saxena, N. (2016). Next generation 5G wireless networks: A comprehensive survey. *IEEE Communications Surveys and Tutorials*, *18*(3), 1617–1655. DOI: 10.1109/COMST.2016.2532458

Akyildiz, I. F., Kak, A., & Nie, S. (2020). 6G and beyond: The future of wireless communications systems. *IEEE Access : Practical Innovations, Open Solutions*, *8*, 133995–134030. DOI: 10.1109/ACCESS.2020.3010896

Andrews, J. G., Buzzi, S., Choi, W., Hanly, S. V., Lozano, A., Soong, A. C. K., & Zhang, J. C. (2014). What will 5G be? *IEEE Journal on Selected Areas in Communications*, *32*(6), 1065–1082. DOI: 10.1109/JSAC.2014.2328098

Bhandari, S., & Chauhan, Y. S. (2021). Nanosheet field-effect transistor device and circuit aspects for future technology nodes. *IEEE Transactions on Electron Devices*, *68*(6), 2796–2803.

Boccardi, F., Heath, R. W., Lozano, A., Marzetta, T. L., & Popovski, P. (2014). Five disruptive technology directions for 5G. *IEEE Communications Magazine*, *52*(2), 74–80. DOI: 10.1109/MCOM.2014.6736746

Chen, S., & Zhao, J. (2014). The requirements, challenges, and technologies for 5G of terrestrial mobile telecommunication. *IEEE Communications Magazine*, *52*(5), 36–43. DOI: 10.1109/MCOM.2014.6815891

Dang, S., Amin, O., Shihada, B., & Alouini, M.-S. (2020). What should 6G be? *Nature Electronics*, *3*(1), 20–29. DOI: 10.1038/s41928-019-0355-6

Fakhimi, M. J., & Akan, O. B. (2024). Nanoantennas and nanoradars: The future of integrated sensing and communication at the nanoscale. *IEEE Communications Magazine*, *62*(1), 68–74.

Kachorovskii, V., Deng, Y., Rumyantsev, S., Lü, J.-Q., Gaska, R., Simin, G., Hu, X., Khan, M. A., & Shur, M. S. (2021). Plasmonic field-effect transistors (TeraFETs) for6G communications. *Nanomaterials (Basel, Switzerland)*, *11*(12), 3301. PMID: 34947650

Knap, W., Kachorovskii, V., Deng, Y., Rumyantsev, S., Lü, J.-Q., Gaska, R., Shur, M., Simin, G., Hu, X., Khan, M. A., & Shur, M. S. (2023). Gigahertz and terahertz transistors for 5G, 6G, and beyond mobile communication systems. *Applied Physics Reviews*, *11*(3), 031318.

Li, X., Zhang, Y., & Wang, P. (2022). Recent advances in new materials for 6G communications. *Advanced Electronic Materials*, *8*(1), 2100978. DOI: 10.1002/aelm.202100978

Osseiran, A., Boccardi, F., Braun, V., Kusume, K., Marsch, P., Maternia, M., Queseth, O., Schellmann, M., Schotten, H., Taoka, H., Tullberg, H., Uusitalo, M. A., Timus, B., & Fallgren, M. (2014). Scenarios for 5G mobile and wireless communications: The vision of the METIS project. *IEEE Communications Magazine*, *52*(5), 26–35. DOI: 10.1109/MCOM.2014.6815890

Rappaport, T. S., Sun, S., Mayzus, R., Zhao, H., Azar, Y., Wang, K., Wong, G. N., Schulz, J. K., Samimi, M., & Gutierrez, F. (2013). Millimeter wave mobile communications for 5G cellular: It will work! *IEEE Access : Practical Innovations, Open Solutions*, *1*, 335–349. DOI: 10.1109/ACCESS.2013.2260813

Rappaport, T. S., Xing, Y., MacCartney, G. R., Molisch, A. F., Mellios, E., & Zhang, J. (2015). Overview of millimeter wave communications for fifth-generation (5G) wireless networks—With a focus on propagation models. *IEEE Transactions on Antennas and Propagation*, *65*(12), 6213–6230. DOI: 10.1109/TAP.2017.2734243

Saad, W., Bennis, M., & Chen, M. (2019). A vision of 6G wireless systems: Applications, trends, technologies, and open research problems. *IEEE Network*, *34*(3), 134–142. DOI: 10.1109/MNET.001.1900287

Shafi, M., Molisch, A. F., Smith, P. J., Haustein, T., Zhu, P., De Silva, P., Tufvesson, F., Benjebbour, A., & Wunder, G. (2017). 5G: A tutorial overview of standards, trials, challenges, deployment, and practice. *IEEE Journal on Selected Areas in Communications*, *35*(6), 1201–1221. DOI: 10.1109/JSAC.2017.2692307

Tsiftsis, T. A., Valagiannopoulos, C., Liu, H., Boulogeorgos, A.-A. A., & Miridakis, N. I. (2020). Metasurface-coated devices: A new paradigm for energy-efficient and secure 6G communications. *IEEE Open Journal of the Communications Society*, *1*, 1446–1455.

Wang, C.-X., Haider, F., Gao, X., You, X.-H., Yang, Y., Yuan, D., Aggoune, H. M., Haas, H., Fletcher, S., & Hepsaydir, E. (2014). Cellular architecture and key technologies for 5G wireless communication networks. *IEEE Communications Magazine*, *52*(2), 122–130. DOI: 10.1109/MCOM.2014.6736752

Yang, K., & Zuo, C. (2022). Nanosheet lithium niobate acoustic resonator for mmWave frequencies. *IEEE Microwave and Wireless Components Letters*, *32*(5), 413–416.

Yoshihisa, K., Satoshi, S., & Satoshi, N. (2021). Trends and Target Implementations for 5G evolution & 6G. *NTT Technical Review*, *19*(11), 18–25. DOI: 10.53829/ntr202111fa1

You, X., Wang, C.-X., Huang, J., Gao, X., Zhang, Z., Wang, M., & Jiang, Y. (2021). Towards 6G wireless communication networks: Vision, enabling technologies, and new paradigm shifts. *Science China. Information Sciences*, *64*(1), 110301. DOI: 10.1007/s11432-020-2955-6

Zhang, Z., Xiao, Y., Ma, Z., Xiao, M., Ding, Z., Lei, X., Karagiannidis, G. K., & Fan, P. (2019). 6G wireless networks: Vision, requirements, architecture, and key technologies. *IEEE Vehicular Technology Magazine*, *14*(3), 28–41. DOI: 10.1109/MVT.2019.2921208

Zong, B., Fan, C., Wang, X., Li, X., Niu, Z., & Li, P. (2019). 6G technologies: Key drivers, core requirements, system architectures, and enabling technologies. *IEEE Vehicular Technology Magazine*, *14*(3), 18–27. DOI: 10.1109/MVT.2019.2921398

Chapter 9

Optimizing Power Efficiency in Wireless Sensor Networks Through Specific Arbitrary Hierarchical Clustering and Enhanced Swastika Routing

Prabira Kumar Sethy
http://orcid.org/0000-0003-3477-6715
Sambalpur University, India

Jagmahon Padhi
http://orcid.org/0000-0001-8080-6597
Sambalpur University, India

A Geetha Devi
PVP Siddharth Institute of Technology, India

Santi Kumari Behera
http://orcid.org/0000-0003-4857-7821
VSSUT Burla, India

Aziz Nanthaamornphong
http://orcid.org/0000-0002-1618-6001
College of Computing, Prince of Songkla University, Phuket, Thailand

DOI: 10.4018/979-8-3373-0852-4.ch009

ABSTRACT

Wireless sensor networks (WSNs) face challenges due to limited human intervention post-deployment and finite battery life at each node, where the failure of one node's battery can jeopardize the entire system. To address this, we proposed a novel routing technique called Specific Arbitrary Hierarchical Clustering Entrenched Advanced Swastika Routing (SAHCEASR). This technique combines clustering with an advanced version of Swastika Routing, employing specific arbitrary clustering for optimal cluster selection. Implemented in MATLAB using a power model, SAHCEASR significantly reduces power consumption at both system-wide and individual node levels. Our adaptive routing approach demonstrated nearly a 50% reduction in battery power consumption, thereby enhancing the network lifetime of WSNs.

1. INTRODUCTION

A WSN is a collection of nodes, where each node acts as an individual functional unit of the system. The node of the WSN is primarily made up of many tiny elements, such as analog-to-digital converters, digital-to-analog converters, batteries, antennas, flash memory, and sensors. The sensor actuates a physical signal, mostly in the form of an analog. The same signal is then converted into a digital format, which further undergoes processing. Finally, the processed data is transferred to the antenna to transmit the signal to a powerful host node. The nodes were distributed in the field to actuate the signal of interest. Sometimes, it is quite difficult to replace the battery or feed an extra electromotive force to the node if the mounted battery is drained out. This implies a malfunctioning of the entire data actuation process; subsequently, the entire WSN fails. Therefore, it is of paramount importance to minimize power dissipation from the battery by minimizing redundant signal transmissions. Typically, multiple deployed nodes actuate the same data and individually transmit the same data to the host node. However, the best thing that can be done is to cluster all related objects into one, which is expected to draw similar data according to the deployed coordinates of the node. Our study's core concept of clustering in the WSN is more significant in maximizing the network lifetime by minimizing the transmission power consumption.

Cluster analysis is the organization of data into groups based on certain parameters and references. Within a cluster, the objects are comparatively similar and differ from those of the other clusters. The linkage metrics can be determined by clustering. For example, in a WSN, finding an object with similar data in multiple nodes can be transmitted once to the host node instead of individual signal transmissions to minimize battery power consumption. The minimization of power will help the system enhance the network lifetime of the entire system. There are many ways to cluster data, and one of the most efficient clustering types is hierarchical data clustering.

Clustering organizes hierarchical data across scales, resulting in a cluster tree or a dendrogram. Instead of a single-cluster collection, the tree is a multilevel hierarchy, with groups at one level interacting with groups at the next. The linkage, cluster, and distance between node functions are all included in the taxonomy data process and may be utilized separately for future in-depth analyses. A dendrogram was used to illustrate the cluster tree.

Hierarchical clustering organizes data of various sizes by constructing a cluster tree or dendrogram. The tree is a multilayered system instead of a unique collection of clusters, with clusters at one level joining clusters at the next level. This enables the selection of the best clustering level or degree of the operation. Agglomerative grouping is supported, and the Statistics Toolbox function cluster data take care of every step. Hierarchical clustering is used for the information set across levels, and the result is a cluster tree or a dendrogram. Despite being a singular aggregate of groups, the tree is a multilevel hierarchy with clusters at one tier interacting with clusters at the following level. Thanks to something like this, you may choose a clustering scale or level that fits perfectly for our operation. The Statistics Toolbox's cluster data function handles each of the necessary tasks. It has capabilities for linkage, clustering, and distance between nodes that can be employed separately for a more thorough investigation. The dendrogram utility, linkage, and cluster functions, which can be individually utilized for further in-depth analysis, depict the cluster structure. The cluster tree was constructed using a dendrogram tool.

The major contributions of this study are as follows:

- The algorithm we formulated can reduce the power consumption of the Wireless Sensor Network by up to 60-70 percent by eliminating redundant data transmission by the node.
- Our model is promising for use in a Wireless Sensor Network where the system has been deployed in a remote area where the node's battery is impossible to replace.
- In this study, we maximized the network lifetime by clustering all deployed nodes, where one of the nodes in each cluster acts as a cluster head that transmits the signal of all likely nodes that are clubbed together.
- Our proposed algorithm uses SWASTIKA routing, which uses the orthogonal path of transmitting the signal from nodes to sink, which can subsequently minimize the total and average power consumption of the Wireless Sensor Network and help in increasing the network lifetime.

- First, any algorithm using the hierarchal clustering of a specific arbitrary cluster number selection according to the application needs to transmit using SWASTIKA routing to maximize the network lifetime.

The remainder of this paper is organized as follows. The literature review is detailed in Section 2. Section 3 describes the proposed methodology, which contains the details of swastika routing, the SAHCEASR algorithm, power modeling, and its implementation. Section 4 presents a mathematical elaboration of the models. The experimental findings are discussed in Section 5. Finally, Section 6 concludes the study.

2. LITERATURE REVIEW

Miodrag Zivkovic et al. (Zivkovic et al., 2021) have proposed another novel metaheuristic approach which they have named the dragonfly algorithm. The dragonfly algorithm is an advanced version of the LEACH and Particle Swarm techniques. They simulated the results of their algorithm and compared them with the pre-existing values from the existing algorithm. They found that the dragonfly algorithm produced better results than the existing algorithm in minimizing power consumption and maximizing network lifetime. Achyut Shankar et al., in the paper (Shankar et al., 2021), have considered wireless sensor networks where human intervention is negligibly low for maintenance and power backup. Insightfully, they proposed a fault-tolerance method by recognizing faults and revitalization approaches for managing faulty systems. This has increased the network lifetime by approximately five times that of conventional methods. The method subsequently improves data transmission and indirectly improves network lifetime. They simulated the proposed method using MATLAB 2018 software. Khushboo Jain et al. (Jain et al., 2022) have proposed cluster-based data collection techniques to minimize the power consumption from the system. SCADA, which is based on scalable clustering, was used in this study. This study states that the cluster head needs to rotate within a specific time interval to minimize the number of reduced packets. They compared their proposed model with the pre-existing protocols model and found that it provides significantly good results and caters to a longer lifetime of the system. J. Padhi et al. (Padhi et al., 2016) proposed a new routing model in which they were able to maximize the total network lifetime of the sensor network by minimizing battery power dissipation. The proposed routing model uses an orthogonal path to transmit signals and reduce the power consumption of the nodes. Patel et al. (Patel et al., 2017) provided advanced routing techniques based on SWASTIKA routing techniques, where they created a sub-group of swastika models to minimize the total network power consumption to enhance the system lifetime. They formulated different layered swastika routing models to enhance the productivity of the traditional swastika routing. Jena et al. (Jena et al., 2016) have minimized the total power consumption by reducing the transmitted distance by adopting an ant colony algorithm instead of the conventional distance counting method. The study claimed to maximize network lifetime using the swastika model associated with ant colony routing. The ant colony algorithm was used in their model to select the path, and for routing, they used the swastika routing model. Using the hybrid model, the sensor network lifetime can be maximized. Amin et al. (Shahraki et al., 2020), in their review paper, have highlighted the clustering, ad-hoc network, and the use of these in IoT. Using these, we can reduce the energy consumption and load balancing in the WSN. They explained grouping techniques that hop from one to another for signal transmission. The papers by Seyyedabbasi et al. (Seyyedabbasi et al., 2020) introduce HEEL, a novel clustering method, enhancing the lifetime of wireless sensor networks (WSN). Kiani (Kiani, 2016) proposes a channel allocation method for time synchronization in WSNs. Kiani et al. (Farzad Kiani, 2013) present EEAR, an energy-effective routing algorithm for WSNs. Seyyedabbasi et al. (Seyyedabbasi et al., 2023) optimize data transmission and pathfinding in WSNs and IoT systems using I-GWO and Ex-GWO algorithms. Nematzadeh et al. (Nematzadeh et al., 2023) focus on maximizing coverage and connectivity through a metaheuristic-based approach for node deployment in WSNs and decentralized IoT, contributing to power efficiency and reliability. In addition to this many authors are report their work on improvement of WSN life time using clustering techniques (Dao et al., 2024; Haghzad Klidbary & Javadian, 2024; Sen Laxminarayan Sahoo & Tiwary, 2024; Vinoth Kumar & Venkatesh, 2024).

3. PROPOSED METHODOLOGY

In this section, swastika routing, SAHCEASR algorithm, power modeling, and its implementation are described in the appropriate sub-section.

3.1. Swastika Routing

In this work, we dubbed our model SWASTIKA because of the resemblance of the routing pattern to the swastika. Swastika, a traditional Hindu religious symbol, is a single word, not an acronym. Hindus hold that the swastika symbol, which commonly resembles a quadrilateral cross with its four legs twisted at a 90-degree angle, directs positivity and makes everything perform effectively in any location, including a home, workplace, cultural region, and society. The swastika can be viewed as a 20-sided polygon from a geometric perspective. The proportions of the Nazi swastika were predetermined using a 5 × 5 diagonal grid. Characteristic features include the existence of two forms of swastikas that are mirror images of one another and 90° rotational symmetry, resulting in the absence of reflection symmetry.

In Figure 1, both counterclockwise and clockwise SWASTIKA have been shown; either design can be used to improve routing and extend the lifespan of a system. Therefore, we computed both patterns in this study.

Figure 1. A right-facing swastika might be described as "clockwise" or "counter-clockwise".

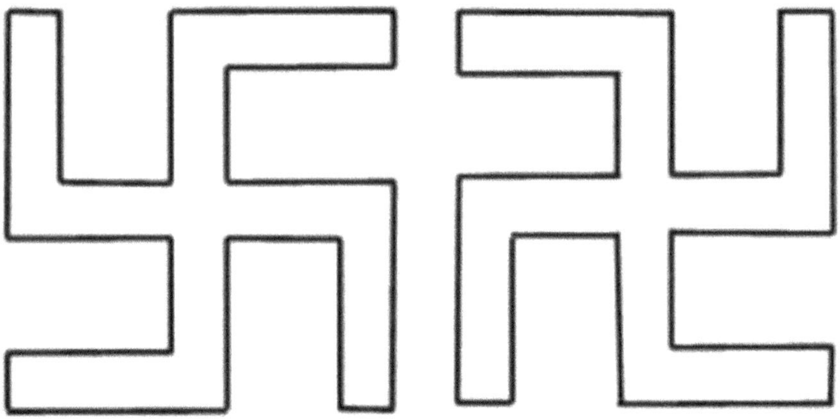

The isomers of swastika can be represented as:

- clockwise ( ) and counter-clockwise ( )
- left-facing ( ) and right-facing ( )
- left-hand ( ) and right-hand ( )

Instead of using the high-energy-consuming Dijkstra path to find the next node until the destination is reached, we develop a route that uses adaptive routing approaches. Unfortunately, this high-energy-consuming Dijkstra path is essentially the diagonal path in any uniformly distributed network, which increases the total power consumption of the network and causes early battery outflow. To address this, we route the diagonal path perpendicular to the endpoint and second-shortest-distance routing for nodes that typically follow the diagonal route.

3.2. Proposed Algorithm of SAHCEASR

SAHCEASR follows the following steps

1. to deploy Nodes in the area of length 'L' and breadth 'B.'
2. First, we determine whether each pair of objects in the node collection is similar. In this phase, users use the distance between the node functions to determine the separation between items. The distance between the node functions supports various methods for calculating this value.
3. A binary hierarchical cluster tree is created from the entities. In this phase, the linkage function is used to connect the adjacent sets of items. First, the linkage function uses the distance data created in step 1 to calculate the closeness of the objects. Then, the freshly generated bunches are combined into larger clusters as objects are coupled to binary clusters, and so on, until a hierarchical tree is constructed.
4. Determine how many groups you want to transmit to cut the hierarchical tree into the required groups accordingly.
5. The group information will be followed the SWASTIKA routing
6. Calculate the total power consumption and individual node power consumption for analysis

3.3. Power Model

As proposed, the total power used by the network is the sum of the power used by each node in the network for radio transmissions. And so is well knowledge; sending and receiving radio signals account for most of the power used by the WSN. Here, we mainly concentrate on radio transmission to study power consumption. The network was set up for an implementation reason that requires each node to submit packets to the sink node. Therefore, the power use of the sensor node has always been closely associated with the rate at which the node transmits traffic and the distance between the sender and the recipient.

Power consumption can be calculated using the following equation:

$$PC = K \times r \times d^{ALPHA} \tag{1}$$

The total power consumption of the WSN system is the summation of the power consumption of all individual nodes. The average power consumption of a node can be determined by dividing the total number of nodes by the total power consumption of the system. Therefore, by calculating the Total Power Consumption (TPC) and Average Power Consumption (APC) and comparing them with the conventional and existing routing models, we can estimate the potential of the proposed network model in terms of its supremacy in increasing network lifetime.

3.4. Implementation

To test our algorithm, we used the SWASTIKA model, where the distance between each node was taken as unity. Assume there are 25 nodes and allow the entire vertical paths to transmit the signal from the sensing node to the sink node, as mentioned in the swastika routing, to minimize the transmission power of the deployed nodes. We clubbed all 25 nodes into four subgroups and aimed to cluster all nodes within those four subgroups. For this purpose, hierarchical clustering was used. It is assumed that the objects coming within a single zone are similar, and that the data coming out of the region will be the same. Instead of sending data from each node, we allow a single node to transmit the signal to the host node to minimize the power consumption. Figure 2 illustrates the object distribution for clustering.

We have taken eight random distinct points that will come within the area we have taken

- Object 1: 3.6000 1.4000
- Object 2: 0.8000 1.2000
- Object 3: 3.6000 3.6000
- Object 4: 3.8000 1.3000
- Object 5: 1.1000 1.3000
- Object 6: 1.4000 3.8000
- Object 7: 1.5000 3.9000
- Object 8: 3.4000 3.3000

Figure 2. Object distribution for clustering

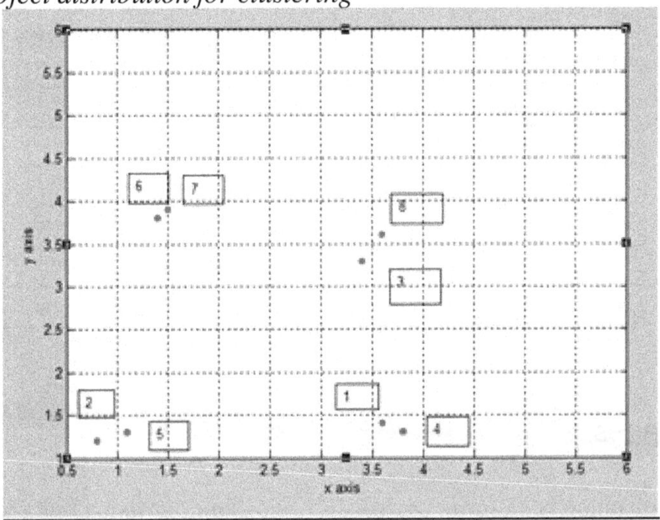

3.4.1. Distance

To determine the distances between pairings, we estimated the distance between items 1 and 2, objects 1 and 3, and so on. The Euclidean distance between objects A and B determines this distance. This distance data are returned by the range between the nodes that function as a vector, D, where each element represents the separation between two entities.

D = 2.8071 2.2000 0.2236 2.5020 3.2558 3.2650 1.9105 3.6878 3.0017 0.3162 2.6683 2.7893 3.3422 2.3087 3.3971 2.2091 2.1213 0.3606 2.7000 3.4655 3.4713 2.0396 2.5179 2.6306 3.0480 0.1414 2.0616 1.9925

To understand the above distance data, the original dataset can be arranged in a square form, where matrix indices indicate the distance between the objects of the respective indices and illustrated in Table 1.

Table 1. Distance Square Form Matrics

0	**2.80**	**2.20**	**0.22**	**2.50**	**3.25**	**3.26**	**1.91**
2.80	0	3.68	3.00	0.31	2.66	2.78	3.34
2.20	3.68	0	2.30	3.39	2.20	2.12	0.36
0.22	3.00	2.30	0	2.70	3.46	3.47	2.03
2.50	0.31	3.39	2.70	0	2.51	2.63	3.04
3.25	2.66	2.20	3.46	2.51	0	0.14	2.06
3.26	2.78	2.12	3.47	2.63	0.14	0	1.99
1.91	3.34	0.36	2.03	3.04	2.06	1.99	0

3.4.2. Linkages

Once the distance is as shown in Table 1, we can establish a linkage among the nodes and generate a linkage table for the assumed dataset. The linkage operator creates a hierarchical cluster tree, for instance, returning the linkage information in a matrix Z, given the distance vector Y produced by the range between the nodes from the data sample set of the x- and y-coordinates.

Table 2. Linkage Matrics 'Z'

6	**7**	**0.141421356237310**
1	4	0.223606797749979
2	5	0.316227766016838
3	8	0.360555127546399
10	12	1.91049731745428
9	13	1.99248588451713
11	14	2.50199920063936

Every row under this output denotes a connection between the objects or groupings. The objects that were linked are listed in the first two columns. The third column presents the separation of these entities. The linkage function starts by grouping items 4 and 5 with the nearest closeness (distance value = 1.0000) for the sample dataset of x- and y-coordinates. Grouping objects 1 and 3, each with a distance value of 1.0000, is the next step in the linkage function.

The third row shows that Objects 6 and 7 were grouped using the linking function. What were objects 6 and 7 if the initial test statistic set had only five objects? Object 6 is a newly formed binary cluster produced when objects 4 and 5 are combined. The linkage function must give each new cluster it creates when two items are combined into one individual index value, starting at m+1, where m is the total number of objects in the initial data collection. (The actual data set essentially uses values 1 through m.) Similar to Item 6, Object 7 is a collection created by combining Objects 1 and 3.

Linkages use distances to choose the order in which items are clustered. Therefore, for example, the original items 1 through 5's distances are represented in the distance vector D. Again, linking also has to calculate the distances between the clusters it makes, like the ones made up of objects 6 and 7. By default, linkage employs a technique called single linkage. However, a variety of approaches can be used.

The linkage function combines object 8, a freshly formed cluster of objects 6 and 7, with object 2 from the initial dataset to form the final cluster. Table 2 presents the linkage metrics.

3.4.3. Dendrograms

With the help of linkage, we can form a hierarchical binary cluster tree, which can easily be understood from the output graph itself. We have a function that reduces the dendrogram plots, which is nothing but the hierarçhal tree. Figure 3 demonstrates how linking organizes items into groups in a hierarchical manner.

Figure 3. Dendrogram of the distance matrix 'D'(x-axis is the object number and y-axis is the magnitude of the linkage)

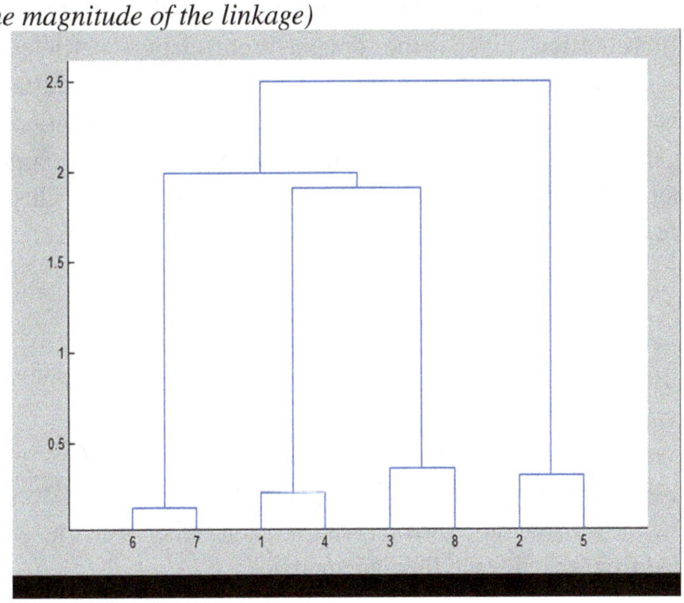

The indices of the objects in the original dataset are shown in the figure values along the x-plane. Positive-side U-shaped bars were used to illustrate the connections between the two things. Height U represents the separation of the objects.

3.4.4. Verify the Cluster Tree

In this subsection, we present the verification of cluster dissimilarity and inconsistency.

3.4.4.1. Verify Dissimilarity

Any two objects from the initial dataset were eventually connected at a certain level in a hierarchical cluster tree. The link height indicates the separation between two clusters home to two objects. The cophenetic distance between two objects was measured at this height. "Comparing the cophenetic distances with the original distance data produced by the distance between the nodes function is one technique to assess how accurately the cluster tree produced by the linkage function matches your data. If the clustering is accurate, there should be a significant correlation between the distances between the objects in the distance vector and the connection of the objects in the cluster tree (Murtagh & Contreras, 2012)." The cophenetic correlation coefficient is the number of cophenet function returns after comparing these two variables and computing their correlations. The cophenetic correlation coefficient increases as it approaches 1. "The cophenet function compares these two sets of variables and calculates their correlation, returning a number known as the cophenetic correlation coefficient. When the cophenetic correlation coefficient value was close to 1, the clustering strategy depicted the data (Contreras & Murtagh, 2015)". "When clustering the same data set using various distance computation techniques or clustering algorithms, the outcomes can be compared using the cophenetic coefficient of correlation (Zhao & Karypis, 2002)." For instance, the clusters created for the sample dataset can be assessed using the cophenet program.

C1 =0.8841

Y is the distance vector produced by the range between node tools, and Z is the matrix produced by the linkage value.

Rerun the same dataset's range between nodes calculation, but this time provides the city block measure. Then, we call the cophenet to assess the clustering solution after executing the linkage function on the new distance between the outputs of the nodes obtained from the average linkage approach.

c =0.8164

The cophenetic correlation coefficient demonstrates that a tree with a varying distance and linkage approach captures the original relationships slightly better.

3.4.4.2. Verify Consistency

"Comparing the height of each link in a cluster tree with the heights of surrounding links under it in the tree is one method of identifying the natural cluster divides in a data collection (Xu & Wunsch, 2005)." There are no clear boundaries between objects coupled at this hierarchy level if a link is at approximately the same height as the connections under it. Therefore, because the distances between the objects being joined and the objects are approximately equal, these links demonstrate a high degree of consistency. However, the objects united at this level in the cluster tree were substantially further away from one another than their components were before they were connected if a link's altitude noticeably changed from the links under it. According to the links under it, this connection is inconsistent. In cluster analysis, inconsistencies in linkages might indicate where a dataset is naturally divided into different categories. The cluster function determines where to split our data into clusters using a quantitative measure of inconsistency. The dendrogram shows inconsistent linkages. In the dendrogram, note how the items were divided into two groups joined by links at a much higher tree level. These connections are inconsistent when matched with the connections in the hierarchy beneath them. We use an inconsistent function to provide a list of inconsistency coefficients for each connection in the cluster tree. The inconsistent function compares each link in the cluster hierarchy to nearby connections with no more than two lower levels. The standard deviations of all links included in the calculation with their inconsistency coefficients are listed in Table 3.

Table 3. The Magnitude of Inconsistency Coefficient

Mean of the heights of all the links included in the calculation	The standard deviation of all the links included in the calculation	Number of links included in the calculation	Inconsistency coefficient
0.20	0	1	0
0.30	0	1	0
0.40	0	1	0
0.50	0	1	0
1.03	1.10	3	1.14
1.23	1.61	3	1.15
3.03	0.70	3	0.94

The fourth column in Table 3 contains the values of inconsistency. It is the difference between the current connection altitude and the mean normalized by the standard deviation:

(1.91049731745428 -1.03333333333333) / 1.10151410945722

=1.14993231207072

3.4.4.3. Create Cluster

After building the hierarchical tree of binary clusters, the clustering utility can be used to trim the tree and divide the data into clusters. There are two ways to build clusters using the cluster function, as discussed in the following sections.

- Find Natural Divisions in Data
- Specify Arbitrary Clusters

In our implementation, we focus on a specified arbitrary cluster, which can be used to determine our cluster number by selecting the number of clusters we want from the process of taxonomy because we are dealing with a WSN system where we are not aware of the similarity zone. The number of clusters might vary from system to system, and depending on the type of actuation signals and the degree of accuracy, we need to cater to the system. Therefore, a specified number of clusters are formed rather than letting the cluster function create clusters based on the natural divisions in the dataset.

4. MATHEMATICAL POWER MODEL

This section describes three power models: the conventional, swastika, and SAHCEASR power models.

4.1. Conventional Power Model

In the conventional model, the individual Node Point (NP) power consumption is expressed as follows:

1 NP=PC 4 μW/Sec, 2 NP=PC4 μW/Sec, 3 NP=PC1, 4 NP=PC4, 5 NP=PC 4, 6 NP=PC4, 7 NP=PC8, 8 NP= PC 4, 9 NP=PC8, 10 NP=PC4, 11 NP=PC1, 12 NP =PC4, 14 NP=PC4, 15 NP=PC1, 16 NP =PC4,17NP = PC8, 18 NP =PC4, 19 NP=PC 8, 20 NP =PC4, 21 NP=*PC* 4, 22 NP=PC4, 23 NP =PC1, 24 NP

=PC 4, 25 NP

= PC4.

The total power consumption is100µW/Sec.

Figure 4. Conventional 5X5 coalition routing model based on shortest path routing (Dijkstra algorithm)

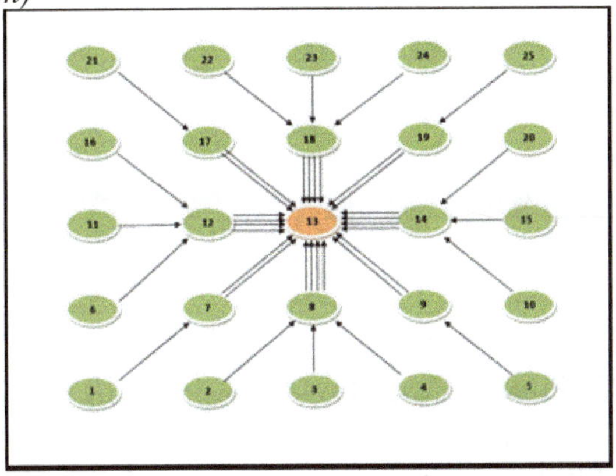

The regular 5X5 conventional model was determined as 100 W/s. Using our adaptive routing, we attempted to minimize the SWASTIKA model, as depicted in Figure 4.

4.2. SWASTIKA Power Model

For the SWASTIKA model, the individual power consumption of the 25 NP points (NP) is as follows:

1 NP= 1 µW/Sec, 2 NP = 2 µW/Sec, 3 NP = 4, 4 NP = 1, 5 NP = 1, 6 NP = 1, 7 NP = 1, 8 NP = 6, 9 NP = 1, 10 NP = 2, 11 NP = 4, 12 NP = 6, 14 NP = 6, 15 NP = 4, 16 NP = 2,17 NP = 1, 18 NP = 6, 19 NP = 1, 20 NP = 1, 21 NP = 1, 22 NP = 1, 23 NP = 4, 24 NP = 2, 25 NP

= 1.

The total power consumption is 60 µW/Sec. The SWASTIKA coalition routing model is shown in Figure 5.

Figure 5. SWASTIKA 5X5 coalition routing model

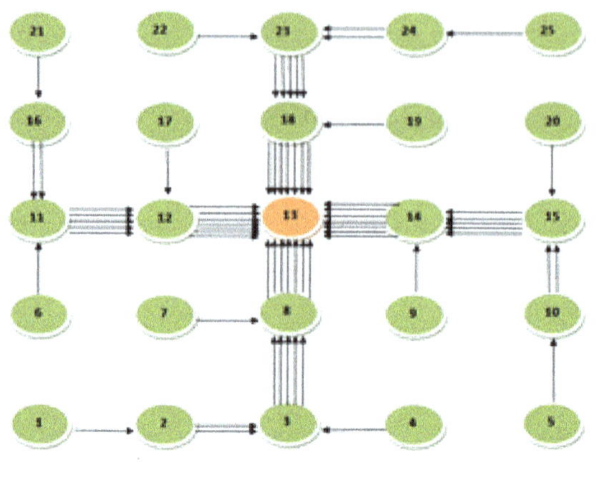

4.3. SAHCEASR Power Model

Figure 6. SAHCEASR Power Model signal routing with four clusters

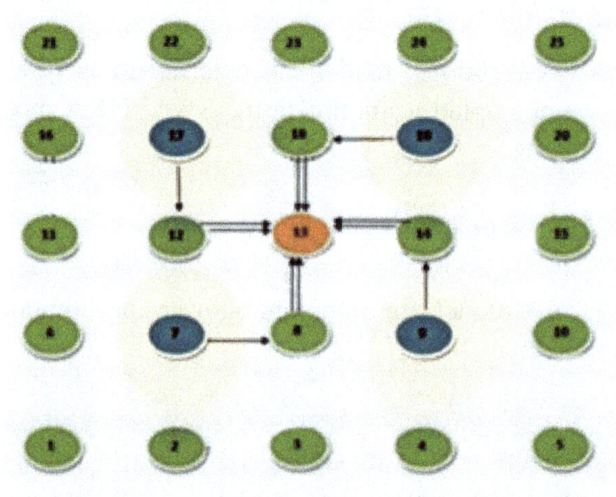

For the SAHCEASR algorithm, we use the same area with the same number of deployed nodes. In our proposed model, we categorize the entire sensor cover area into four parts using clustering. The clustering used here is based on likeliness to one another, which is very clear from Figure 6. From the figure, we deduce the outcome noted in Table 4.

Table 4. Cluster Detail

Cluster Number	Objects Included	Coordinates	Node included	Event likeliness
1	6,7	[1.4,3.8] [1.5,3.9]	11,12,13,16,17 18,21,22,23	Closest
2	1,4	[3.6,1.4] [3.8,1.3]	3,4,5,8,9,10 13,14,15	Close but comparatively less than cluster 1
3	3,8	[3.6,3.6] [3.4,3.3]	13,14,15,18,19, 20,23,24,25	Close but comparatively less than clusters 1,2
4	2,5	[0.8,1.2] [1.1, 1.3]	1,2,3,6,7, 8,11,12,13	Close but comparatively less than clusters 1,2,3

The object was clustered into four groups, which allowed the conventional and traditional swastika models to be reformed, as shown in fig. 6. Each cluster was allocated a cluster head. Thus, the signal is communicated to the sink node. Node numbers 7,9,17 &19 are the cluster heads, and node number 13 is the sink node, also called the sink node of the system. The power consumption can be calculated using the power model such as 7 Node Point= 1 unit, 8 Node Point = 2 unit, 9 Node Point = 1unit, 12 Node Point =2 unit, 14 Node Point = 2 unit, 17 Node Point = 1 unit, 18 Node Point =2 unit, 19 Node Point = 1 unit. The total power consumption is 12μW/Sec, which is 1/5 that of the total power consumption of the swastika model and 1/8th of the conventional routing model.

5. EXPERIMENTAL RESULTS

The models were implemented in MATLAB, where we simulated the conventional and swastika model power consumption and compared the results with our proposed algorithm. We found a significant difference in the total power consumption of the WSN system with 25 nodes deployed uniformly with a separation of distance unity. The total power consumption in the SAHCEAR, conventional, and SWASTIKA models is listed in Table 5.

Table 5. Total Power Consumption of SAHCEASR vs. Conventional model and SWASTIKA model

Total Power Consumption= Sum of gross power consumption of a WSN			
MODEL	SAHCEASR	Conventional	SWASTIKA
TPC	12 μW/Sec	100 μW/Sec	60 μW/Sec

Based on Table 5, Figure 6 has been plotted, where it is quite clear that the power consumption in the proposed model is quite low compared to the other conventional models. This ensures a longer network lifetime. Similarly, the average power consumption is tabulated in Table 6, and the plot has been deduced from the MATLAB simulation in Figure 7, which indicates that the proposed model power consumption varies significantly compared to the conventional network routing model.

Figure 7. Total power consumption of SAHCEASR vs. Conventional model and SWASTIKA model

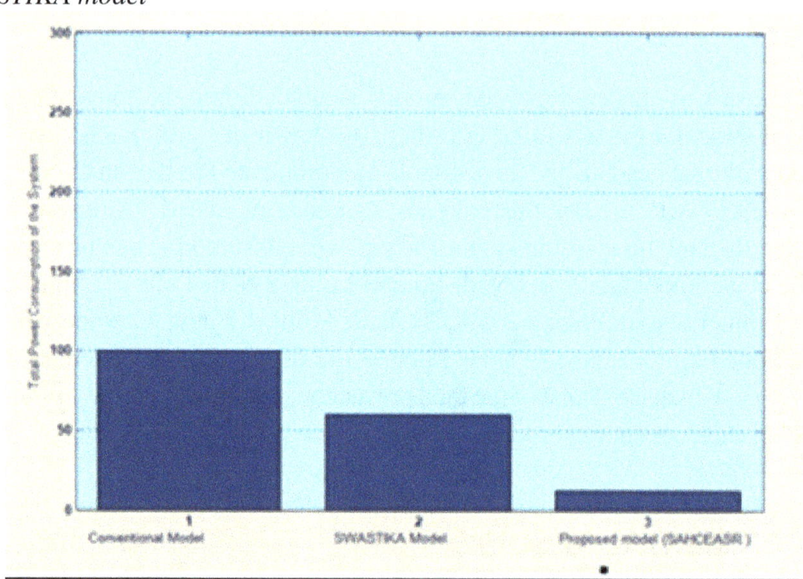

Table 6. Average Power Consumption of SAHCEASR vs. Conventional model and SWASTIKA model

Average Power Consumption= gross power/ no of object deployed			
MODEL	**SAHCEASR**	**Conventional**	**SWASTIKA**
APC	1.4 μW/Sec	4 μW/Sec	2.5 μW/Sec

Figure 8. Average power consumption of SAHCEASR vs. Conventional model and SWASTIKA model

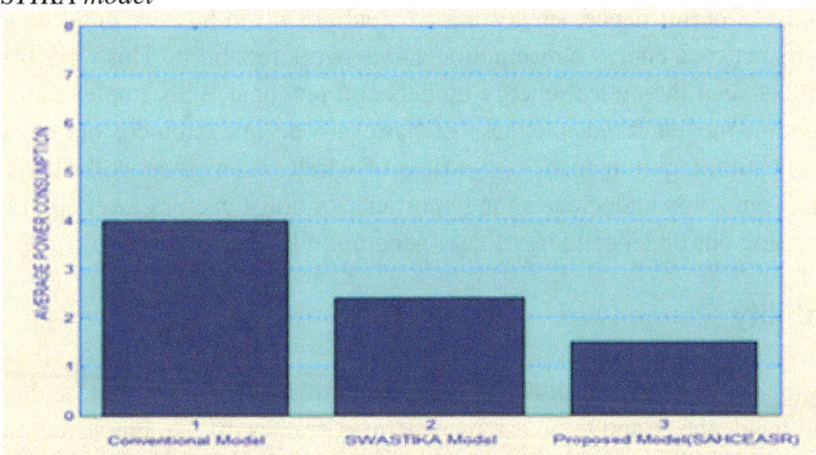

6. IMPLICATIONS, TRENDS, AND FUTURE DIRECTIONS IN POWER-EFFICIENT WSNS

Practical Relevance

The findings of this study have significant implications for real-world applications in wireless sensor network deployments and Internet of Things (IoT) systems. Our results demonstrate that optimized power management techniques can be leveraged to address [battery life limitations in remote sensing applications. For instance, our adaptive duty cycling algorithm could extend the operational lifetime of environmental monitoring sensors by up to 40%. This has the potential to improve long-term data collection in hard-to-reach areas by reducing maintenance visits and battery replacements. Industry professionals and policymakers can utilize these insights to design more sustainable and cost-effective IoT solutions. Moreover, our findings suggest that [the proposed energy harvesting integration] could lead to self-sustaining sensor networks, which may result in expanded deployment of WSNs in smart cities and precision agriculture for urban planners and farmers.

Broader Trends

Our research contributes to the growing body of knowledge on energy-efficient protocols in wireless communications. The results align with the emerging trend of green IoT and sustainable computing, as observed in recent studies. However, our findings also challenge the conventional wisdom that power efficiency always comes at the cost of performance, suggesting a potential shift in how we understand the trade-offs between energy consumption and network reliability. This work bridges the gap between theoretical energy models and practical WSN implementations, highlighting the interconnectedness of hardware design, protocol optimization, and application requirements. As the field of wireless sensor networks continues to evolve, our study underscores the importance of holistic, cross-layer approaches in advancing our understanding of next-generation IoT ecosystems.

Scalability

The scalability of our findings presents both opportunities and challenges. On one hand, the adaptive power management framework we employed can be readily adapted to larger datasets or diverse populations, potentially expanding its applicability to heterogeneous sensor networks and multi-modal IoT deployments. This scalability is particularly promising for smart city infrastructure monitoring. However, scaling up this research may face limitations such as increased network complexity and potential interference in dense deployments, which would need to be addressed through advanced spectrum management and machine learning-based optimization techniques. Future studies could explore the effectiveness of our approach in large-scale industrial IoT settings, which would further validate its scalability. Additionally, developing an open-source power optimization toolkit based on our findings could facilitate broader implementation across various IoT application domains, from healthcare to smart manufacturing.

7. CONCLUSION

The WSN system, which is deployed remotely, can be considered with the proposed SAHCEASR algorithm for less power dissipation from the system and to keep the node's battery power running longer. From the results of the power model, it is clear that our proposed routing model SAHCASR uses less battery power, as it takes less power in the transmission of the data by adopting a novel routing method that has been fetched by aggregating similar nodes by hierarchal clustering and by adopting an advanced form of swastika clustering. The minimum power consumption indicates better network life. Thus, we can maximize the lifetime of the wireless sensor network.

AUTHORS' CONTRIBUTIONS

Jagmohan Padhi: Developed the main idea, designed the hierarchical clustering algorithm, and wrote the first draft of the paper. Geetah Devi: Ran simulations, worked on optimizing the Swastika Routing algorithm, and helped revise the manuscript. Prabira Kumar Sethy (Corresponding Author): Supervised the research, contributed to the clustering algorithm, and revised the manuscript for publication.Santi Kumari Behera: Helped improve the Swastika Routing, provided insights for combining it with hierarchical clustering, and assisted in editing the final manuscript.

REFERENCES

Contreras, P., & Murtagh, F. (2015). Hierarchical clustering. Handbook of cluster analysis, pages 103–123.

Dao, T. C., Tam, N. T., & Huynh, T. T. B. (2024). Node depth representation-based evolutionary multitasking optimization for maximizing the network lifetime of wireless sensor networks. *Engineering Applications of Artificial Intelligence, 128*, 107463. DOI: 10.1016/j.engappai.2023.107463

Farzad Kiani, A. (2013). An energy effective-accuracy routing algorithm for wireless sensor networks. *Life Science Journal, 10*(2), 39–45.

Haghzad Klidbary, S., & Javadian, M. (2024). Improvement of low energy adaptive clustering hierarchical protocol based on genetic algorithm to increase network lifetime of wireless sensor network. *International Journal of Engineering, 37*(9), 1800–1811. DOI: 10.5829/IJE.2024.37.09C.10

Jain, K., Mehra, P. S., Dwivedi, A. K., & Agarwal, A. (2022). Scada: Scalable cluster-based data aggregation technique for improving network lifetime of wireless sensor networks. *The Journal of Supercomputing, 78*(11), 13624–13652. DOI: 10.1007/s11227-022-04419-1

Jena, S. R., Mallick, M. R., Padhi, J., & Panda, M. (2016). Ant-colony algorithm based routing using swastika clustering in wsns. [IJCTA]. *International Journal of Control Theory and Applications, 9*(16), 8181–8192.

Kiani, F. (2016). A novel channel allocation method for time synchronization in wireless sensor networks. *International Journal of Numerical Modelling, 29*(5), 805–816. DOI: 10.1002/jnm.2143

Murtagh, F., & Contreras, P. (2012). Algorithms for hierarchical clustering: An overview. *Wiley Interdisciplinary Reviews. Data Mining and Knowledge Discovery, 2*(1), 86–97. DOI: 10.1002/widm.53

Nematzadeh, S., Torkamanian-Afshar, M., Seyyedabbasi, A., & Kiani, F. (2023). Maximizing coverage and maintaining connectivity in wsn and decentralized iot: An efficient metaheuristic-based method for environment-aware node deployment. *Neural Computing & Applications, 35*(1), 611–641. DOI: 10.1007/s00521-022-07786-1

Padhi, J., Biswal, R. R., Das, K., & Bishwal, P. K. (2016). Novel routing to maximize network life time using swastika model. *International Journal of Control Theory and Applications, 9*(16), 8171–8180.

Patel, S., Pradhan, S. S., & Patra, K. C. (2017). Advance scale-3 swastika model to improve network lifetime in wireless sensor network. [IJRSI]. *International Journal of Research and Scientific Innovation, IV*(VI).

Sen Laxminarayan Sahoo, S., & Tiwary, K. S. (2024). Sarbast Moslem, and Tapan Senapati. Improvement of wireless sensor network lifetime via intelligent clustering under uncertainty. *IEEE Access : Practical Innovations, Open Solutions.*

Seyyedabbasi, A., Dogan, G., & Kiani, F. (2020). Heel: A new clustering method to improve wireless sensor network lifetime. *IET Wireless Sensor Systems, 10*(3), 130–136. DOI: 10.1049/iet-wss.2019.0153

Seyyedabbasi, A., Kiani, F., & Allahviranloo, T. (2023). Unai FernandezGamiz, and Samad Noeiaghdam. Optimal data transmission and pathfinding for wsn and decentralized iot systems using i-gwo and ex-gwo algorithms. *Alexandria Engineering Journal, 63*, 339–357. DOI: 10.1016/j.aej.2022.08.009

Shahraki, A., Taherkordi, A., Haugen, Ø., & Eliassen, F. (2020). A survey and future directions on clustering: From wsns to iot and modern networking paradigms. *IEEE Transactions on Network and Service Management, 18*(2), 2242–2274. DOI: 10.1109/TNSM.2020.3035315

Shankar, A., Sivakumar, N. R., Sivaram, M., Ambikapathy, A., Nguyen, T. K., & Dhasarathan, V. (2021). Increasing fault tolerance ability and network lifetime with clustered pollination in wireless sensor networks. *Journal of Ambient Intelligence and Humanized Computing, 12*(2), 2285–2298. DOI: 10.1007/s12652-020-02325-z

Vinoth Kumar, P., & Venkatesh, K. (2024). Hybrid seagull and whale optimization algorithm-based dynamic clustering protocol for improving network longevity in wireless sensor networks. *China Communications.*

Xu, R., & Wunsch, D.II. (2005). Survey of clustering algorithms. *IEEE Transactions on Neural Networks, 16*(3), 645–678. DOI: 10.1109/TNN.2005.845141 PMID: 15940994

Zhao, Y., & Karypis, G. (2002). Evaluation of hierarchical clustering algorithms for document datasets. In *Proceedings of the eleventh international conference on Information and knowledge management*, pages 515–524, 2002. DOI: 10.21236/ADA439551

Zivkovic, M., & Zivkovic, T., Venkatachalam, K., & Bacanin, N. (2021). Enhanced dragonfly algorithm adapted for wireless sensor network lifetime optimization. In Data Intelligence and Cognitive Informatics: Proceedings of ICDICI 2020, pages 803–817. Springer.

Chapter 10
Traffic Shaping and Flow Management in Wide Area Networks

Selçuk Göçer
Ahmet Yesevi University, Turkey

Gurkan Tuna
http://orcid.org/0000-0002-6466-4696
Trakya University, Turkey

ABSTRACT

Wide area networks, which are the basis of access to information and communication today, form the backbone of the internet. The ever-increasing data traffic, different application types, and service quality requirements of different systems used in local area networks cause problems such as congestion and loss of performance in networks due to the limited bandwidth received from the internet service provider. This study focuses on the effects of QoS functions and critical network management tools in wide area networks with limited bandwidth, such as traffic shaping, traffic control, bandwidth management, and data traffic prioritization methods, on network performance. The results obtained from this study show that traffic shaping methods are used according to the need to adapt to the limited bandwidth within the framework of the designed rules and are important in improving the performance of the managed network and controlling sudden increases and traffic fluctuations.

DOI: 10.4018/979-8-3373-0852-4.ch010

INTRODUCTION

Nowadays, network traffic density is increasing day by day. It is of great importance to transfer information flow uninterruptedly and securely in wide area networks. The increase in data amounts negatively affects the operating performance and stability of wide area networks with limited bandwidth. Frequent network traffic congestion or delays in network traffic due to intense data flow both reduce efficiency and interrupt critical information flow.

In wide area networks, it is difficult to manage network traffic smoothly due to hardware costs, balancing network load and ensuring network performance. Due to the increase in the number of users, applications, different systems and network devices, effective management of the limited bandwidth received from the internet service provider has become a necessity. It is important to find out solutions to how to use traffic shaping and flow management methods in wide area networks.

Network communication quality of service (QoS) is a technology used to manage traffic in a network and prioritize planned data types. Traffic shaping and flow management methods within the scope of QoS aim to solve these problems. When QoS methods are not applied to network devices, serious disruptions can occur in data traffic that needs to be transmitted, especially in emergency situations. However, faced with increasing Internet traffic and diverse QoS requirements, traditional centralized network architectures encounter challenges (She et al., 2025). The Internet best-effort service led to the lacking deployment of network QoS on the Internet architecture since there are problems associated with currently used short path routing (Elbasheer et al., 2021).

In this study the main purpose is to limit the bandwidth of existing systems used in local area networks so that they do not exceed the limited capacity defined by the Internet service provider when going to the wide area network and to determine the important network traffic and to ensure uninterrupted wide area network connection. This way, the replies will be given to the following questions.

- Is managing the network by traffic shaping in wide area networks with limited bandwidth sufficient for network performance?
- Which method is useful in using the QoS model to prioritize critical traffic, optimize network performance, create traffic classes, and minimize data loss?

In the application phase of this study, a network structure consisting of a Cisco 4451 router, a Cisco 9300 network switch, and a group of computers was designed. One of the computers was used to carry out performance and bandwidth tests for traffic coming from other systems in the network. Limiting bandwidth was not the only purpose of the study. Another purpose was to determine the important network traffic in the limited capacity and to perform traffic shaping and traffic control to ensure uninterrupted wide area network connection and to reveal the importance of service quality in wide area networks. Organizations can guarantee critical data flow, network traffic performance, and network service quality with basic applications and services by using QoS methods in their networks.

RELATED WORK

In today's Internet infrastructure, data traffic on wide area networks is increasing rapidly. This increase causes a decrease in network performance, an increase in delays, and disruptions in critical applications. QoS is a set of methods developed to find solutions to these problems. QoS methods such as traffic shaping, traffic flow management, and traffic control provide the desired quality of service for specified applications by classifying, prioritizing, and shaping network traffic. Therefore, the application of traffic shaping methods allows efficient use of network resources and regulates data flow.

By centralizing control and decoupling the data plane from the control plane, software-defined networking started to revolutionize network administration. In spite of its advantages, the escalating volume of network traffic induces congestion at nodes, adversely affecting routing quality and overall performance. Therefore, addressing congestion has become imperative because of its emergence as a fundamental challenge in network management (Prasanth & Uma, 2024). In recent years, deep learning algorithms and software-defined networking technologies started to enable traffic management to be more controllable in IP networking and mobile Internet to yield quality services to subscribers (Wassie, Ding, & Wondie, 2024).

In real-time applications such as voice and video, latency and data loss are unacceptable. Network traffic can sometimes increase suddenly. These sudden and uncontrolled fluctuations can negatively affect the performance of the network. Especially in corporate or security-critical networks, it is necessary to prevent certain users, applications or systems from overloading the network. This means limiting unauthorized access or malicious traffic. Efficient use of bandwidth in the network makes it possible to provide faster and more stable service to more users at lower costs.

Hasenoot, Rellermeyer & Uta (2023) discussed the effects of traffic shaping methods on common distributed applications. Their results revealed several implications on how to distribute certain applications with traffic shaping protocols to minimize performance impact.

Bai, Atiquzzaman & Ivancic (2002) examined the QoS guarantees to be obtained from end applications when integrated services are run over decoupled services and evaluated that lowering the priority of low priority packets before they are transmitted over the network ensures stable QoS operation.

In the research conducted by Kassim, Ismail, Yusof, & Abdullah (2017), instead of the traditional traffic control and shaping algorithm in bandwidth management in campus networks, a time-based control and shaping algorithm was developed and simulated. The results obtained in the simulation environment showed that the new algorithm reduced the sudden traffic in the traffic control phase, while it delayed the traffic flow in the traffic shaping phase. It is evaluated that this method will help to increase the quality of service in limited bandwidth usage, traffic flow rate and network performance.

This study is important in terms of ensuring that traffic shaping methods, traffic control methods and QoS protocols are used effectively in wide area networks to use limited bandwidth resources more efficiently, to keep the network performance under control, to prioritize critical traffic, to ensure uninterrupted data transfer and to improve the user experience. It provides a contribution to network administrators in terms of improving network performance in limited bandwidth and preventing losses in the network, and facilitates more efficient network management planning.

CONCEPTUAL FRAMEWORK

In this section, the conceptual framework of the research on the definition of wide area network, definition of network service quality, definition and methods of traffic shaping, definition and methods of traffic flow management, definition of traffic control, traffic shaping and traffic control application areas and the effect of traffic shaping on wide area network performance is established.

A wide area network is simply a collection of local area networks or other networks that communicate with each other. A wide area network can communicate over very long distances through an Internet service provider. A wide area network is essentially a network of networks, and the Internet can be described as the world's largest wide area network.

The first known wide area network was created by the US Air Force in the 1950s to connect buildings in the Semi-Automatic Ground Environment (SAGE) radar defense system. Buildings were connected to each other by a network of dedicated telephone lines, telephones, and modems. This was followed in the late 1960s by the Advanced Research Projects Agency Network (ARPANET), which became the first wide-area packet switching network and the first network to implement the TCP/IP suite, and became the first wide-area network that laid the foundations of today's Internet (Hellige, 1994).

Based on the fact that ARPANET is the basis of the Internet, we can define wide area networks as networks that provide data, voice and image communication in large geographical areas. These networks are generally used to establish connections between institutions, offices, factories or facilities scattered in different cities, countries or continents. Wide area networks, unlike local area networks, provide data transmission over very long distances and generally use telecommunication infrastructures and large-scale networks such as the Internet. Wide area networks can be created with a wide variety of technologies and can be used for military, commercial or academic purposes (Forouzan, 2007).

Wide area networks can be seen as one of the fundamental building blocks of the Internet and other large-scale network infrastructures. They are designed to provide reliable and fast data transmission, while eliminating geographical barriers and forming the basis of global communication. In addition, large-scale enterprises, government agencies and military forces use wide area networks to bring units in different locations together, providing centralized management and control.

One of the main functions of wide area networks is to enable users and devices in different geographic areas to communicate with each other. One of the most important features of wide area networks is their wide coverage and the optimization of data transmission speeds in these areas. For example, while a local area network provides high-speed data transmission between computers in the same building, a wide area network can perform similar data transmission over much greater distances (Taner, 2018).

Wide area networks are of critical importance, especially for military and government institutions. These networks are equipped with secure communication protocols and ensure the secure transfer of sensitive data. In military operations, wide area networks are indispensable for information sharing between various locations, operational coordination, and remote command-control systems. Especially in critical operations and emergency situations, establishing an uninterrupted and reliable communication network over wide area networks is the key to operational success.

In addition to the advantages of wide area networks, there are also some challenges. One of the biggest challenges is that wide area networks require a more complex infrastructure due to their spread over a large geographical area. This can increase the maintenance costs and management difficulties of the network. In addition, wide area networks have to deal with longer distances, which can cause delays in data transmission (Stallings, 2014). In addition, data security is one of the biggest concerns of wide area networks. Especially in military wide area networks, various crypto devices, encryption techniques and security protocols are used to ensure the secure transfer of data.

The development of wide area network technologies is parallel to the needs in the field of data communication and network management. Technologies such as fiber optic networks, high-speed internet connections and 5G increase the capacity and performance of wide area networks. For example, routing technologies such as multiprotocol label switching (MPLS) ensure that data reaches its destination more quickly and reliably (Kar, Kodialam, & Lakshman, 2000). In addition, new technologies such as software-defined wide area networking increase the efficiency of wide area networks by making network management more flexible and dynamic.

Network QoS is the definition or measurement of overall performance, especially the performance seen by network users. QoS provides the ability to prioritize end-to-end traffic with the ability to provide different priorities to different applications, users, or data streams, or to guarantee a certain level of performance in a data stream. It ensures that the most important traffic is transmitted first in order to use the available bandwidth more efficiently. Factors such as delay, jitter, packet loss, and capacity are taken into account when determining important traffic for QoS (Cisco, 2022).

Although file transfer services such as File Transfer Protocol (FTP) are very important, data packets are not as sensitive to delay as voice or video packets. If a web page we are trying to open opens a second later, it does not cause a problem. Although one-second interruptions are acceptable in uses such as video conferencing, teleconferencing, music broadcasting, and IPTV, in corporate uses, for example, the delay of telemetry data belonging to unmanned aerial vehicles (UAVs) can pose a national security risk. QoS is important and necessary in network management. Common definitions used in QoS configurations are presented in Table 1.

Table 1. Important definitions used in QoS configurations

Definition	Explanation
Committed Information Rate (CIR)	Committed data rate
Peak Information Rate (PIR)	Peak data rate
Committed Burst Rate (CBR)	Committed data block size (Guaranteed size of data packets sent or received)
Maximum Burst Rate (MBR)	Maximum size of data packets allowed to be processed
Delay	Delay between each voice packet reaching the receiver
Jitter	Time difference between delays

QoS Models

QoS models have some rules developed to optimize data flow on the network and prioritize certain types of traffic. Applying QoS models increass network performance, manages traffic, and minimizes problems such as delay, jitter, and packet loss (Fgee et al., 2005). QoS models aim to provide access to the resources each application requires, especially in large and complex networks where different applications operate simultaneously. There are three main models that should be preferred for network requirements in order to provide QoS in networks. A comparison of these three models is presented in Table 2.

- **Best Effort:** No QoS mechanism is implemented in this model, so it does not provide data delivery guarantee in network traffic. Many data applications such as FTP are transmitted with the best effort model.
- **Integrated Services (IntServ):** In IntServ, one of the first implementations of QoS models, a dedicated resource is planned for each data flow. This model can be defined as the most stringent QoS model used to obtain service guarantees from the network for end applications that require guaranteed services (Harju & Kivimaki, 2000). However, IntServ is difficult and costly to implement in large-scale networks due to the necessity of reserving resources for each data.
- **Differentiated Services (DiffServ):** This model provides a bridge between integrated services that provide network service guarantees and best effort services that provide basic QoS requirements (Harju & Kivimaki, 2000). It can be defined as a flexible and effective QoS model that can prioritize and mark traffic classes.

Table 2. Comparison of QoS models

QoS Model	Network Traffic Management	Resource Consumption	Ease of Application	Scalability	Advantage/ Disadvantage
Best Effort	None	Low	Easy	Very High	No guaranteed service quality is provided, packet delay or loss may occur.
Integrated Services	Detailed	High	Complicated	Low	It provides end-to-end QoS guarantee, making it an ideal solution for critical applications.
Differentiated Services	Classification	Middle	Middle	High	It does not provide end-to-end QoS guarantee, but provides prioritization by creating traffic classes.

Within the differentiated services model, the prioritization of important traffic is related to the 3 bits in the Type of Service (ToS) field in the IP packet header. This 3-bit field determines a priority level according to how urgent and critical the data is. IP priority levels are numbered from 0 to 7, with each level determined by different types of traffic. Figure 1 shows the IP priority levels taken from the CLI screen of the Cisco 4451 router. In addition, an access control list (ACL) can be used in the differentiated services model. ACLs are used to classify network traffic according to certain criteria and apply appropriate QoS rules to this traffic. ACLs can be used as a source IP address, a destination IP address, or a combination of both.

Figure 1. IP priority levels

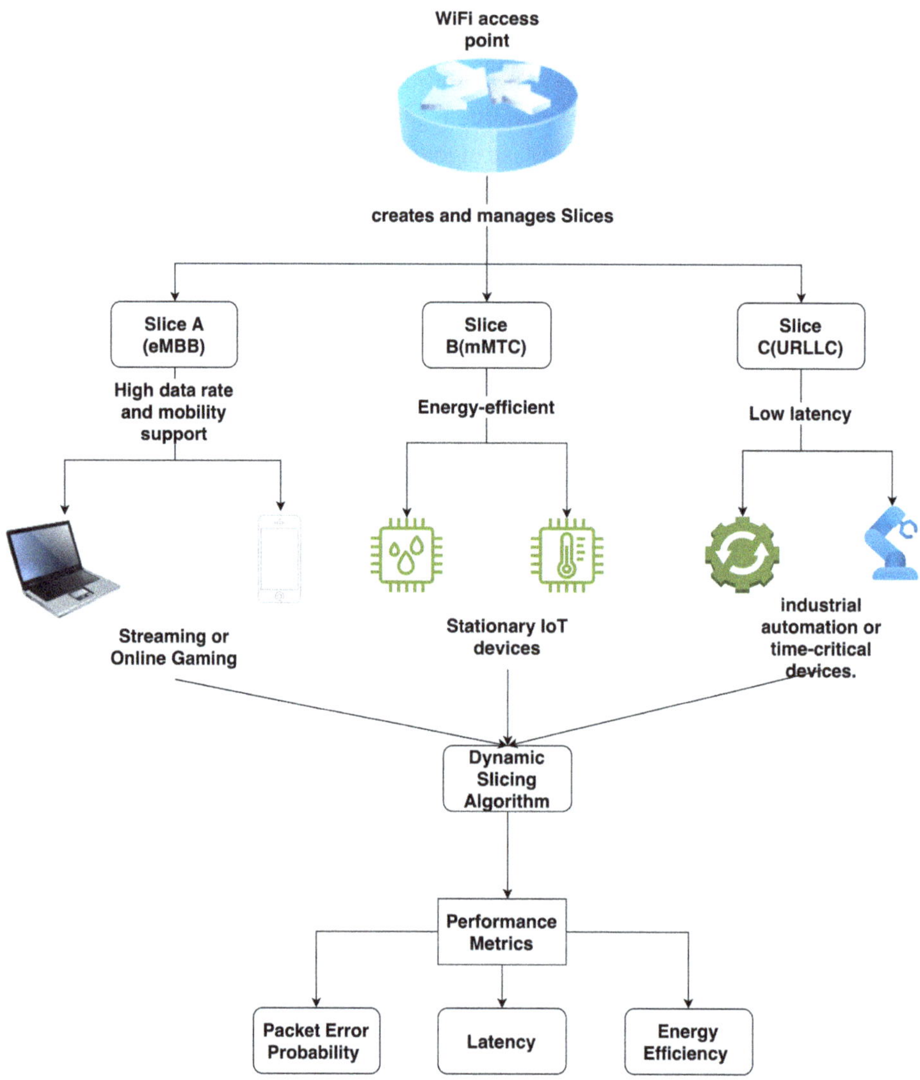

Traffic Flow Management

Traffic flow management is a set of techniques and tactics that enable effective, efficient and secure organization and management of data traffic on a network. This process is carried out in order to optimize network resources and minimize delays in data transmission (White, 1997). Data traffic, especially in networks such as data centers, wide area networks and cloud infrastructures, can have a complex structure and therefore effective traffic flow management is of great importance. In addition, communication and data sharing are of vital importance during military operations. In these networks, data packets must be transmitted in a timely and secure manner. Traffic is prioritized and security is kept at the highest level for the execution of critical tasks. In addition, an infrastructure that is resistant to cyber attacks is required. In this context, strategies such as traffic shaping, prioritization and load balancing provide reliable communication in networks (Stoica & Zhang, 1999).

Traffic flow management must meet the following four important characteristics.

- **Reliability:** The aim of the network is for data packets to reach the correct destination, in the correct order and at the desired speed. This is a critical feature for the managed network and the services offered over this network.
- **Delay:** It refers to the time it takes for a data packet to reach the destination from the source.
- **Jitter:** It can be defined as the differences that occur in packet delays. High jitter means that the difference between delays is large, while low jitter means that this difference is small.
- **Bandwidth:** It refers to the maximum amount of data that a network can carry in a certain period of time.

The success of network management depends on the accuracy and effectiveness of network traffic classification in line with the determined plans. The methods used in flow management are as follows:

- **Traffic classification:** Data traveling over the network is classified according to its type and priority. For example, traffic for real-time applications such as video conferencing or voice over IP takes precedence over less time-sensitive traffic such as file downloads. This classification determines which traffic is processed first.
- **Traffic shaping:** It is used to optimize bandwidth usage on the network and make traffic more controlled. Traffic shaping limits the amount of data sent to ensure that a network operates at a certain data rate and manages the flow rate

of data according to the prioritization status of the data in overload situations. This prevents network congestion and increases overall network performance.

- **Traffic control (traffic policing):** It is a method of dropping or rerouting data packets when the specified bandwidth is exceeded. In this method, when the data rate on the network exceeds the specified limit, excess data packets are either deleted or resent.
- **Load balancing:** A method that aims to distribute traffic on a network evenly across multiple connections or servers. This improves network performance by preventing a resource from becoming overloaded. For example, when a data center encounters traffic coming to multiple servers at the same time, the load balancing mechanism distributes this traffic evenly.
- **Traffic engineering:** The process of designing and structuring network infrastructure to direct and optimize traffic flows. It aims to minimize congestion on the network and maximize performance by providing control over network resources.
- **Reserved traffic method:** It aims to reserve a certain bandwidth of the network for a specific application or data flow. In this way, the bandwidth required by critical applications is guaranteed at all times.
- **Traffic routing:** A method that determines which paths data packets will follow on the network. This method aims to transmit the data flow in the best way by choosing the most efficient and least congested path.
- **Fair queuing:** This method aims to ensure a balanced transmission of all traffic on the network by allocating equal resources to each data flow.
- **Traffic monitoring and analysis:** Network administrators can identify potential problems by monitoring real-time traffic on the network. This is important for monitoring network performance, identifying problem areas, and ensuring the efficient operation of the network. Monitoring and analysis also play a critical role in network security. Threats such as Distributed Denial of Service (DDoS) can be quickly detected this way.

Traffic Shaping and Traffic Control

Traffic shaping and traffic control are QoS methods used to manage and optimize network traffic according to certain rules. These methods have been developed specifically to improve network performance and balance bandwidth usage.

Depending on the needs and goals of the network administrator, traffic shaping can be used in a variety of network scenarios. For example, an enterprise network can utilize traffic shaping to reduce or delay bandwidth for non-essential activities like social networking or gaming while allocating greater bandwidth to business-critical apps like email or video conferencing. By limiting the bandwidth available for peer to peer (P2P) traffic during periods of high demand or peak hours, an Internet service provider (ISP) network can also utilize traffic shaping to control its users' bandwidth usage. Similar to this, a cloud network might block or delay traffic for its free or basic users, who pay less or nothing for lower bandwidth and higher latency, while prioritizing traffic for its premium customers, who pay more for more bandwidth and lower latency. Employee productivity and efficiency can be increased, all clients can receive fair and consistent service quality, customer happiness and loyalty can be raised, and service level agreements and regulations can be followed.

Traffic shaping provides more efficient network management by controlling the speed and direction of data flow through the network in a certain period of time (Walrand, 2023). Packets that exceed the traffic speed limit are queued and sent at the allowed speed. Examples of traffic shaping include giving priority to important business applications, managing ISP data caps or throttling during peak hours, guaranteeing smooth video streaming, prioritizing Voice over IP (VoIP) calls for clear audio quality, maintaining low latency for online gaming, and adhering to service level agreements by prioritizing particular traffic or meeting latency and jitter requirements. These examples demonstrate how traffic shaping can improve user experience, network speed, and adherence to rules or agreements.

Traffic control is used to prevent a specified traffic speed from being exceeded on the network. Packets that exceed the traffic speed limit are dropped without being transmitted. Areas of use of traffic shaping and traffic control are as follows:

- **Provision of QoS:** Traffic shaping minimizes latency, jitter, and packet loss, especially for real-time applications (such as VoIP, video conferencing). It allows network traffic to be managed by categorizing it according to its priorities. For example, prioritizing UAV telemetry data over voice or video calls, and leaving some voice or data transfers in the background.
- **Bandwidth optimization:** Organizations or ISPs use traffic shaping to prevent users or specific applications from overusing the bandwidth on the network, thus ensuring that the network operates more balanced and stable.
- **Network security:** Traffic shaping, which is used especially in defense against cyber attacks, helps protect the network against abnormal traffic increases such as DDoS.
- **Traffic management:** Traffic shaping methods are used to manage traffic during periods of high network congestion. This prevents performance degradation due to heavy traffic.

Main differences between traffic control and traffic shaping methods are listed in Table 3.

Table 3. Main differences between traffic shaping and traffic control

Traffic Control	Traffic Shaping
It ensures that traffic does not exceed the speed limit, and drops or marks the parts that exceed it.	It checks the speed of traffic, queues the excess and sends it later.
It usually drops excess traffic or manages it through flagging.	It queues excess traffic and sends it according to the speed limit.
Data loss may occur due to dropping excess traffic.	There is no data loss, but packets may be delayed.
There is no delay as it reduces excess traffic.	There is a delay because excess traffic is queued.
It limits bandwidth by dropping excess traffic.	It regulates bandwidth by queuing excess traffic.
Consumes less resources.	Consumes more resources and memory.

Traffic shaping includes methods used to manage, optimize and organize data traffic on the network within the framework of certain rules. These methods have been developed to improve network performance, increase service quality and ensure more efficient use of network resources. Commonly used methods in traffic shaping are:

- **Leaky bucket algorithm:** This method is used to flow traffic at a certain speed. Like a bucket, incoming data packets are poured into the bucket, and this bucket releases packets at a certain speed. If the bucket accumulates, that is, if the amount of incoming data is higher than the specified speed, excess packets are discarded or accumulated (Tanenbaum, Feamster, & Wetherall, 2021). It is especially useful in military networks to prevent unexpected traffic increases from crashing the systems. The downside is that it does not provide flexibility for applications that require high bandwidth at times. It can cause data loss, especially when the bucket is full, incoming packets are lost.
- **Token bucket algorithm:** This method provides flexibility while controlling the amount of data sent at a certain speed. The user is given a certain amount of tokens and each token represents a data packet (Mathews, Kramer, & Gotzhein, 2017). The user has the right to send as much data as the number of tokens. Over time, tokens can be accumulated and then used when a high traffic flow is needed. This algorithm allows flexible use of bandwidth. For example, during a military operation, when there is a short-term need for high bandwidth, this technique comes into play and meets the need. The downside is that when token accumulation is limited, packet losses may occur if sudden traffic is prolonged.
- **Fair queuing:** Fair queuing is a method used to provide fair traffic management. In this method, a certain bandwidth is assigned to each flow in the network (Tanenbaum, Feamster, & Wetherall, 2021). A balanced traffic

distribution is made between systems or applications. A separate queue is created for each data flow. The system receives data packets from each queue in order and transmits them over the network. Each system or application is assigned equal bandwidth, thus preventing any system or application from affecting the performance of others by using excessive bandwidth. Its disadvantages are that it requires a complex structure and is difficult to implement. Low priority traffic may be delayed in this method.

- **Weighted fair queuing:** Weighted fair queuing is similar to fair queuing, but instead of giving equal bandwidth to each stream, certain streams are prioritized by allocating more bandwidth. This method allows critical applications to be delivered first (Tanenbaum, Feamster, & Wetherall, 2021). Each stream is assigned a weight. High bandwidth-consuming, priority applications are given more bandwidth, while less important or low-priority traffic is allocated less resources. This allows high-priority services, such as voice and video, to operate without interruption. The downside is that it requires complex configuration. Low-priority streams may have to wait for a long time.
- **Traffic policing:** Traffic control is a method used to control traffic that exceeds specified speed limits and reduce it when necessary. This method is generally preferred by ISPs. A certain speed limit is set on the network. If this limit is exceeded, excess traffic is reduced or blocked completely. This prevents network resources from being overused and reduces the impact of traffic that harms the network. The negative aspect is that it can cause packet losses. It does not provide sudden traffic flexibility.
- **Priority queuing:** Prioritized queuing manages traffic by prioritizing different types of traffic (Cormen, Leiserson, Rivest, & Stein, 2022). Real-time data (voice and video) traffic can be prioritized, while low-priority data (such as email) can be processed in the background. Different priority queues are set for different types of traffic. High-priority traffic is always processed first, while low-priority traffic is processed later. This ensures uninterrupted operation, especially for real-time applications. The downside is that it can cause low-priority traffic to lag.
- **Class-based queuing:** Class-based queuing separates data traffic into classes and provides separate bandwidth management for each class. This method provides more efficient traffic management by processing traffic classified according to priority. Network traffic is divided into specific classes. Each class has a predefined bandwidth percentage. Priority classes use more bandwidth, while low priority classes receive less bandwidth. This method allows certain types of traffic to be processed faster on the network. The downside is that it requires complex configuration. Low priority classes may be delayed.

The abovementioned traffic shaping methods allow network traffic to be managed according to different needs. Considering the advantages and disadvantages of each method, the right method should be selected for network performance and QoS.

Traffic control is a set of methods used to control network traffic and enforce bandwidth limits. These methods are used to reduce unnecessary traffic on the network and use resources more efficiently. Traffic control methods are as follows (Cisco, 2022):

- **Rate limiting:** It limits traffic to a specified speed, ensuring that packets exceeding this speed are discarded or marked. This method is generally used to control bandwidth and prevent unnecessary traffic.
- **Token bucket algorithm:** As abovementioned, this method uses a token bucket that adds tokens at a certain rate. When data is to be transmitted, it is checked whether there are tokens in the bucket. If there are enough tokens in the bucket, the packet is sent, if there are no tokens, the packet is thrown away or put on hold. The Token bucket method tries to keep sudden traffic increases under control.
- **Leaky bucket algorithm:** As abovementioned, t his method uses a bucket that tries to keep the data flowing at a steady pace. If the data rate exceeds the bucket capacity, excess packets are not allowed. Leaky bucket generally tries to make the network traffic more stable and keep sudden traffic fluctuations under control.
- **Two-rate, three-color marker (TRTCM):** With this method, network traffic is marked with two different speed limits and three colors (green, yellow, red). If the traffic remains below a certain threshold, it is marked green; when it reaches an upper limit, it is marked yellow; when it exceeds these limits, it is marked red. Red packets are discarded, yellow packets are transmitted with low priority, while green packets are transmitted normally (Heinanen, Finland, & Guerin, 1999a).
- **Single-rate, three-color marker (SRTCM):** This method is similar to TRTCM, except that it has a single rate limit. Traffic is marked green, yellow or red according to the specified limit. Packets exceeding the capacity are discarded or marked with low priority (Heinanen, Finland, & Guerin, 1999b).
- **Committed access rate (CAR):** The CAR method developed by Cisco is a traffic control technique used to control traffic with rate limits. If traffic exceeds the specified rate, packets are discarded or marked.
- **Hierarchical traffic policing**: This method organizes different rate limiting rules in a hierarchical structure. A connection with a lower priority is subject to stricter rules, while a connection with a higher priority can be more flexible. This method is used to limit specified groups.

The abovementioned traffic control methods provide flexibility according to the needs and plans of the network management. Traffic control is very important to effectively manage bandwidth and balance network traffic.

Traffic shaping is a method that increases the efficiency of network management. It is used especially in the following areas:

- **Use in military networks:** Military networks have high security and performance requirements because they provide critical information and data communication. Traffic shaping is used in these networks to control traffic on the network and ensure the smooth transmission of critical operational data. For example, a military unit in the field must share information with its command center without experiencing delays or data loss. In such cases, traffic shaping minimizes operational disruptions by prioritizing critical communication channels.
- **Use in enterprise networks:** Large companies and organizations rely on large networks to ensure business continuity. Traffic shaping ensures that critical applications (such as customer relationship management or database applications) operate uninterrupted across these networks. It also regulates the bandwidth needed for services such as video conferencing or data backup.
- **ISPs:** ISPs frequently use traffic shaping and traffic control to improve the quality of service they provide to their customers. This allows users to watch videos, play games, or download data in a stable manner, even during peak network hours.

Traffic shaping plays a critical role in optimizing performance, especially in complex networks such as wide area networks. Managing data flow on the wide area network serves many important goals, such as reducing latency, using bandwidth efficiently, and preventing congestion. The effects of traffic shaping on wide area network performance are as follows:

- **Bandwidth management:** Limited bandwidth in wide area networks has to accommodate different types of traffic (voice, video, file transfer, etc.). Traffic shaping optimizes the available bandwidth to make more efficient use of certain traffic (priority traffic).
- **Reducing latency:** In wide area networks, especially long-distance connections are vulnerable to latency problems. Traffic shaping keeps latency to a minimum by ensuring that data packets arrive in the correct order and on time. Traffic shaping plays a critical role, especially in applications that are sensitive to latency (voice/video conversation, live broadcast).

- **Traffic prioritization:** Traffic shaping provides prioritized traffic management. Latency is reduced by prioritizing real-time traffic so that delays to lower priority traffic (such as file transfers) do not impact critical workflows.
- **Reducing packet loss:** When there is a high load on the wide area network, packet loss may occur due to this congestion. Traffic shaping reduces packet loss by balancing the data flow on the network.
- **Congestion management:** Excessive traffic can cause congestion in routers. Traffic shaping staggers packet sending to manage this congestion, which minimizes data loss.
- **Fair bandwidth sharing:** Traffic shaping ensures that all users use their bandwidth fairly. This increases the efficient use of bandwidth, especially during peak times, and reduces data loss.
- **QoS support:** Traffic shaping enables efficient enforcement of QoS policies. QoS optimizes the performance of critical applications by giving different priorities to different types of traffic.
- **Prioritize critical traffic types:** With QoS, specific traffic (for example, video conferencing or VoIP) can be prioritized on the wide area network. Traffic shaping enforces these priorities to prevent latency or data loss.
- **Traffic classification:** Traffic shaping classifies traffic to determine which types of traffic will use more resources. Applications like voice and video are given high priority, while large file transfers are given low priority.
- **Preventing network congestion:** When too many users on the wide area network send or receive data simultaneously, this can cause congestion. Traffic shaping directs data flow in a controlled manner to prevent these congestions.
- **Managing traffic overload:** Traffic shaping improves overall performance on the wide area network by limiting excessive traffic. This means more balanced use of network resources, with no performance degradation even during peak times.
- **Efficient bandwidth utilization:** In situations where data flow is intense in wide area networks, efficient bandwidth utilization minimizes network congestion. Traffic shaping reduces the impact of congestion by preventing bandwidth overload.
- **Security and control:** Traffic shaping also provides significant security benefits. Traffic shaping rules can prevent network threats by limiting malicious or security-threatening traffic on wide area networks.
- **Limiting suspicious traffic:** Traffic shaping, working in conjunction with security policies, can detect and reduce hostile traffic. Especially in attacks such as DoS or DDoS, shaping can be a critical measure to prevent network collapse.

- **Abuse/misuse prevention:** Ensures balanced and fair use of wide area networks by limiting users from excessively using bandwidth.
- **Cost savings:** Cost advantages can be achieved in wide area networks through efficient traffic management. Especially in large wide area network structures connected to cloud services, efficient use of bandwidth reduces data transmission costs.
- **Optimized data transfer:** Traffic shaping ensures more efficient data transmission with less data loss and latency. This not only reduces network resource consumption but also reduces the costs of connections to data centers or cloud providers.
- **Reduced need for hardware investment:** Bandwidth and performance issues in wide area networks are often solved by investing in additional hardware. However, traffic shaping reduces the need for such investments because it enables efficient use of existing hardware.
- **Improved user experience:** Traffic shaping directly affects end-user network performance. Balancing traffic density, reducing latency and packet loss, ensures users experience uninterrupted and fast connection.
- **Uninterrupted connectivity:** Especially during critical business processes (for example, financial transactions or healthcare applications), traffic shaping prevents performance issues by ensuring an uninterrupted data flow.
- **Faster and more efficient connectivity:** Thanks to traffic shaping, applications providing services over wide area networks can respond faster, improving overall user satisfaction.

METHODOLOGY

In this study, services separated from QoS models, traffic shaping and traffic control methods will be used to manage the limited bandwidth in networks. These methods will be configured and applied to the network device that will act as a router in the network. QoS methods in network networks generally divide the capacity between applications in order to use the existing bandwidth efficiently. When performing QoS, a limited capacity is given to systems/applications so that they do not fall below that capacity. If the capacity is sufficient, the idle bandwidth is used according to the priority values. In this study, unlike general QoS methods, it is aimed to assign a standard capacity to systems/applications and not exceed that capacity limit. Even if there is idle capacity, systems/applications will not be able to exceed their determined capacity.

The architecture to be used in the application is shown in Figure 2. As can be seen in the architecture, while there is a limited bandwidth of 1 Gbps capacity received from the ISP for the wide area network, the total bandwidth on the local area network side is 2 Gbps. In the study to be carried out, it is aimed to adapt the high bandwidth on the LAN side to the limited bandwidth received from the ISP on the wide area network side, so that the network traffic, where critical data also flows, is not interrupted, and packet losses and delays in the network are at a minimum level.

The implementation phase of this study was carried out in a test environment with real devices. In the study to be carried out, a network structure consisting of 1 Cisco 4451 router, 1 Cisco 9300 network switching device and 4 Windows 10 installed computers was designed. Secure CRT ver 8.0.4, OpenSpeedTest Server ver 2.1.8 and FileZilla Server ver 3.68.0 programs were used on the computers.

The router, and Secure CRT terminal emulation program was used to manage the Cisco 4451 router. On the local area network side, the other three computers connected to the Cisco 9300 network switch device were used to generate traffic for the study in question towards the server address with the IP addresses allocated by representing the systems used behind. No configuration was made on the Cisco 9300 network switch device, which was used as a standard layer 2 device.

Figure 2. Network infrastructure

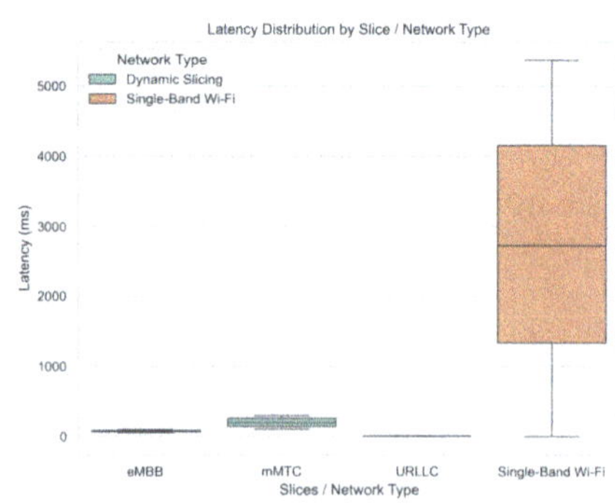

In order to perform QoS on the LAN side and to separate packets of different systems, different methods such as IP address, port number, protocols and ToS bits on the IP address header can be used. Since the system used in the application is a network closed to the Internet (assuming that the entire IP address space can be used), different network addresses are given to each system.

After marking according to IP addresses, each system will be assigned to a separate class. By assigning priority values to these classes created for each system, it will be determined which one will be forwarded first if packets fall into the buffer in the router during high capacity usage. By applying traffic control to these classes, capacity usage will be limited and traffic will not exceed 1 Gbps. If the total capacity exceeds the specified capacity, packets will be discarded without being transmitted. However, by performing traffic shaping within the capacity allocated for each system, traffic will be used more effectively and transmitted more stably from the buffer memory.

To perform differentiated traffic management, traffic flowing on the network can be separated from each other using different methods such as IP addresses, ToS bits on the IP address header, port numbers and protocols. Generally, protocols or ToS bits are used. Since the network architecture used in the study is closed to the internet, the entire IP space can be used. For this reason, systems are classified and separated according to IP addresses for traffic flow management. The classification list by IP addresses is shown in Table 4.

Table 4. Classification table according to IP addresses

172.16.0.0/21 IP address block	
172.16.0.0/24	**MGMT**
172.16.1.0/24	VOIP
172.16.2.0/24	VOIP
172.16.3.0/24	FTP
172.16.4.0/24	UAV
172.16.5.0/24	OTHER
172.16.6.0/24	OTHER
172.16.7.0/24	SPARE

After classifying network traffic by IP address, ACL rules will be applied on the router to apply appropriate QoS policies to this traffic.

The following shows the access control list configuration:

```
Router#configure terminal
Router(config)#
Router(config)#ip access-list extended ACL_DIGER
Router(config-ext-nacl)#10 permit ip 172.16.5.0 0.0.0.255
any
Router(config-ext-nacl)#20 permit ip 172.16.6.0 0.0.0.255
any
Router(config-ext-nacl)#30 permit ip 172.16.7.0 0.0.0.255
any
Router(config-ext-nacl)#40 permit ip any 172.16.5.0
0.0.0.255
```

```
Router(config-ext-nacl)#50 permit ip any 172.16.6.0
0.0.0.255
Router(config-ext-nacl)#60 permit ip any 172.16.7.0
0.0.0.255
Router(config-ext-nacl)#exit
Router(config)#
Router(config)#ip access-list extended ACL_FTP
Router(config-ext-nacl)#10 permit ip 172.16.3.0 0.0.0.255
any
Router(config-ext-nacl)#20 permit ip any 172.16.3.0
0.0.0.255
Router(config-ext-nacl)#exit
Router(config)#
Router(config)#ip access-list extended ACL_IHA
Router(config-ext-nacl)#10 permit ip 172.16.4.0 0.0.0.255
any
Router(config-ext-nacl)#20 permit ip any 172.16.4.0
0.0.0.255
Router(config-ext-nacl)#exit
Router(config)#
Router(config)#ip access-list extended ACL_MGMT
Router(config-ext-nacl)#10 permit ip 172.16.0.0 0.0.0.255
any
Router(config-ext-nacl)#20 permit ip any 172.16.0.0
0.0.0.255
Router(config-ext-nacl)#exit
Router(config)#
Router(config)#ip access-list extended ACL_VOIP
Router(config-ext-nacl)#10 permit ip 172.16.1.0 0.0.0.255
any
Router(config-ext-nacl)#20 permit ip 172.16.2.0 0.0.0.255
any
Router(config-ext-nacl)#30 permit ip any 172.16.1.0
0.0.0.255
Router(config-ext-nacl)#40 permit ip any 172.16.2.0
0.0.0.255
Router(config-ext-nacl)#exit
Router(config)#
```

A summary of the access control list configured in the router is shown in Figure 3.

Figure 3. Access control list (Please note that in the figure and the configuration below, DIGER represents OTHER and IHA represents UAV)

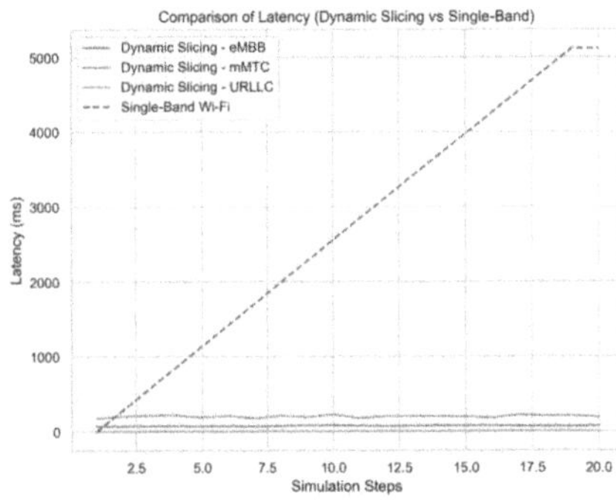

IP addresses marked with the access control list are divided into different classes. The classification configuration is given below.

```
Router#configure terminal
Router(config)#
Router(config)#class-map match-all class_MGMT
Router(config-cmap)#match access-group name ACL_MGMT
Router(config-cmap)#exit
Router(config)#
Router(config)#class-map match-all class_VOIP
Router(config-cmap)#match access-group name ACL_VOIP
Router(config-cmap)#exit
Router(config)#
Router(config)#class-map match-all class_DIGER
Router(config-cmap)#match access-group name ACL_DIGER
Router(config-cmap)#exit
Router(config)#
Router(config)#class-map match-all class_IHA
Router(config-cmap)#match access-group name ACL_IHA
Router(config-cmap)#exit
Router(config)#
Router(config)#class-map match-all class_FTP
Router(config-cmap)#match access-group name ACL_FTP
Router(config-cmap)#exit
Router(config)#
```

A summary of the classification configured in the router is shown in Figure 4.

Figure 4. Traffic classification

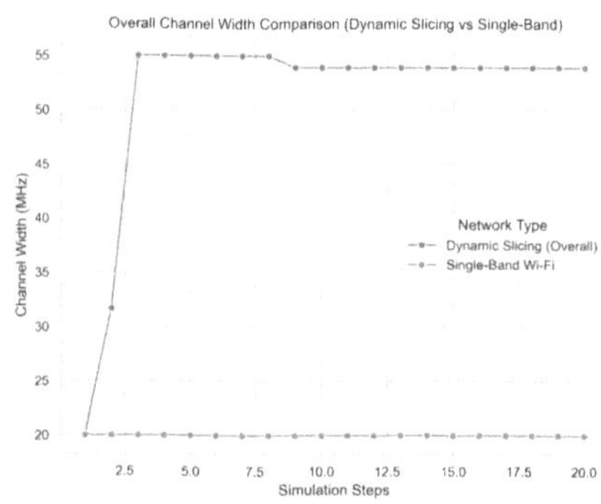

Classified IP address blocks must be assigned a priority level. According to the IP priority levels given above, which priority level should be used for which purpose is explained in Table 5.

Table 5. IP priority levels and purpose of use

Priority		**Purpose of Use**
Level	**Ordinary**	
0	Normal traffic	Default level
1	Priority traffic	Data applications
2	Urgent traffic to be forwarded	Data applications
3	Traffic that needs to be delivered immediately	Call Signal
4	Traffic that needs to be forwarded immediately as a priority	Video conferencing and video streaming
5	Critical (sensitive) traffic	Sound
6	Internet traffic	Network traffic control
7	Network traffic	Network traffic control

The following shows the IP prioritization level configuration:

```
Router#configure terminal
Router(config)#
Router(config)#class-map match-all class_MGMT
Router(config-cmap)#match access-group name ACL_MGMT
```

```
Router(config-cmap)#match precedence 1
Router(config-cmap)#exit
Router(config)#
Router(config)#class-map match-all class_VOIP
Router(config-cmap)#match access-group name ACL_VOIP
Router(config-cmap)#match precedence 5
Router(config-cmap)#exit
Router(config)#
Router(config)#class-map match-all class_DIGER
Router(config-cmap)#match access-group name ACL_DIGER
Router(config-cmap)#match precedence 0
Router(config-cmap)#exit
Router(config)#
Router(config)#class-map match-all class_IHA
Router(config-cmap)#match access-group name ACL_IHA
Router(config-cmap)#match precedence 3
Router(config-cmap)#exit
Router(config)#
Router(config)#class-map match-all class_FTP
Router(config-cmap)#match access-group name ACL_FTP
Router(config-cmap)#match precedence 2
Router(config-cmap)#exit
Router(config)#
```

A summary of IP priority levels according to the classes configured on the router is shown in Figure 5.

Figure 5. IP priority value assignment according to classification

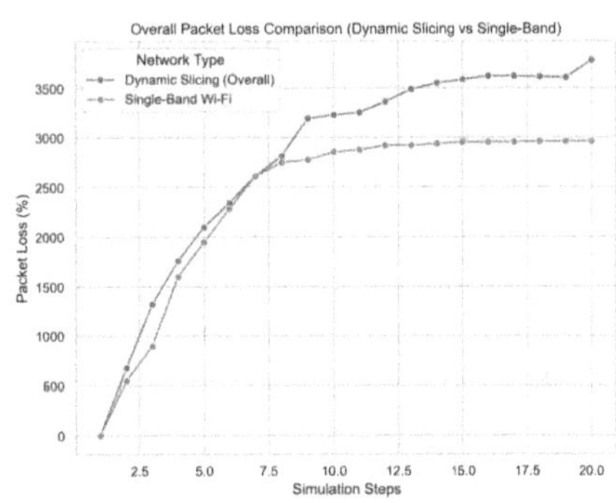

At this stage, the traffic control method will be applied and the existing bandwidth will be limited according to the total bandwidth received from the ISP. Packets exceeding the capacity will be discarded. By performing traffic shaping on classes within the traffic control rules, the traffic will be used more effectively and transmitted more stably by the buffer memory. Below is the configuration where classes are subjected to traffic shaping, limiting their capacity as a percentage of the total bandwidth:

```
Router(config)#
Router(config)#policy-map policing_shaping
Router(config-pmap)#class class_MGMT
Router(config-pmap-c)#shape average percent 5
Router(config-pmap-c)#exit
Router(config)#
Router(config)#policy-map policing_shaping
Router(config-pmap)#class class_VOIP
Router(config-pmap-c)#shape average percent 10
Router(config-pmap-c)#exit
Router(config)#
Router(config)#policy-map policing_shaping
Router(config-pmap)#class class_DIGER
Router(config-pmap-c)#shape average percent 30
Router(config-pmap-c)#exit
Router(config)#
Router(config)#policy-map policing_shaping
Router(config-pmap)#class class_IHA
Router(config-pmap-c)#shape average percent 40
Router(config-pmap-c)#exit
Router(config)#
Router(config)#policy-map policing_shaping
Router(config-pmap)#class class_FTP
Router(config-pmap-c)#shape average percent 15
Router(config-pmap-c)#exit
```

A summary of the traffic shaping percentage process within the traffic control configured on the router is given in Figure 6.

Figure 6. Limiting classes with traffic shaping

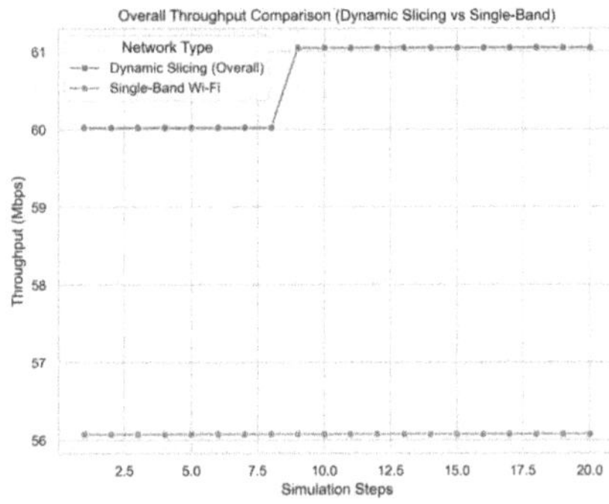

A configuration that allows traffic that exceeds capacity to be discarded by applying traffic control is given below.

```
Router(config)#
Router(config)#policy-map policing_shaping
Router(config-pmap)#class class_default
Router(config-pmap-c)#police 1000000000 conform-action
transmit exceed-action drop
Router(config-pmap-c)#exit
```

A summary of the process configured on the router is given in Figure 7.

Figure 7. Applying traffic control to traffic exceeding capacity

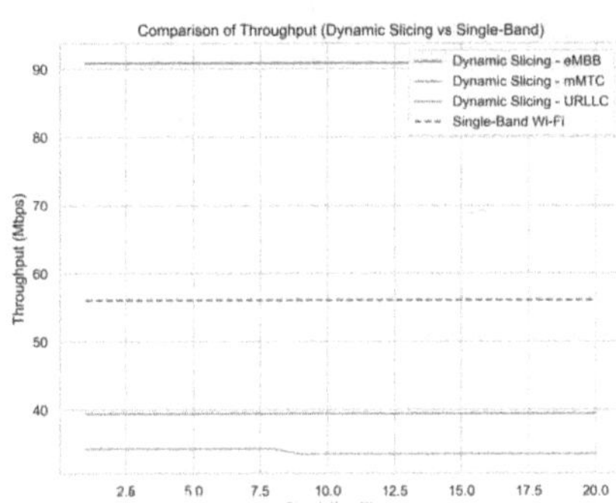

The configuration commands for the interfaces used in the router are given below.

```
Router(config)#
Router(config)#interface GigabitEthernet0/0/0
Router(config-if)#ip address 192.168.1.1 255.255.255.0
Router(config-if)#service-policy output policing_shaping
Router(config-if)#no shutdown
Router(config-if)#exit
Router(config)#
Router(config)#interface GigabitEthernet0/0/3
Router(config-if)#ip address 172.16.0.1 255.255.248.0
Router(config-if)#no shutdown
Router(config-if)#service-policy output policing_shaping
Router(config-if)#exit
Router(config)#
Router(config)#interface GigabitEthernet0
Router(config-if)#vrf forwarding Mgmt-intf
Router(config-if)#service-policy output policing_shaping
Router(config-if)#exit
```

IP and interface information of the devices used in the application architecture are shown in Figure 8.

Table 6 lists the IP address information of the computers used in this architecture. These IP addresses were defined for the computers.

Table 6. IP addresses of computers used in the application

Interface	Server Computer	FTP Computer	UAV Computer	Other Computer
IP address	192.168.1.10	172.16.3.10	172.16.4.10	172.16.5.10
Subnet mask	255.255.255.0	255.255.248.0	255.255.248.0	255.255.248.0
Default gateway	192.168.1.1	172.16.0.1	172.16.0.1	172.16.0.1

Figure 8. Application architecture

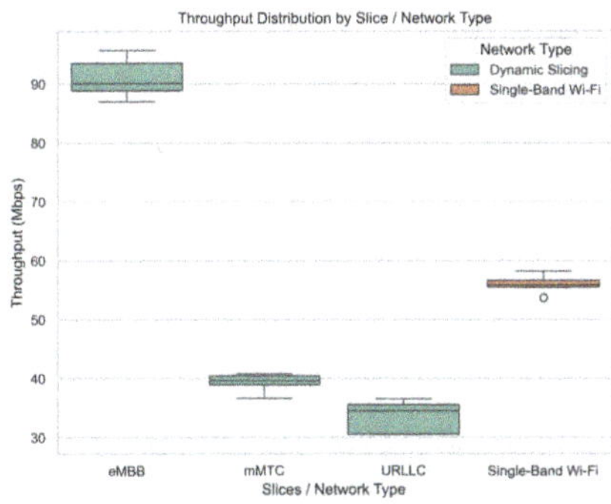

Final configuration of the Cisco 4451 router is given below.

```
!
class-map match-all class_MGMT
match access-group name ACL_MGMT
match precedence 1
class-map match-all class_VOIP
match access-group name ACL_VOIP
match precedence 5
class-map match-all class_DIGER
match access-group name ACL_DIGER
match precedence 0
class-map match-all class_IHA
match access-group name ACL_IHA
match precedence 3
class-map match-all class_FTP
match access-group name ACL_FTP
match precedence 2
!
policy-map policing_shaping
class class_MGMT
shape average percent 5
class class_VOIP
shape average percent 10
class class_FTP
shape average percent 15
class class_IHA
```

```
shape average percent 40
class class OTHER
shape average percent 30
class class-default
police 1000000000 conform-action transmit exceed-action drop
!
!
interface GigabitEthernet0/0/0
ip address 192.168.1.1 255.255.255.0
negotiation auto
service-policy output policing_shaping
!
interface GigabitEthernet0/0/1
no ip address
shutdown
negotiation auto
!
interface GigabitEthernet0/0/2
no ip address
shutdown
negotiation auto
!
interface GigabitEthernet0/0/3
ip address 172.16.0.1 255.255.248.0
negotiation auto
service-policy output policing_shaping
!
interface GigabitEthernet0
vrf forwarding Mgmt-intf
no ip address
shutdown
negotiation auto
!
ip http server
ip http authentication local
ip http secure-server
ip forward-protocol nd
ip tftp source-interface GigabitEthernet0
!
!
```

```
!
ip access-list extended ACL_DIGER
10 permit ip 172.16.5.0 0.0.0.255 any
20 permit ip 172.16.6.0 0.0.0.255 any
30 permit ip 172.16.7.0 0.0.0.255 any
40 permit ip any 172.16.5.0 0.0.0.255
50 permit ip any 172.16.6.0 0.0.0.255
60 permit ip any 172.16.7.0 0.0.0.255
ip access-list extended ACL_FTP
10 permit ip 172.16.3.0 0.0.0.255 any
20 permit ip any 172.16.3.0 0.0.0.255
ip access-list extended ACL_IHA
10 permit ip 172.16.4.0 0.0.0.255 any
20 permit ip any 172.16.4.0 0.0.0.255
ip access-list extended ACL_MGMT
10 permit ip 172.16.0.0 0.0.0.255 any
20 permit ip any 172.16.0.0 0.0.0.255
ip access-list extended ACL_VOIP
10 permit ip 172.16.1.0 0.0.0.255 any
20 permit ip 172.16.2.0 0.0.0.255 any
30 permit ip any 172.16.1.0 0.0.0.255
40 permit ip any 172.16.2.0 0.0.0.255
!
```

After the configuration of the devices, the adaptation of the high bandwidth of 2 Gbps on the local area network side to the limited 1 Gbps bandwidth received from the ISP on the wide area network side was tested in line with the purpose of this study. In order to generate high bandwidth in the application, upload and download values were examined using the OpenSpeed speed test application. Figure 9 shows the upload and download values taken from the UAV computer before and after configuration.

Figure 9. Test result from UAV computer

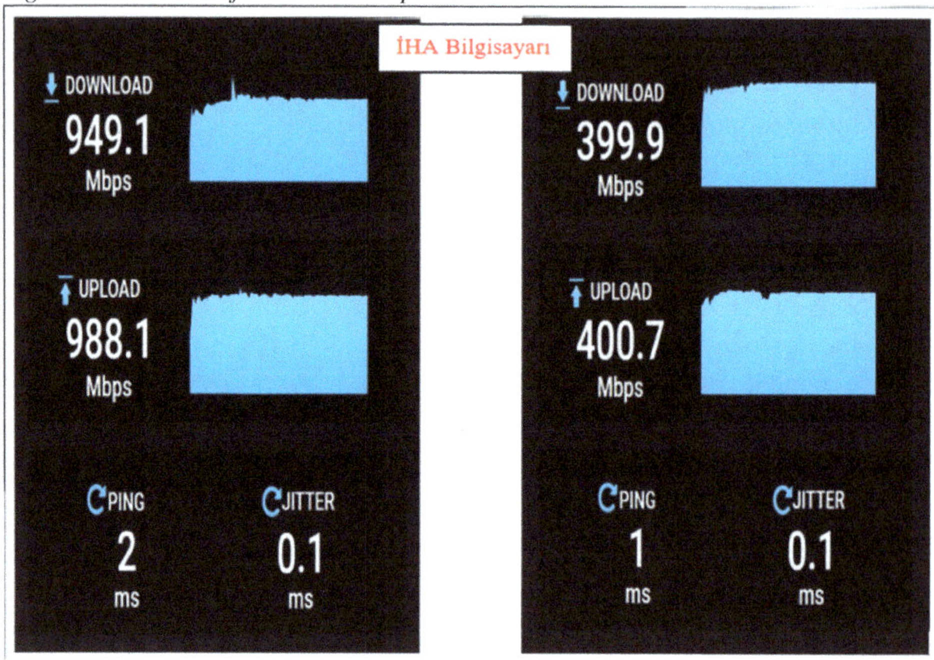

Figure 10 shows the upload and download values obtained from the FTP computer before and after configuration.

Figure 10. Test result from FTP computer

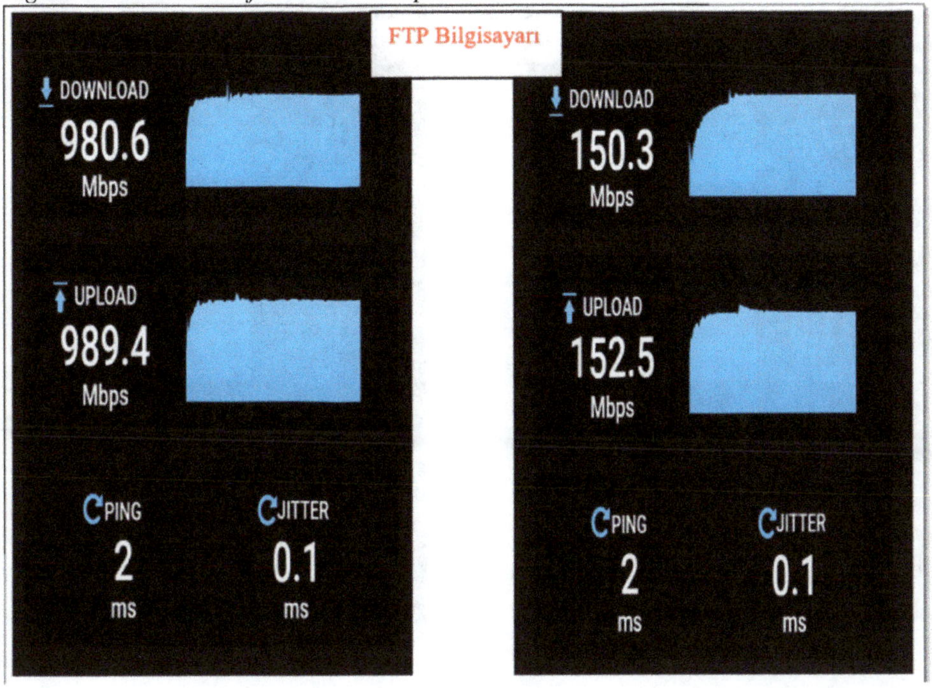

Figure 11 shows the upload and download values taken from the Other computer before and after the configuration.

Figure 11. Test result from the other computer

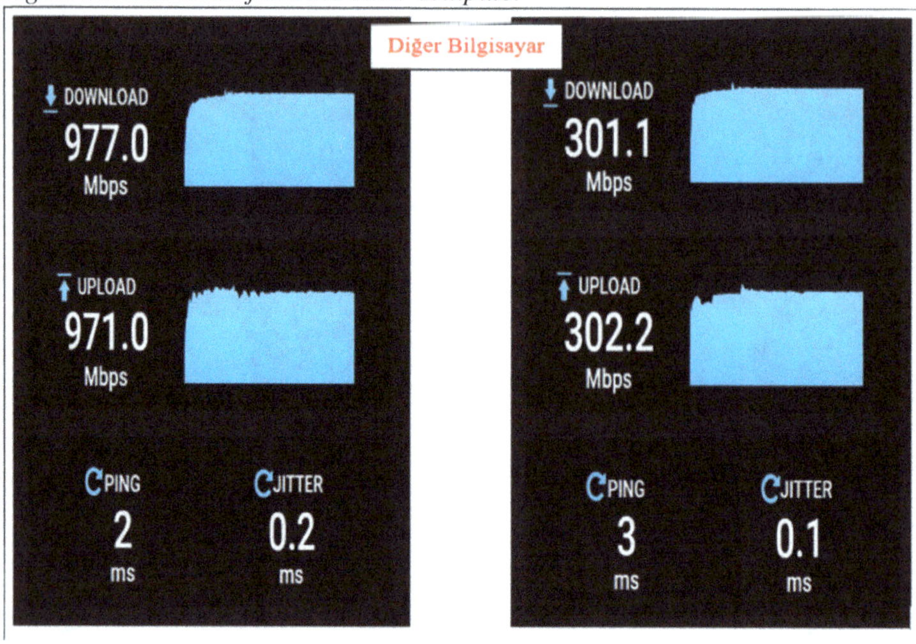

Figure 12 shows the photos of the devices used in this study. Figure 13 shows how to connect to the Cisco 4451 router with SecureCRT. Figure 14 shows Open-SpeedTest program being used as a test server. Finally, Figure 15 shows the access control list of the router.

Figure 12. Devices used in this study

Figure 13. Setting up the connection to the Cisco 4451 router with SecureCRT

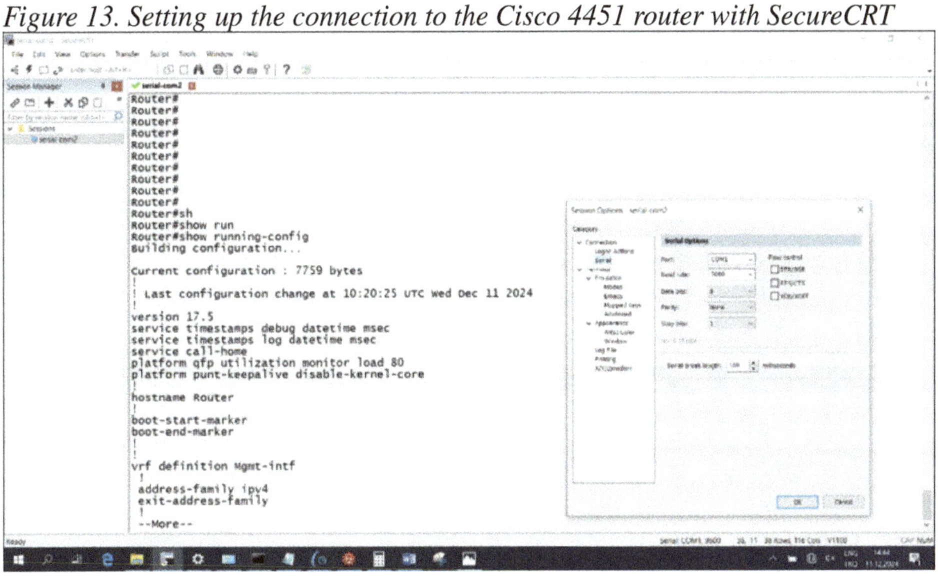

Figure 14. OpenSpeedTest program being used as a test server

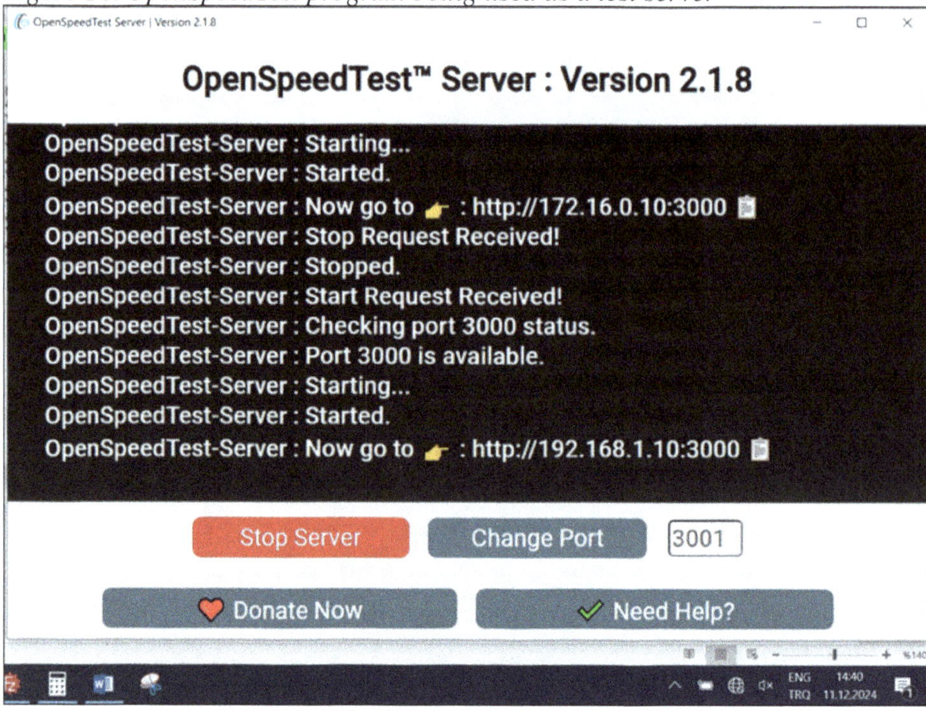

Figure 15. Access control list of the router

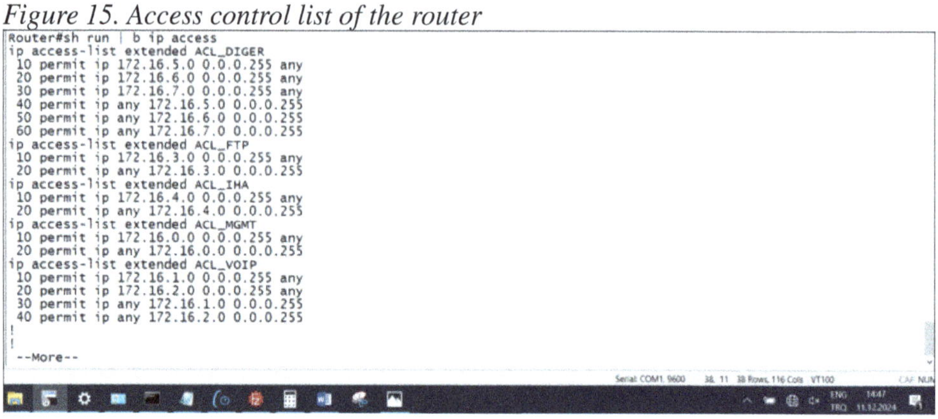

CONCLUSION

This study focused on the need to manage the bandwidth constraints that high-bandwidth systems in the local area network encounter when going out to the wide area network. Exceeding this limited capacity defined by the Internet service providers negatively affects the network performance and interrupts the critical network traffic. This undesirable situation requires an effective network management plan. In this context, QoS models, traffic shaping and traffic control methods were examined in detail in order to manage the network more stably and effectively.

During the implementation phase, traffic shaping methods were tested and the focus was on prioritizing critical data and managing bandwidth effectively, which is one of the issues we are looking for a solution for. In this study, traffic control and traffic shaping methods were used to manage traffic flow, along with access control lists and traffic classes, and decoupled traffic management, one of the three basic QoS models, was used for prioritization and marking. It was observed that creating traffic classes was effective in allocating special resources to critical applications.

The results obtained show that in wide area networks, in limited bandwidth, prioritizing critical traffic by limiting the capacity to be used by systems and applications with traffic shaping provides more stable operation of the network, increases network performance and provides more efficient use of network resources. In this study, instead of Cisco 4451 router, Cisco 9300 network device supporting layer 3 was intended to be used, but it could not be used because a separate license had to be purchased to use QoS features on these devices. It is recommended that researchers test this study on a larger scale network and with other QoS and traffic shaping methods, taking into account issues such as hardware and cost effectiveness.

ACKNOWLEDGMENT

Some parts of this chapter were produced from the first author's master's project. This research received no specific grant from any funding agency in the public, commercial, or not-for-profit sectors.

REFERENCES

Bai, H., Atiquzzaman, M., & Ivancic, W. A. (2002). Running integrated services over differentiated service networks: quantitative performance measurements. In *Quality of Service over Next-Generation Internet* (Vol. 4866, pp. 11–22). SPIE. DOI: 10.1117/12.473015

Cisco Systems, Inc. (2022). *Quality of service configuration guide, Cisco IOS XE 17.x*. Cisco Systems. https://www.cisco.com

Cormen, T. H., Leiserson, C. E., Rivest, R. L., & Stein, C. (2022). *Introduction to algorithms* (4th ed.). The MIT Press.

Elbasheer, M. O., Aldegheishem, A., Lloret, J., & Alrajeh, N. (2021). A QoS-based routing algorithm over Software Defined Networks. *Journal of Network and Computer Applications*, *194*, 103215. DOI: 10.1016/j.jnca.2021.103215

Fgee, E.-B., Kenney, J. D., Phillips, W. J., Robertson, W., & Sivakumar, S. (2005). Comparison of QoS performance between IPv6 QoS management model and IntServ and DiffServ QoS models. *3rd Annual Communication Networks and Services Research Conference (CNSR'05)*, Halifax, NS, Canada, 2005, pp. 287-292. DOI: 10.1109/CNSR.2005.28

Forouzan, B. A. (2007). *Data communications and networking* (4th ed.). McGraw-Hill.

Harju, J., & Kivimaki, P. (2000). Co-operation and comparison of DiffServ and IntServ: performance measurements. *Proceedings 25th Annual IEEE Conference on Local Computer Networks* (pp. 177-186). Tampa, FL, USA. DOI: 10.1109/LCN.2000.891025

Hasenoot, J. A., Rellermeyer, J. S., & Uta, A. (2023). The performance of distributed applications: a traffic shaping perspective. *In Proceedings of the 2023 ACM/SPEC International Conference on Performance Engineering* (pp. 207-220). DOI: 10.1145/3578244.3583733

Heinanen, J., Finland, T., & Guerin, R. (1999a). A Two Rate Three Color Marker. https://datatracker.ietf.org/doc/html/rfc2698

Heinanen, J., Finland, T., & Guerin, R. (1999b). A Single Rate Three Color Marker. https://www.rfc-editor.org/rfc/rfc2697.html

Hellige, H. D. (1994). From SAGE via ARPANET to ETHERNET: Stages in Computer Communications Concepts between 1950 and 1980. *History and Technology, an International Journal, 11*(1), 49-75.

Kar, K., Kodialam, M., & Lakshman, T. V. (2000). Minimum interference routing of bandwidth guaranteed tunnels with MPLS traffic engineering applications. *IEEE Journal on Selected Areas in Communications, 18*(12), 2566–2579. DOI: 10.1109/49.898737

Kassim, M., Ismail, M., Yusof, M. I., & Abdullah, M. A. (2017). Time based traffic policing and shaping algorithms on campus network internet traffic. *Journal of Telecommunication, Electronic and Computer Engineering*, *9*(2-3), 135–140.

Mathews, K., Kramer, C., & Gotzhein, R. (2017). Token bucket based traffic shaping and monitoring for WLAN-based control systems. 2017 IEEE 28th Annual International Symposium on Personal, Indoor, and Mobile Radio Communications (PIMRC), Montreal, QC, Canada. DOI: 10.1109/PIMRC.2017.8292201

Prasanth, L. L., & Uma, E. (2024). A computationally intelligent framework for traffic engineering and congestion management in software-defined network (SDN). EURASIP Journal on Wireless Communications and Networking, 2024(1). DOI: 10.1186/s13638-024-02392-2

She, H., Yan, L., Mao, C., Bu, Q., & Guo, Y. (2025). Service-driven dynamic QoS on-demand routing algorithm. *Future Generation Computer Systems*, *166*, 107685. DOI: 10.1016/j.future.2024.107685

Stallings, W. (2014). Data and computer communications (Tenth edition) India: Pearson Education.

Stoica, I., & Zhang, H. (1999). Providing guaranteed service without per flow management. *Computer Communication Review*, *29*(4), 81–94. DOI: 10.1145/316194.316208

Tanenbaum, A. S., Feamster, N., & Wetherall, D. (2021). *Computer networks* (6th ed.). Pearson.

Taner, C. (2018). *Fundamentals of network administration* (3rd ed.). Abaküs Publications.

Walrand, J. (2023). A concise tutorial on traffic shaping and scheduling in time-sensitive networks. *IEEE Communications Surveys and Tutorials*, *25*(3), 1941–1953. DOI: 10.1109/COMST.2023.3275038

Wassie, G., Ding, J., & Wondie, Y. (2024). Detecting and predicting models for QoS optimization in SDN. *Journal of Computer Networks and Communications, 2024*(1). DOI: 10.1155/2024/3073388

White, P. P. (1997). RSVP and integrated services in the Internet: A tutorial. *IEEE Communications Magazine*, *35*(5), 100–106. DOI: 10.1109/35.592102

ADDITIONAL READING

Barreiros, M., & Lundqvist, P. (2016). *QOS-enabled networks: tools and foundations* (2nd ed.). Wiley. DOI: 10.1002/9781119109136

Joseph, V., & Chapman, B. (2009). *Deploying QoS for Cisco IP and next-generation networks: the definitive guide*. Morgan Kaufmann.

Liu, X., Chen, J., & Yang, Y. (2012). *Temporal QoS management in scientific cloud workflow systems*. Elsevier.

Szigeti, T., Hattingh, C., & Barton, R. (2013). *End-to-end QoS network design: quality of service for rich-media & cloud networks* (2nd ed.).

KEY TERMS AND DEFINITIONS

Bandwidth: It refers to the maximum amount of data that the network can transmit in a certain time period.

Buffer Memory: It provides temporary storage of data packets in routers and switching devices and regulates fluctuations in network traffic.

Data Volume (Throughput): It refers to the amount of data successfully transmitted over the network.

Conclusion

The journey through the 30 chapters of *Emerging Technologies Shaping New Communication and Electronic Infrastructures* has taken me across the vast and dynamic terrain of modern communication systems, showcasing an interdisciplinary mosaic of breakthroughs, methodologies, and applications. This concluding section aims to draw together the threads woven throughout this volume, reflect on their collective significance, and project their implications for the future of technology, society, and global innovation ecosystems.

At the heart of this collection is a central theme: the convergence of communication, computation, and control is not just transforming the physical infrastructure of our world, but also reshaping the way we interact with it. Whether it's a quantum channel enabling secure communications between continents, a microstrip antenna embedded in a wearable health monitor, or a real-time edge analytics engine managing traffic in a smart city, these technologies are actively constructing the digital fabric of the 21st century.

One of the strongest messages emerging from this book is the role of intelligent systems in shaping future infrastructures. Chapters on machine learning-optimized biotelemetry, robotic automation, and AI-enhanced routing protocols in mobile ad hoc networks illustrate how decision-making and adaptability are being integrated into communication technologies. These innovations reduce human intervention, increase system efficiency, and allow for dynamic optimization in ever-changing environments. They are not only technically impressive but also point toward a future where infrastructure is responsive, predictive, and increasingly autonomous.

Equally significant is the interplay between hardware innovation and system-level intelligence. Chapters focused on antenna design, bandpass filters, and nanosheet devices for 5G/6G applications illustrate how advancements at the device level feed directly into broader communication capabilities. These developments are essential not only for meeting the ever-growing demands for bandwidth, speed, and reliability but also for achieving miniaturization, energy efficiency, and environmental sustainability. As we move towards the widespread adoption of 6G, innovations in hardware will become even more critical.

The emphasis on quantum technologies and post-quantum cryptography marks another major frontier addressed in this book. The chapters by Kisalaya Chakrabarti, Dr. Rajkumar S, Dr. Kavitha K, Dr. Shyam Sihare, and others provide crucial insights into the challenges and solutions emerging in the realm of quantum communication. The use of blockchain in secure key distribution, discussions on quantum coherence, and cryptographic primitives for the post-quantum era collectively underscore the urgent need to prepare our communication systems for the quantum age. These insights are not theoretical abstractions; they have practical relevance for national security, financial systems, and critical infrastructure worldwide.

Cybersecurity and privacy emerged as cross-cutting themes, underscoring the importance of designing communication systems that are not only fast and efficient but also secure and trustworthy. From industrial control system vulnerability assessments to IoT-specific threat detection using machine learning, the chapters in this area highlight that security must be a foundational pillar rather than an afterthought in system design. As more devices connect to the internet and as data becomes an even more valuable asset, ensuring the integrity and privacy of communication will remain a paramount concern.

Another dimension explored in this volume is the social and economic impact of communication technologies. The discussions on 5G's influence on financial systems, smart city infrastructure, and healthcare connectivity reflect the broader ripple effects of technological innovation. These technologies are no longer confined to the labs or the tech sector; they are deeply integrated into the social fabric, influencing everything from how patients receive care to how cities manage energy and transportation.

A key strength of this book is its global perspective. Contributors from the United Kingdom, India, Nigeria, Turkey, and Thailand bring a wide range of cultural, academic, and industrial experiences to the table. This geographic diversity enriches the book's content, illustrating how regional priorities, infrastructural realities, and research traditions shape the development and application of emerging technologies. It reinforces the idea that building future communication infrastructures is a global endeavour, requiring cross-border collaboration, inclusive innovation, and mutual learning.

Importantly, the book also serves as a platform for early-career researchers and educators who are shaping the next generation of communication technology experts. By documenting state-of-the-art research and offering practical solutions, the chapters contribute to building academic curricula and industry training programs that reflect current and future demands.

From an academic standpoint, the book contributes to filling knowledge gaps in rapidly evolving areas. It brings together fragmented research areas under one umbrella, making connections between quantum encryption and blockchain, between microwave components and machine learning, between NLP and materials science. This integrative approach is vital for generating new research questions, fostering interdisciplinary collaboration, and driving innovation at the intersections of different fields.

From an industrial perspective, the book provides real-world use cases, performance evaluations, and design methodologies that can inform the development and deployment of commercial communication systems. Engineers, product designers, and R&D professionals will find actionable insights and proven frameworks that can accelerate the development cycle and reduce risks.

As the pace of technological change continues to accelerate, books like this are critical in providing a moment of reflection, synthesis, and projection. They offer a snapshot of where we are today while providing a roadmap for where we might go next. The chapters in this volume not only describe emerging technologies but also model the kind of multidisciplinary, collaborative, and forward-thinking approach that will be essential for tackling the complex challenges of the future.

Looking ahead, I envision several key trajectories where the contributions in this book will have lasting impact:

1. **Adaptive, Resilient Networks**: Future communication systems will need to respond dynamically to shifting demands, environmental changes, and security threats. The research presented here lays the groundwork for systems that can self-optimize and self-heal.
2. **Sustainable Infrastructure**: With the global focus on sustainability, innovations that reduce energy consumption, use recyclable materials, and minimize electromagnetic pollution will become increasingly valuable. This book highlights early efforts in that direction.
3. **Ubiquitous Intelligence**: As AI and machine learning become foundational components of communication systems, the distinction between 'communication' and 'computation' will blur. Many chapters anticipate this convergence and offer frameworks for its implementation.
4. **Quantum-Ready Systems**: Preparing for a quantum future is no longer optional. The foundational concepts and applied frameworks discussed in this book can help guide organizations and governments in making informed investments and policy decisions.
5. **Democratized Innovation**: Finally, by showcasing research from a variety of institutions and regions, the book promotes a vision of inclusive, globally

distributed innovation. This democratization of knowledge and capability is essential for building communication systems that serve all of humanity.

In conclusion, *Emerging Technologies Shaping New Communication and Electronic Infrastructures* is more than just a collection of academic chapters. It is a collective vision of a technological future that is intelligent, secure, inclusive, and sustainable. By bringing together diverse voices and disciplines, it serves as both a scholarly resource and a practical guide for anyone invested in shaping the future of communication. I hope it inspires further research, sparks meaningful collaboration, and contributes to the ongoing quest to build a more connected, capable, and equitable world.

Eugene Ogbodo
University of Hertfordshire, UK

Compilation of References

Abbas, M., Khan, S. U., Aamir, M., Khan, N. U., Ali, U., Ullah, S., Basir, A., & Björninen, T. (2024). Compact implantable antenna for enhanced Bio-Telemetry communication in the MICS band. *IEEE Access : Practical Innovations, Open Solutions*, *1*, 1. Advance online publication. DOI: 10.1109/ACCESS.2024.3512950

Abhivarun, S. N., Sankar, A., Harshan, K., Priharini, S., & Natarajamani, S. (2023, November). Design and Performance Evaluation of MIMO Antenna for 5G Smartphones. In 2023 IEEE 3rd International Conference on Applied Electromagnetics, Signal Processing, & Communication (AESPC) (pp. 1-4). IEEE.

Aboualalaa, M., Abdel-Rahman, A. B., Allam, A., Elsadek, H., & Pokharel, R. K. (2017). Design of a dual-band microstrip antenna with enhanced gain for energy harvesting applications. *IEEE Antennas and Wireless Propagation Letters*, *16*, 1622–1626.

Agiwal, M., Roy, A., & Saxena, N. (2016). Next generation 5G wireless networks: A comprehensive survey. *IEEE Communications Surveys and Tutorials*, *18*(3), 1617–1655. DOI: 10.1109/COMST.2016.2532458

Ahmad, A., Choi, D., & Ullah, S. (2022). A compact two elements MIMO antenna for 5G communication. *Scientific Reports*, *12*(1), 3608. DOI: 10.1038/s41598-022-07579-5 PMID: 35246591

Akkole, S., & Vasudevan, N. (2021). "Microstrip Fractal MultiBand Antenna Design and Optimization by using DGS Technique for Wireless Communication," 2021 6th International Conference on Inventive Computation Technologies (ICICT), pp. 72-77, Coimbatore, India DOI: 10.1109/ICICT50816.2021.9358712

Akyildiz, I. F., Kak, A., & Nie, S. (2020). 6G and Beyond: The Future of Wireless Communications Systems. *IEEE Access : Practical Innovations, Open Solutions*, *8*, 133995–134030. DOI: 10.1109/ACCESS.2020.3010896

Al-Sehemi, A., Al-Ghamdi, A., Dishovsky, N., Atanasova, G., & Atanasov, N. (2020). A flexible broadband antenna for IoT applications. *International Journal of Microwave and Wireless Technologies*, *12*(6), 531–540. DOI: 10.1017/S1759078720000161

Altufaili, M. M. S., Najaf, A. N., & Idan, Z. S. (2022). Design of circular-shaped microstrip patch antenna for 5G applications. *Telkomnika*, *20*(1), 19–26. DOI: 10.12928/telkomnika.v20i1.21019

Anchidin, L., Lavric, A., Mutescu, P., Petrariu, A. I., & Popa, V. (2023). The design and development of a microstrip antenna for internet of things applications. *Sensors (Basel)*, *23*(3), 1062. DOI: 10.3390/s23031062 PMID: 36772099

Andrews, C. J. M., Narayanan, A. S. K., & Sunil, A. M., (2024). Compact Metamaterial based Antenna for 5G Applications. Volume 24, 103269 .DOI: 10.1016/j.rineng.2024.103269

Andrews, J. G., Buzzi, S., Choi, W., Hanly, S. V., Lozano, A., Soong, A. C. K., & Zhang, J. C. (2014). What will 5G be? *IEEE Journal on Selected Areas in Communications*, *32*(6), 1065–1082. DOI: 10.1109/JSAC.2014.2328098

Anwar, U., Ajijola, O. A., Shivkumar, K., & Markovic, D. (2022). Towards a Leadless Wirelessly Controlled Intravenous Cardiac Pacemaker. *IEEE Transactions on Biomedical Engineering*, *69*(10), 3074–3086. DOI: 10.1109/TBME.2022.3161415 PMID: 35320081

Aptiv Corporation (Ed.). (2022) Placement of sensing technologies, Satellite architecture. Available at: https://www.aptiv.com/docs/default-source/infographics/aptivsatellitearch_infographic.pdf?sfvrsn=4804163d_2

Arif, A., Zubair, M., Ali, M., Khan, M. U., & Mehmood, M. Q. (2019, May). A Compact, Low-Profile Fractal Antenna for Wearable On-Body WBAN Applications. *IEEE Antennas and Wireless Propagation Letters*, *18*(5), 981–985. DOI: 10.1109/LAWP.2019.2906829

Attia, H. A., El-Metwally, M., & Fahmy, O. M. (2010). Harmonic distortion effects and mitigation in distribution systems. *The Journal of American Science*, *6*(10), 173–183.

Azari, A., Skrivervik, A., & Aliakbarian, H. (2024). A mmWave Low complexity and Low-Cost Super Wideband Dual-Polarized Aperture Coupled Antenna for 5G Applications. *IEEE Access : Practical Innovations, Open Solutions*, *12*, 85601–85607. DOI: 10.1109/ACCESS.2024.3415167

Bai, H., Atiquzzaman, M., & Ivancic, W. A. (2002). Running integrated services over differentiated service networks: quantitative performance measurements. In *Quality of Service over Next-Generation Internet* (Vol. 4866, pp. 11–22). SPIE. DOI: 10.1117/12.473015

Balanis, C. A. (2016). *Antenna theory: analysis and design*. John Wiley & Sons.

Bhandari, S., & Chauhan, Y. S. (2021). Nanosheet field-effect transistor device and circuit aspects for future technology nodes. *IEEE Transactions on Electron Devices*, *68*(6), 2796–2803.

Bhatia, S. S., Sahni, A., & Rana, S. B. (2018, January). A Novel Design of Compact Monopole Antenna with Defected Ground Plane for Wideband Applications. *Progress in Electromagnetics Research M. Pier M*, *70*, 21–31. DOI: 10.2528/PIERM18050201

Bisdikian, C. (2002). An overview of the Bluetooth wireless technology. *IEEE Communications Magazine*, *39*(12), 86–94. DOI: 10.1109/35.968817

Bluetooth, S. I. G. Inc. (n.d.). *Tech overview*. Bluetooth Technology Website. https://www.bluetooth.com/learn-about-bluetooth/tech-overview/

Boccardi, F., Heath, R. W., Lozano, A., Marzetta, T. L., & Popovski, P. (2014). Five disruptive technology directions for 5G. *IEEE Communications Magazine*, *52*(2), 74–80. DOI: 10.1109/MCOM.2014.6736746

Bose, M., & Karuppiah, V. (2024). Metamaterial inspired superstrate loaded miniaturized quad port MIMO antenna for 5G C-Band applications. *Optics Communications, 131123*. Advance online publication.

Bukhsh, Z. A., Molegraaf, H., & Jansen, N. (2023, April). A Maintenance Planning Framework Using Online and Offline Deep Reinforcement Learning. *Neural Computing & Applications*. PMID: 40584427

Cai, P., Zhang, Y., Tang, Z., & Gao, S. (2022). Chameleon Swarm algorithm enhanced by differential evolution. *2022 Joint 12th International Conference on Soft Computing and Intelligent Systems and 23rd International Symposium on Advanced Intelligent Systems (SCIS&Amp;ISIS)*, 1–6.

Cao, Y., Brubaker, M. A., Fleet, D. J., & Hertzmann, A. (2015). Efficient Optimization for Sparse Gaussian Process Regression. *IEEE Transactions on Pattern Analysis and Machine Intelligence*, *37*(12), 2415–2427. DOI: 10.1109/TPAMI.2015.2424873 PMID: 26539847

Carmo, M., Jardim, S., Neto, A., Aguiar, R., Corujo, D., & Rodrigues, J. J. P. C. (2018). Slicing WiFi WLAN Sharing Access Infrastructures to Enhance Ultra-Dense 5G Networking. In: *IEEE International Conference on Communications (ICC)* DOI: 10.1109/ICC.2018.8422769

Chahbar, M., Diaz, G., Dandoush, A., C'erin, C., & Ghoumid, K. (2021). A Comprehensive Survey on the E2E 5G Network Slicing Model. *IEEE Transactions on Network and Service Management, 18*(1), 49–62. DOI: 10.1109/TNSM.2020.3044626

Challoo, R., Oladeinde, A., Yilmazer, N., Ozcelik, S., & Challoo, L. (2012). An overview and assessment of wireless technologies and co-existence of ZigBee, Bluetooth and Wi-Fi devices. *Procedia Computer Science, 12*, 386–391. DOI: 10.1016/j.procs.2012.09.091

Chaudhari, Q. (n.d.). FMCW Radar Part 1 – Ranging. Wireless Pi. https://wirelesspi.com/fmcw-radar-part-1-ranging/

Chen, S., & Zhao, J. (2014). The requirements, challenges, and technologies for 5G of terrestrial mobile telecommunication. *IEEE Communications Magazine, 52*(5), 36–43. DOI: 10.1109/MCOM.2014.6815891

Christian Wolff Radar basics, Radar tutorial. Available at: https://www.radartutorial.eu/02.basics/Pulse%20Radar.en.html

Cisco Systems, Inc. (2022). *Quality of service configuration guide, Cisco IOS XE 17.x.* Cisco Systems. https://www.cisco.com

Commission, F. C. (2023) Federal Communications Commission FCC 23-35, New opportunities for unlicensed field disturbance sensors. Available at: https://docs.fcc.gov/public/attachments/FCC-23-35A1.pdf

Commission, F. C. (2023a) Federal Communications Commission, FCC Empowers Short-Range Radars in the 60 GHz Band. Available at: https://www.federalregister.gov/documents/2023/07/24/2023-15367/fcc-empowers-short-range-radars-in-the-60-ghz-band

Commission, F. C. (2025) The Federal Communications Commission, 15.255 Operation within the band 57-71 GHz. Available at: https://www.ecfr.gov/current/title-47/chapter-I/subchapter-A/part-15/subpart-C/subject-group-ECFR2f2e5828339709e/section-15.255

Contreras, P., & Murtagh, F. (2015). Hierarchical clustering. Handbook of cluster analysis, pages 103–123.

Cormen, T. H., Leiserson, C. E., Rivest, R. L., & Stein, C. (2022). *Introduction to algorithms* (4th ed.). The MIT Press.

Dangi, R., Jadhav, A., Choudhary, G., Dragoni, N., Mishra, M. K., & Lalwani, P. (2022). ML-Based 5G Network Slicing Security: A Comprehensive Survey. *Future Internet, 14*(4), 116. DOI: 10.3390/fi14040116

Dang, S., Amin, O., Shihada, B., & Alouini, M.-S. (2020). What should 6G be? *Nature Electronics, 3*(1), 20–29. DOI: 10.1038/s41928-019-0355-6

Dao, T. C., Tam, N. T., & Huynh, T. T. B. (2024). Node depth representation-based evolutionary multitasking optimization for maximizing the network lifetime of wireless sensor networks. *Engineering Applications of Artificial Intelligence, 128*, 107463. DOI: 10.1016/j.engappai.2023.107463

Darboe, O., Konditi, D. B. O., & Manene, F. (2019). A 28 GHz Rectangular Microstrip Patch Antenna for 5G Applications. *International Journal of Engineering Research & Technology (Ahmedabad), 12*(6), 854–857.

Das, S. C., Paul, L. C., Hossain, M. N., Mahmud, M. Z., Azim, R., (2021). A Dual Band Miniaturized Spiral-shaped Patch Antenna for 5G and WiFi-5/6 Applications, DOI: 10.1109/SPICSCON54707.2021.9885246

Dash, R. K., Saha, P. B., & Ghoshal, D. (2022, June). Slotted Patch Based Multiband Antenna with Multiple DGS Effect to Suppress Cross Polarized Radiation. *Progress in Electromagnetics Research C. Pier C, 120*, 179–193. DOI: 10.2528/PIERC22031707

Dhaliwal, B. S., Kaur, G., Kaur, S., & Pattnaik, S. (Feb 2020). Low-Cost Fractal MIMO Antenna Design for ISM Band Using ANN. In *Advances in Computational Intelligence Techniques* (pp. 219–226). Springer. DOI: 10.1007/978-981-15-2620-6_15

Dhaliwal, B. S., Kaur, G., Saini, N., Pattnaik, S. S., & Josan, S. K. (2019, August). A Modified Two-Step ANN Ensemble Approach to Improve Generalization and its Application in Fractal Antenna Design. *Journal of Circuits, Systems, and Computers, 29*(07), 2050106. DOI: 10.1142/S0218126620501066

Dhaliwal, B. S., & Pattnaik, S. S. (2017, September). Development of PSO-ANN Ensemble Hybrid Algorithm and Its Application in Compact Crown Circular Fractal Patch Antenna Design. *Wireless Personal Communications, 96*(1), 135–152. DOI: 10.1007/s11277-017-4157-8

Elabd, R. H., & Al-Gburi, A. J. A. (2024, June). Design and Optimization of a Circular-Ring Shaped UWB Fractal Antenna for Wireless Multiband Applications Using Particle Swarm Optimization. *Progress in Electromagnetics Research B. Pier B, 106*, 101–112. DOI: 10.2528/PIERB24033002

Elbasheer, M. O., Aldegheishem, A., Lloret, J., & Alrajeh, N. (2021). A QoS-based routing algorithm over Software Defined Networks. *Journal of Network and Computer Applications, 194*, 103215. DOI: 10.1016/j.jnca.2021.103215

Elijah, A. A., & Mokayef, M. (2020). Miniature microstrip antenna for IoT application. *Materials Today: Proceedings, 29*, 43–47. DOI: 10.1016/j.matpr.2020.05.678

Elmezughi, M. K., Salih, O., Afullo, T. J., & Duffy, K. J. (2023, September). Path Loss Modeling Based on Neural Networks and Ensemble Method for Future Wireless Networks. *Heliyon, 9*(9), e19685. DOI: 10.1016/j.heliyon.2023.e19685 PMID: 37809436

Esmail, B. F., & Koziel, S. (2022). Design and optimization of Metamaterial-Based Dual-Band 28/38 GHz 5G MIMO antenna with modified ground for isolation and bandwidth improvement. *IEEE Antennas and Wireless Propagation Letters, 22*(5), 1069–1073. DOI: 10.1109/LAWP.2022.3232622

Essa, A., Almajali, E., Mahmoud, S., Amaya, R. E., Alja'Afreh, S. S., & Ikram, M. (2024, June). "Wireless Power Transfer for Implantable Medical Devices: Impact of Implantable Antennas on Energy Harvesting,". *IEEE Open Journal of Antennas and Propagation, 5*(3), 739–758. DOI: 10.1109/OJAP.2024.3392160

European regulation (2021) Delegated regulation - 2022/30 - en - EUR-lex, COMMISSION DELEGATED REGULATION (EU) 2022/30. Available at: https://eur-lex.europa.eu/eli/reg_del/2022/30/oj

Fakhimi, M. J., & Akan, O. B. (2024). Nanoantennas and nanoradars: The future of integrated sensing and communication at the nanoscale. *IEEE Communications Magazine, 62*(1), 68–74.

Falkner, B. J., Zhou, H., Zhang, W., Metaxas, G., Mehta, A., Ziegler, V., Multerer, T., Mirshekar-Syahkal, D., & Nakano, H. (2022). A low-profile wideband compressed single-arm spiral antenna array for mid-band 5G beam steering applications. *Scientific Reports, 12*(1), 13790. DOI: 10.1038/s41598-022-16423-9 PMID: 35963882

Farzad Kiani, A. (2013). An energy effective-accuracy routing algorithm for wireless sensor networks. *Life Science Journal, 10*(2), 39–45.

Fei, P., Shen, Z., Wen, X., & Nian, F. (2015). A Single-Layer circular polarizer based on hybrid meander line and loop configuration. *IEEE Transactions on Antennas and Propagation*, *63*(10), 4609–4614. DOI: 10.1109/TAP.2015.2462128

Ferdous, N., Hock, G. C., Hamid, S. H. A., Raman, M. N. A., Kiong, T. S., & Ismail, M. (2019). Design of a small patch antenna at 3.5 GHz for 5G application. *IOP Conference Series. Earth and Environmental Science*, *268*(1), 012152. DOI: 10.1088/1755-1315/268/1/012152

Fgee, E.-B., Kenney, J. D., Phillips, W. J., Robertson, W., & Sivakumar, S. (2005). Comparison of QoS performance between IPv6 QoS management model and IntServ and DiffServ QoS models. *3rd Annual Communication Networks and Services Research Conference (CNSR'05)*, Halifax, NS, Canada, 2005, pp. 287-292. DOI: 10.1109/CNSR.2005.28

Forouzan, B. A. (2007). *Data communications and networking* (4th ed.). McGraw-Hill.

Frontoni, E., Romeo, L., Bernardini, M., Moccia, S., Migliorelli, L., Paolanti, M., Ferri, A., Misericordia, P., Mancini, A., & Zingaretti, P. (2020). A decision support system for diabetes chronic care models based on general practitioner engagement and EHR data sharing. *IEEE Journal of Translational Engineering in Health and Medicine*, *8*, 1–12. DOI: 10.1109/JTEHM.2020.3031107 PMID: 33150095

FURUNO Radar basics, FURUNO. Available at: https://www.furuno.com/en/technology/radar/basic/

Garroppo, R. G., Gazzarrini, L., Giordano, S., & Tavanti, L. (2011, June). Experimental assessment of the coexistence of Wi-Fi, ZigBee, and Bluetooth devices. In *2011 IEEE International Symposium on a World of Wireless, Mobile and Multimedia Networks* (pp. 1-9). IEEE. DOI: 10.1109/WoWMoM.2011.5986182

Gokul, N., & Sankaran, S. (2021, December). Modeling and defending against resource depletion attacks in 5g networks. In 2021 IEEE 18th India Council International Conference (INDICON) (pp. 1-7). IEEE.

Gupta, A., Bansal, S., Al-Gburi, A. J. A., Forsyth, D. I., & Ismail, M. M. (2024). Design of an elliptical arc-shaped antenna at 37 GHz and performance analysis for 5G on-body application https://pe.org.pl/articles/2024/9/52

Gupta, A., Joshi, H. D., & Khanna, R. (2017, June). An X-shaped Fractal antenna with DGS for Multiband Applications. *International Journal of Microwave and Wireless Technologies*, *9*(5), 1075–1083. DOI: 10.1017/S1759078716000994

Gupta, M., & Mathur, V. (2017, September). Wheel Shaped Modified Fractal Antenna Realization for Wireless Communications. *AEÜ. International Journal of Electronics and Communications*, *79*, 257–266. DOI: 10.1016/j.aeue.2017.06.017

Haghzad Klidbary, S., & Javadian, M. (2024). Improvement of low energy adaptive clustering hierarchical protocol based on genetic algorithm to increase network lifetime of wireless sensor network. *International Journal of Engineering*, *37*(9), 1800–1811. DOI: 10.5829/IJE.2024.37.09C.10

Hannan, M., Ghosh, S., Shanmuganantham, T., & Shafi, F. (2024). Compact Slotted Modified Triangular Patch Antenna For 5G Applications in 28 GHz Band, DOI: 10.21203/rs.3.rs-3985130/v1

Harju, J., & Kivimaki, P. (2000). Co-operation and comparison of DiffServ and IntServ: performance measurements. *Proceedings 25th Annual IEEE Conference on Local Computer Networks* (pp. 177-186). Tampa, FL, USA. DOI: 10.1109/LCN.2000.891025

Hasenoot, J. A., Rellermeyer, J. S., & Uta, A. (2023). The performance of distributed applications: a traffic shaping perspective. *In Proceedings of the 2023 ACM/SPEC International Conference on Performance Engineering* (pp. 207-220). DOI: 10.1145/3578244.3583733

Heinanen, J., Finland, T., & Guerin, R. (1999a). A Two Rate Three Color Marker. https://datatracker.ietf.org/doc/html/rfc2698

Heinanen, J., Finland, T., & Guerin, R. (1999b). A Single Rate Three Color Marker. https://www.rfc-editor.org/rfc/rfc2697.html

Hellige, H. D. (1994). From SAGE via ARPANET to ETHERNET: Stages in Computer Communications Concepts between 1950 and 1980. *History and Technology, an International Journal, 11*(1), 49-75.

Heo, J., Jang, D., Choo, H., Na, D.-Y., & Bae Park, Y. (2025). Real-Time Duct Prediction and Propagation Analysis Using Genetic Algorithm-Based RFC and Parabolic Equation Method. *IEEE Access : Practical Innovations, Open Solutions*, *13*, 471–479. DOI: 10.1109/ACCESS.2024.3522376

He, S., Zhang, B., Luo, J., & Qu, M. (2021, May). Research on coexistence and mutual interference of Wi-Fi and Bluetooth. In *2021 International Conference on Microwave and Millimeter Wave Technology (ICMMT)* (pp. 1-3). IEEE. DOI: 10.1109/ICMMT52847.2021.9617988

Hiertz, G. R., Denteneer, D., Stibor, L., Zang, Y., Costa, X. P., & Walke, B. (2010). The IEEE 802.11 universe. *IEEE Communications Magazine*, *48*(1), 62–70. DOI: 10.1109/MCOM.2010.5394032

Hui, W., Guo, Y., Xie, K., & Zhao, X. (2023). Investigation of circularly polarized microstrip antenna with Tri-Band polarization conversion metamaterials. *IEEE Antennas and Wireless Propagation Letters*, *22*(12), 2831–2835. DOI: 10.1109/LAWP.2023.3300298

Hussain, M., Jarchavi, S. M. R., Naqvi, S. I., Gulzar, U., Khan, S., Alibakhshikenari, M., & Huynen, I. (2021, October 31). Design and Fabrication of a Printed Tri-Band Antenna for 5G Applications Operating across Ka-, and V-Band Spectrums. *Electronics (Basel)*, *10*(21), 2674. DOI: 10.3390/electronics10212674

Iqbal, A., Al-Hasan, M., Mabrouk, I. B., & Denidni, T. A. (2024). A compact Self-Triplexing antenna for head implants. *IEEE Transactions on Antennas and Propagation*, *72*(11), 8207–8214. Advance online publication. DOI: 10.1109/TAP.2024.3450332

Islam, M. A., Alam, F., Rahman, M. A., Hossain, M. R., & Roy, S. C., (2023). Design and analysis of the performance of triangular patch antenna for THz applications .DOI: 10.31893/multiscience.2024043

Islam, M. S., Islam, M. T., Ullah, M. A., Beng, G. K., Amin, N., & Misran, N. (2019). A modified Meander Line microstrip patch antenna with enhanced bandwidth for 2.4 GHz ISM-Band internet of things (IoT) applications. *IEEE Access : Practical Innovations, Open Solutions*, *7*, 127850–127861. DOI: 10.1109/ACCESS.2019.2940049

ISO 26262 series of standards. Available at: https://www.iso.org/obp/ui/en/#iso:std:iso:26262:-1:ed-2:v1:en

Jacob, S., Bose, M., & Karuppiah, V. (2024). Artificial Magnetic Conductor-Backed Microstrip-Fed Dual-Band antenna for 5G wearable sensor nodes and C Band application in WBAN network. In *Advances in information security, privacy, and ethics book series* (pp. 221–241).

Jacob, S., Bose, M., & Karuppiah, V. (2024). Artificial Magnetic Conductor-Backed Microstrip-Fed Dual-Band antenna for 5G wearable sensor nodes and C Band application in WBAN network. In *Advances in information security, privacy, and ethics book series* (pp. 221–241). DOI: 10.4018/979-8-3693-2786-9.ch009

Jain, K., Mehra, P. S., Dwivedi, A. K., & Agarwal, A. (2022). Scada: Scalable cluster-based data aggregation technique for improving network lifetime of wireless sensor networks. *The Journal of Supercomputing*, *78*(11), 13624–13652. DOI: 10.1007/s11227-022-04419-1

Jamil, A., Rauf, M., Sami, A., Ansari, A., & Idrees, M. D. (2022, February). "A Wideband Hybrid Fractal Ring Antenna for WLAN Applications," International Journal of Antennas and Propagation, Vol. 2022. *Article ID*, *6136916*, 1–8.

Jena, S. R., Mallick, M. R., Padhi, J., & Panda, M. (2016). Ant-colony algorithm based routing using swastika clustering in wsns. [IJCTA]. *International Journal of Control Theory and Applications*, *9*(16), 8181–8192.

Kachorovskii, V., Deng, Y., Rumyantsev, S., Lü, J.-Q., Gaska, R., Simin, G., Hu, X., Khan, M. A., & Shur, M. S. (2021). Plasmonic field-effect transistors (TeraFETs) for6G communications. *Nanomaterials (Basel, Switzerland)*, *11*(12), 3301. PMID: 34947650

Kaloxylos, A. (2018). A Survey and an Analysis of Network Slicing in 5G Networks. *IEEE Communications Standards Magazine*, *2*(1), 60–65. DOI: 10.1109/MCOMSTD.2018.1700072

Kar, K., Kodialam, M., & Lakshman, T. V. (2000). Minimum interference routing of bandwidth guaranteed tunnels with MPLS traffic engineering applications. *IEEE Journal on Selected Areas in Communications*, *18*(12), 2566–2579. DOI: 10.1109/49.898737

Kassim, M., Ismail, M., Yusof, M. I., & Abdullah, M. A. (2017). Time based traffic policing and shaping algorithms on campus network internet traffic. *Journal of Telecommunication, Electronic and Computer Engineering*, *9*(2-3), 135–140.

Kaur, A., & Malik, P. K. (2021, March). Multiband Elliptical Patch Fractal and Defected Ground Structures Microstrip Patch Antenna for Wireless Applications. *Progress in Electromagnetics Research B. Pier B*, *91*, 157–173. DOI: 10.2528/PIERB20102704

Kaur, M. (2024). Multifunctional and multistandard compact antenna design using hybridization of fractals and artificial neural network. In *Bio-Inspired Intelligence for Smart Decision-Making* (pp. 26–45). IGI Global Scientific Publishing.

Kaur, M., Krishan, R., Kaur, N., & Sivia, J. S. (2024). Invasive weed optimization based compact hybridized fractal antenna design for RF energy harvesting and multifunctional wireless applications. *AEÜ. International Journal of Electronics and Communications*, *184*, 155428.

Kaur, M., Krishan, R., Sivia, J. S., & Kaur, N. (2024). Shuffled frog Leaping algorithm based circular patterns loaded ring shaped fractal antenna for multistandard wireless applications. *AEÜ. International Journal of Electronics and Communications*, *176*, 155123.

Kaur, M., & Sivia, J. S. (2019, February). Minkowski, Giuseppe Peano and Koch Curves based Design of Compact Hybrid Fractal Antenna for Biomedical Applications using ANN and PSO. *AEÜ. International Journal of Electronics and Communications*, *99*, 14–24. DOI: 10.1016/j.aeue.2018.11.005

Kaur, M., & Sivia, J. S. (2019, March). ANN-based Design of Hybrid Fractal Antenna for Biomedical Applications. *International Journal of Electronics*, *106*(8), 1184–1199. DOI: 10.1080/00207217.2019.1582712

Kaur, M., & Sivia, J. S. (2019, November). Giuseppe Peano and Cantor Set Fractals based Miniaturized Hybrid Fractal Antenna for Biomedical Applications using Artificial Neural Network and Firefly Algorithm. *International Journal of RF and Microwave Computer-Aided Engineering*, *30*(1), 1–11.

Kaur, N., Singh, J., & Kumar, M. (2020, August). Hexagonal Ring Shaped Dual Band Antenna Using Staircase Fractal Geometry for Wireless Applications. *Wireless Personal Communications*, *113*(4), 2067–2078. DOI: 10.1007/s11277-020-07307-0

Kerkhoff, A. J., & Ling, H. (2007). Design of a Band-Notched Planar Monopole antenna using genetic algorithm optimization. *IEEE Transactions on Antennas and Propagation*, *55*(3), 604–610. DOI: 10.1109/TAP.2007.891563

Khan, M. M., Alhakami, W., Rahman, H. M. A., Shovon, M. N. A., Hadjouni, M., Elmannai, H., & Bourouis, S. (2022). Design and Analysis of a 5G Wideband Antenna for Wireless Body-Centric Network. *Wireless Communications and Mobile Computing*, *2022*(1), 1–16. DOI: 10.1155/2022/1558791

Khattak, R. Y., Ahmed, Q. Z., Shoaib, S., Lazaridis, P. I., & Alade, T. (2024). Recent advances in antennas for biotelemetry and healthcare applications. *IEEE Open Journal of Antennas and Propagation*, *1*(6), 1499–1522. Advance online publication. DOI: 10.1109/OJAP.2024.3462289

Kiani, F. (2016). A novel channel allocation method for time synchronization in wireless sensor networks. *International Journal of Numerical Modelling*, *29*(5), 805–816. DOI: 10.1002/jnm.2143

Kim, I.-K., Yook, J.-G., & Park, H.-K.II-K. Kim. (2002, July). J-G. Yook and H-K. Park, "Fractal-Shape Small Size Microstrip Patch Antenna,". *Microwave and Optical Technology Letters*, *34*(1), 15–17. DOI: 10.1002/mop.10359

Kiran, T., Mounisha, N., Mythily, C., Akhil, D., & Phani Kumar, T. V. B. (2018). Design of Microstrip Patch Antenna for 5G Applications. *IOSR Journal of Electronics and Communication Engineering*, *13*(1), 14–17. DOI: 10.9790/2834-1301011417

Kishore, S., & Abdul Rajak, A. R. (2022). Microstrip Patch Antenna with C Slot for 5G Communication at 30 GHz. *Emerging Science Journal*, *6*(6), 1315–1327. Advance online publication. DOI: 10.28991/ESJ-2022-06-06-06

Knap, W., Kachorovskii, V., Deng, Y., Rumyantsev, S., Lü, J.-Q., Gaska, R., Shur, M., Simin, G., Hu, X., Khan, M. A., & Shur, M. S. (2023). Gigahertz and terahertz transistors for 5G, 6G, and beyond mobile communication systems. *Applied Physics Reviews*, *11*(3), 031318.

Korea (2022) Radio Waves Act. Available at: https://elaw.klri.re.kr/eng_service/lawView.do?hseq=7162&lang=ENG

Kumari, J. J. J., Thangam, S., & Raja, A. S. (2023, December). An Adaptive on Demand Modified Ant Colony Optimization Routing for VANETs. In International Conference on Intelligent Systems Design and Applications (pp. 480-492). Cham: Springer Nature Switzerland.

Kumari, J. J. J., Thangam, S., & Raja, A. S. (2024, March). QoS CBSC: An Enhanced Metaheuristic Strategy on QoS-Cloud-based Service in Cloud. In 2024 5th International Conference on Intelligent Communication Technologies and Virtual Mobile Networks (ICICV) (pp. 685-691). IEEE.

Kumari, J. J. J., Thangam, S., & Raja, A. S. (2023, October). Efficacious Routing Approaches in Vehicular Ad hoc Network: An Empirical Study. In *2023 First International Conference on Advances in Electrical, Electronics and Computational Intelligence (ICAEECI)* (pp. 1-7). IEEE.

Kumar, S. A., & Dileepan, D. (2017, June). Design and Development of CPW Fed Monopole Antenna at 2.45 GHz and 5.5 GHz for Wireless Applications. *Alexandria Engineering Journal*, *56*(2), 231–234. DOI: 10.1016/j.aej.2016.12.018

Kumar, S., Dixit, A. S., Malekar, R. R., Raut, H. D., & Shevada, L. K. (2020). Fifth Generation Antennas: A Comprehensive Review of Design and Performance Enhancement Techniques. *IEEE Access : Practical Innovations, Open Solutions*, *8*, 163568–163593. DOI: 10.1109/ACCESS.2020.3020952

Kyama, R., Sanapureddy, S. D., Gyadapaka, J., K N V, K. K., (2023). Design and Analysis of Microstrip Rectangular Patch Antenna for 5G Applications. E3S Web of Conferences, 391, 01113 .DOI: 10.1051/e3sconf/202339101113

Lieto, I., Malanchini, I., Mandelli, S., Moro, E., & Capone, A. (2020). Malanchini, S. Mandelli, E. Moro, A. Capone: Strategic Network Slicing Management in Radio Access Networks. *IEEE Transactions on Mobile Computing*, *21*(4), 1434–1448. DOI: 10.1109/TMC.2020.3025027

Lin, Q., Alkaraki, S., Wong, H., & Kelly, J. R. (2022). A wideband circularly polarized antenna based on anisotropic metamaterial. *IEEE Transactions on Antennas and Propagation*, *71*(2), 1254–1262. DOI: 10.1109/TAP.2022.3225157

Liu, B., & El Kamel, A. (2015). V2X-based decentralized cooperative adaptive cruise control in the vicinity of intersections. *IEEE Transactions on Intelligent Transportation Systems*, *17*(3), 644–658. DOI: 10.1109/TITS.2015.2486140

Liu, L., Liu, T., Zheng, Y., Chernogor, L. F., Jin, Z., & Sun, Z. (2022). Archimedean spiral antenna based on metamaterial structure with wideband circular polarization. *AEÜ. International Journal of Electronics and Communications*, *152*, 154257. DOI: 10.1016/j.aeue.2022.154257

Liu, Y., Chen, Y., Lin, H., & Juwono, F. H. (2016). A novel differentially fed compact Dual-Band implantable antenna for biotelemetry applications. *IEEE Antennas and Wireless Propagation Letters*, *15*, 1791–1794. DOI: 10.1109/LAWP.2016.2536735

Li, X., Zhang, Y., & Wang, P. (2022). Recent advances in new materials for 6G communications. *Advanced Electronic Materials*, *8*(1), 2100978. DOI: 10.1002/aelm.202100978

Li, Y. (2023). Li. Y, A microstrip patch antenna for 5G mobile communications. *Journal of Physics: Conference Series*, *2580*(1), 012063. DOI: 10.1088/1742-6596/2580/1/012063

Lorincz, J., Kukuruzovi'c, A., & Blažević, Z. (2024). Z. Bla˘zevi'c: A Comprehensive Overview of Network Slicing for Improving the Energy Efficiency of Fifth-Generation Networks. *Sensors (Basel)*, *24*(10), 3242. DOI: 10.3390/s24103242 PMID: 38794095

Maheswari, T., Uma Sankar, V., Sai Reddy, S., Pradeep Kumar, B., & Rushendrababu, K. (2019). Design of a Microstrip patch Antenna for 5G Wireless Applications. *International Journal of Research and Analytical Reviews*, *6*(1).

Males-Ilic, N., Milovanovic, B., & Budimir, D. (2001, December). Low intermodulation amplifiers for RF and microwave wireless systems. In *APMC 2001. 2001 Asia-Pacific Microwave Conference (Cat. No. 01TH8577)* (Vol. 3, pp. 984-987). IEEE. DOI: 10.1109/APMC.2001.985282

Malik, J., Kim, S., Seo, J. M., Cho, Y. M., & Bien, F. (2022). Minimally invasive implant type electromagnetic biosensor for continuous glucose monitoring System: In vivo evaluation. *IEEE Transactions on Biomedical Engineering*, *70*(3), 1000–1011. DOI: 10.1109/TBME.2022.3207240 PMID: 36112548

Manimegalai, A., & Kalphana, I. (2024). Design and Analysis of Microstrip Patch Antenna for Satellite Communication. *International Journal of Creative Research Thoughts*, *12*(2).

Mark, R., Mishra, N., Mandal, K., Sarkar, P. P., & Das, S. (2018, September). Hexagonal Ring Fractal Antenna with Dumb Bell Shaped Defected Ground Structure for Multiband Wireless Applications. *AEÜ. International Journal of Electronics and Communications*, *94*, 42–50. DOI: 10.1016/j.aeue.2018.06.039

Masroor, I., Ansari, J. A., Pandey, A., & Mishra, P. K. (2022, June). Design and Analysis of Multiband Fractal Reconfigurable Antenna Using PIN Diodes for Smart Wireless Communications. *Progress in Electromagnetics Research C. Pier C*, *120*, 209–222. DOI: 10.2528/PIERC22041402

Mathews, K., Kramer, C., & Gotzhein, R. (2017). Token bucket based traffic shaping and monitoring for WLAN-based control systems. 2017 IEEE 28th Annual International Symposium on Personal, Indoor, and Mobile Radio Communications (PIMRC), Montreal, QC, Canada. DOI: 10.1109/PIMRC.2017.8292201

Matthiesen, B., Aydin, O., & Jorswieck, E. A. (2018). Throughput and Energy-Efficient Network Slicing. In: WSA 2018; 22nd International ITG Workshop on Smart Antennas, pp. 1–6

Meenambal, B., Jacob, S., & Vasudevan, K. (2024). Metamaterial-Inspired concatenated dual ring antenna design for 5G NR C-Band applications. In *Advances in information security, privacy, and ethics book series* (pp. 200–220).

Meenambal, B., Jacob, S., & Vasudevan, K. (2024). Metamaterial-Inspired concatenated dual ring antenna design for 5G NR C-Band applications. In *Advances in information security, privacy, and ethics book series* (pp. 200–220). DOI: 10.4018/979-8-3693-2786-9.ch008

Meenambal, B., Vasudevan, K., Maheshwari, G. U., & Sherene, J. (2024). Offset-Fed Defected Ground Structure (DGS)-Based Metamaterial-Inspired Asymmetric SRR for 5G terminal applications. In *Lecture notes in networks and systems* (pp. 163–178).

MIPI Alliance. (2024) Camera serial interface 2 (MIPI CSI-2) v4.1, MIPI Alliance. Available at: https://www.mipi.org/specifications/csi-2

Misala, S., & Mosa, S. A. (2023, December). A Novel Stacked Rectangular with Surface Mounted Short Rectangle Dielectric Resonator Antenna in C-Band Applications. *Progress in Electromagnetics Research Letters, 115*, 81–89. DOI: 10.2528/PIERL23070902

Miucic, R., Sheikh, A., Medenica, Z., & Kunde, R. (2018, August). V2X applications using collaborative perception. In 2018 IEEE 88th Vehicular Technology Conference (VTC-Fall) (pp. 1-6). IEEE. doi:DOI: 10.1109/VTCFall.2018.8690818

Mohammed, N., Cluff, K., Griffith, J., & Loflin, B. (2019). "A Non-invasive Wearable Readout System for Bio-fluid Phenomena Detection," *2019 IEEE 16th International Conference on Wearable and Implantable Body Sensor Networks (BSN)*, Chicago, IL, USA, pp. 1-4, DOI: 10.1109/BSN.2019.8771042

Murtagh, F., & Contreras, P. (2012). Algorithms for hierarchical clustering: An overview. *Wiley Interdisciplinary Reviews. Data Mining and Knowledge Discovery, 2*(1), 86–97. DOI: 10.1002/widm.53

Mwang'amba, R., Mei, P., Akinsolu, M. O., Liu, B., & Zhang, S. (2024). Gain bandwidth enhancement and sidelobe level stabilization of MM-Wave lens antennas using AI-driven optimization. *IEEE Antennas and Wireless Propagation Letters, 1–5*. Advance online publication.

Nagaraju, P., Kumaraswamy, H. V., Sudhindra, K. R., & Lokesh Kumar, M, (2022). Design of Compact Sized Frequency Reconfigurable Spiral Antenna for 5G Applications

Nematzadeh, S., Torkamanian-Afshar, M., Seyyedabbasi, A., & Kiani, F. (2023). Maximizing coverage and maintaining connectivity in wsn and decentralized iot: An efficient metaheuristic-based method for environment-aware node deployment. *Neural Computing & Applications, 35*(1), 611–641. DOI: 10.1007/s00521-022-07786-1

Nguyen, D. C., Ding, M., Pathirana, P. N., Seneviratne, A., Li, J., Niyato, D., Dobre, O., & Poor, H. V. (2021). 6G Internet of Things: A Comprehensive survey. *IEEE Internet of Things Journal, 9*(1), 359–383. DOI: 10.1109/JIOT.2021.3103320

O'Donnell, R. M. (2001) Radar: Introduction to radar systems - online course, Radar: Introduction to Radar Systems - Online Course | MIT Lincoln Laboratory. Available at: https://www.ll.mit.edu/outreach/radar-introduction-radar-systems-online-course

Oladejo, S., & Falowo, O. (2019). An Energy-Efficient Resource Allocation Scheme for 5G Slice Networks. In: *Proc. Southern Afr. Telecommun. Netw. Appl. Conf.*, pp. 1–4

Olimid, R. F., & Nencioni, G. (2020). 5G Network Slicing: A Security Overview. *IEEE Access : Practical Innovations, Open Solutions*, *8*, 99999–100009. DOI: 10.1109/ACCESS.2020.2997702

Ophir, L., Bitran, Y., & Sherman, I. (2004, September). Wi-Fi (IEEE 802.11) and Bluetooth coexistence: Issues and solutions. In *2004 IEEE 15th International Symposium on Personal, Indoor and Mobile Radio Communications (IEEE Cat. No. 04TH8754)* (Vol. 2, pp. 847-852). IEEE.

Osseiran, A., Boccardi, F., Braun, V., Kusume, K., Marsch, P., Maternia, M., Queseth, O., Schellmann, M., Schotten, H., Taoka, H., Tullberg, H., Uusitalo, M. A., Timus, B., & Fallgren, M. (2014). Scenarios for 5G mobile and wireless communications: The vision of the METIS project. *IEEE Communications Magazine*, *52*(5), 26–35. DOI: 10.1109/MCOM.2014.6815890

Overview of ARIB Standards (STD-T48) Association of Radio Industries and Businesses. (n.d.). https://www.arib.or.jp/english/std_tr/telecommunications/desc/std-t48.html

Padhi, J., Biswal, R. R., Das, K., & Bishwal, P. K. (2016). Novel routing to maximize network life time using swastika model. *International Journal of Control Theory and Applications*, *9*(16), 8171–8180.

Pahlavan, K., & Krishnamurthy, P. (2021). Evolution and impact of Wi-Fi technology and applications: A historical perspective. *International Journal of Wireless Information Networks*, *28*(1), 3–19. DOI: 10.1007/s10776-020-00501-8

Pallavi, M., Kumar, P., Ali, T., Shenoy, S. B., Mane, P. R., & Prabhu, G. (2024). Design and experimental study of a coupled U and L-Shape negative index metamaterial for aircraft navigation applications. *IEEE Access : Practical Innovations, Open Solutions*, *12*, 16127–16150. DOI: 10.1109/ACCESS.2024.3358898

Panda, P. K., & Ghosh, D. (2019, November). Wideband and High Gain Tuning Fork Shaped Monopole Antenna Using High Impedance Surface. *AEÜ. International Journal of Electronics and Communications*, *111*(1), 152920. DOI: 10.1016/j.aeue.2019.152920

Pant, M., & Malviya, L. (2023). Design, developments, and applications of 5G antennas: A review. *International Journal of Microwave and Wireless Technologies*, *15*(1), 156–182. DOI: 10.1017/S1759078722000095

Patel, S., Pradhan, S. S., & Patra, K. C. (2017). Advance scale-3 swastika model to improve network lifetime in wireless sensor network. [IJRSI]. *International Journal of Research and Scientific Innovation*, *IV*(VI).

Patel, Y. K. (2022). Design of Microstrip Patch Antenna for 5G Applications. *International Research Journal of Modernization in Engineering Technology and Science*, *4*(5).

Patil, S., Patil, V., & Sharma, M. (2022). A Review on 5G Antenna: Challenges and Parameter Enhancement Techniques. International Journal of Electrical and Electronics Research, 12(2), 545-556, (2024), .DOI: 10.37391/ijeer.120229

Peng, N. J., Li, N. K., & Irwin, G. (2008). A new Jacobian matrix for optimal learning of Single-Layer neural networks. *IEEE Transactions on Neural Networks*, *19*(1), 119–129. DOI: 10.1109/TNN.2007.903150 PMID: 18269943

Prasanth, L. L., & Uma, E. (2024). A computationally intelligent framework for traffic engineering and congestion management in software-defined network (SDN). EURASIP Journal on Wireless Communications and Networking, 2024(1). DOI: 10.1186/s13638-024-02392-2

Pucci, L., Bacchielli, T., & Giorgetti, A. (2025). "Cooperative Maximum Likelihood Target Position Estimation for MIMO-ISAC Networks," in *IEEE Wireless Communications Letters*, DOI: 10.1109/LWC.2025.3548446

Punith, S., Praveenkumar, S. K., Abhinandan, A. J., & Ahmed, M.R. (2019). A Novel Multiband Microstrip Patch Antenna for 5G Communications. (2019), .DOI: 10.1016/j.procs.2020.04.224

Qian, K., He, Z., & Zhang, X. (2020). 3D point cloud generation with millimeter-wave radar. *Proceedings of the ACM on Interactive, Mobile, Wearable and Ubiquitous Technologies*, *4*(4), 1–23. DOI: 10.1145/3432221

Raj, T., Mishra, R., Kumar, P., & Kapoor, A. (2023). Advances in MIMO Antenna Design for 5G: A Comprehensive Review. *Sensors (Basel)*, *23*(14), 6329. DOI: 10.3390/s23146329 PMID: 37514623

Ramya, C. M., Shanmugaraj, M., & Prabakaran, R. (2011, April). Study on ZigBee technology. In *2011 3rd international conference on electronics computer technology* (Vol. 6, pp. 297-301). IEEE. DOI: 10.1109/ICECTECH.2011.5942102

Rana, M. S., & Rahman, M. M. (2022). Study of Microstrip Patch Antenna for Wireless Communication System, *International Conference for Advancement in Technology (ICONAT)*, DOI: 10.1109/ICONAT53423.2022.9726110

Rao, S. & Texas Instruments Incorporated. (2018b). MIMO Radar. https://www.ti .com/lit/an/swra554a/swra554a.pdf?ts=1739548406903

Rappaport, T. S., Sun, S., Mayzus, R., Zhao, H., Azar, Y., Wang, K., Wong, G. N., Schulz, J. K., Samimi, M., & Gutierrez, F. (2013). Millimeter wave mobile communications for 5G cellular: It will work! *IEEE Access : Practical Innovations, Open Solutions*, *1*, 335–349. DOI: 10.1109/ACCESS.2013.2260813

Rappaport, T. S., Xing, Y., MacCartney, G. R., Molisch, A. F., Mellios, E., & Zhang, J. (2015). Overview of millimeter wave communications for fifth-generation (5G) wireless networks—With a focus on propagation models. *IEEE Transactions on Antennas and Propagation*, *65*(12), 6213–6230. DOI: 10.1109/TAP.2017.2734243

Rashmitha, R., Niran, N., Jugale, A. A., & Ahmed, M. R. (2020). Microstrip Patch Antenna Design for Fixed, Mobile, and Satellite 5G Communications, *Third International Conference on Computing and Network Communications (CoCoNet'19)*, DOI: 10.1016/j.procs.2020.04.223

Rasilainen, K., Phan, T. D., Berg, M., Pärssinen, A., & Soh, P. J. (2023). Hardware aspects of Sub-THZ antennas and reconfigurable intelligent surfaces for 6G communications. *IEEE Journal on Selected Areas in Communications*, *41*(8), 2530–2546. DOI: 10.1109/JSAC.2023.3288250

Ratnam, D. V., Pasha, A., Swathi, P., & Rao, M. V. G. "Acquisition of GPS L1 signals using Cooley-tukey FFT algorithm," *2013 IEEE International Conference on Signal Processing, Computing and Control (ISPCC)*, Solan, India, 2013, pp. 1-4, DOI: 10.1109/ISPCC.2013.6663397

Razavi, B., & Behzad, R. (2012). *RF microelectronics* (Vol. 2). Prentice hall.

Regulagadda, N. R., & Kumari, U. V. R. (2023, July). A Multiband Flexible Wideband CPW Wearable Slot Antenna for Biomedical and IoT Applications. *Progress in Electromagnetics Research C. Pier C*, *135*, 131–144. DOI: 10.2528/PIERC23052102

Renesas Electronics. (2023) Automotive radar, Millimeter wave automotive radars. Available at: https://www.renesas.com/en/blogs/how-connect-antennas

Richart, M., Baliosian, J., Serrat, J., Gorricho, J.-L., & Ag¨uero, R. (2020). Slicing With Guaranteed Quality of Service in WiFi Networks. *IEEE Transactions on Network and Service Management*, *17*(3), 1822–1837. DOI: 10.1109/TNSM.2020.3005594

Rodriguez-Soto, M., Lopez-Sanchez, M., & Rodriguez-Aguilar, J. A. (2025). Multi-objective reinforcement learning for designing ethical multi-agent environments. *Neural Computing & Applications*, *37*(31), 25619–25644.

Ruppel, C. D., Javali, S. L., & Ltd, A. T. (2019, May 14). *US10938487B2 - Monitoring a vehicle-to-everything communication system - Available on Google Patents*. https://patents.google.com/patent/US10938487B2/en

Saad, W., Bennis, M., & Chen, M. (2019). A vision of 6G wireless systems: Applications, trends, technologies, and open research problems. *IEEE Network, 34*(3), 134–142. DOI: 10.1109/MNET.001.1900287

Sampath, K. J., Kumar, N. K., Yeswanth, K., Snehith, K., & Anooja, B. (2021, June). An efficient channel coding architecture for 5G wireless using high-level synthesis. In 2021 5th International Conference on Trends in Electronics and Informatics (ICOEI) (pp. 674-680). IEEE.

Sarker, N., Podder, P., Mondal, M. R. H., Shafin, S. S., & Kamruzzaman, J. (2023). Applications of Machine Learning and Deep Learning in antenna Design, Optimization, and Selection: A review. *IEEE Access : Practical Innovations, Open Solutions, 11*, 103890–103915. DOI: 10.1109/ACCESS.2023.3317371

Sathishkumar, N., Natarajan, R., Divya, D., Divya, S., (2023). A broadband dual-polarized antenna for 5G wireless communication systems, .DOI: 10.1016/j.rio.2024.100659

Semiconductors, N. X. P.Automotive and Industrial Radar Systems, NXP Semiconductors. Available at: https://www.nxp.com/applications/RADAR-SYSTEMS

Sen Laxminarayan Sahoo, S., & Tiwary, K. S. (2024). Sarbast Moslem, and Tapan Senapati. Improvement of wireless sensor network lifetime via intelligent clustering under uncertainty. *IEEE Access : Practical Innovations, Open Solutions.*

Seyyedabbasi, A., Dogan, G., & Kiani, F. (2020). Heel: A new clustering method to improve wireless sensor network lifetime. *IET Wireless Sensor Systems, 10*(3), 130–136. DOI: 10.1049/iet-wss.2019.0153

Seyyedabbasi, A., Kiani, F., & Allahviranloo, T. (2023). Unai FernandezGamiz, and Samad Noeiaghdam. Optimal data transmission and pathfinding for wsn and decentralized iot systems using i-gwo and ex-gwo algorithms. *Alexandria Engineering Journal, 63*, 339–357. DOI: 10.1016/j.aej.2022.08.009

Shafi, M., Molisch, A. F., Smith, P. J., Haustein, T., Zhu, P., De Silva, P., Tufvesson, F., Benjebbour, A., & Wunder, G. (2017). 5G: A tutorial overview of standards, trials, challenges, deployment, and practice. *IEEE Journal on Selected Areas in Communications, 35*(6), 1201–1221. DOI: 10.1109/JSAC.2017.2692307

Shahraki, A., Taherkordi, A., Haugen, Ø., & Eliassen, F. (2020). A survey and future directions on clustering: From wsns to iot and modern networking paradigms. *IEEE Transactions on Network and Service Management, 18*(2), 2242–2274. DOI: 10.1109/TNSM.2020.3035315

Shankar, A., Sivakumar, N. R., Sivaram, M., Ambikapathy, A., Nguyen, T. K., & Dhasarathan, V. (2021). Increasing fault tolerance ability and network lifetime with clustered pollination in wireless sensor networks. *Journal of Ambient Intelligence and Humanized Computing*, *12*(2), 2285–2298. DOI: 10.1007/s12652-020-02325-z

Sharma, K., & Pandey, G.P. (2021). "Efficient modelling of compact microstrip antenna using machine learning", *International Journal of Electronics and Communications* DOI: 10.1016/j.aeue.2021.153739

Sharma, M. S. S. S., Sai, M. T. V., & Sankaran, S. (2024, August). Deep Learning based approach for detection of Resource Depletion Attacks in 5G Networks. In 2024 IEEE 5th India Council International Subsections Conference (INDISCON) (pp. 1-6). IEEE.

Sharma, N., & Bhatia, S. S. (2021, March). Ultra-wideband Fractal Antenna using Rhombus Shaped Patch with Stub loaded Defected Ground Plane: Design and Measurement. *AEÜ. International Journal of Electronics and Communications*, *131*, 153604. DOI: 10.1016/j.aeue.2021.153604

Sharma, N., & Sharma, V. (2018, December). A Design of Microstrip Patch Antenna using Hybrid Fractal Slot for Wideband Applications. *Ain Shams Engineering Journal*, *9*(4), 2491–2497. DOI: 10.1016/j.asej.2017.05.008

Sharma, N., Sharma, V., & Bhatia, S. S. (2018, September). A Novel Hybrid Fractal Antenna for Wireless Applications. *Progress in Electromagnetics Research M. Pier M*, *73*, 25–35. DOI: 10.2528/PIERM18052403

Sharma, S., Dudeja, R. K., Aujla, G. S., Bali, R. S., & Kumar, N. (2020, September). DeTrAs: Deep Learning-Based Healthcare Framework for IoT-Based Assistance of Alzheimer Patients. *Neural Computing & Applications*.

She, H., Yan, L., Mao, C., Bu, Q., & Guo, Y. (2025). Service-driven dynamic QoS on-demand routing algorithm. *Future Generation Computer Systems*, *166*, 107685. DOI: 10.1016/j.future.2024.107685

Sherene, J., Vasudevan, K., Maheshwari, G. U., & Meenambal, B. (2024). Design and analysis of AMC-Based anti-symmetric Dual L-Shaped antenna for WBAN Application. In *Lecture notes in networks and systems* (pp. 179–194).

Sherene, J., Vasudevan, K., Maheshwari, G. U., & Meenambal, B. (2024). Design and analysis of AMC-Based anti-symmetric Dual L-Shaped antenna for WBAN Application. In *Lecture notes in networks and systems* (pp. 179–194). DOI: 10.1007/978-981-97-1943-3_16

Shi, W., Tong, P., & Bi, X. (2025). Moving-Least-Squares-Enhanced 3D Object Detection for 4D Millimeter-Wave Radar. *Remote Sensing (Basel)*, *17*(8), 1465. DOI: 10.3390/rs17081465

Shi, Y., & Wang, B. (2022, October). Cartoon Art Style Rendering Algorithm Based on Deep Learning. *Neural Computing & Applications*.

Shorey, R., & Miller, B. A. (2000, December). The Bluetooth technology: merits and limitations. In *2000 IEEE international conference on personal wireless communications. Conference proceedings (Cat. No. 00TH8488)* (pp. 80-84). IEEE. DOI: 10.1109/ICPWC.2000.905777

Sim, C., Lo, J., & Chen, Z. N. (2022). Design of a broadband Millimeter-Wave array antenna for 5G applications. *IEEE Antennas and Wireless Propagation Letters*, *22*(5), 1030–1034. DOI: 10.1109/LAWP.2022.3231358

Singh, G., Kumar, A., Singh, M. J., & Singh, A. P. (2024, March). Modified Pentagonal Fractal Antenna for Ultra-Wideband Applications in Communication Systems. *International Journal of Electronics Letters*, *12*(4), 413–421. DOI: 10.1080/21681724.2024.2327091

Singh, U., Rathor, S., & Kumar, M. (2024, November). Hybrid Deep Learning and Machine Learning Approach for Detecting Spatial and Temporal Forgeries in Videos. *Neural Computing & Applications*.

Sivia, J. S., Pharwaha, A. P. S., & Kamal, T. S. (2013, November). Analysis and Design of Circular Fractal Antenna using Artificial Neural Networks. *Progress in Electromagnetics Research B. Pier B*, *56*, 251–267. DOI: 10.2528/PIERB13091611

Skolnik, M. I. (2008). Radar handbook. *IEEE Aerospace and Electronic Systems Magazine*, *23*(5), 41–41.

Son, J., & Cha, H. (2020). An implantable neural stimulator IC with anodic current pulse modulation based active charge balancing. *IEEE Access : Practical Innovations, Open Solutions*, *8*, 136449–136458. DOI: 10.1109/ACCESS.2020.3012028

Sran, S. S., & Sivia, J. S. (2020, August). ANN and IFS based Wearable Hybrid Fractal Antenna with DGS for S, C and X band Application. *AEÜ. International Journal of Electronics and Communications*, *127*, 153425. DOI: 10.1016/j.aeue.2020.153425

Sran, S. S., & Sivia, J. S. (2021, February). PSO and IFS Techniques for the Design of Wearable Hybrid Fractal Antenna. *International Journal of Electronics*, *108*(12), 2039–2057. DOI: 10.1080/00207217.2021.1885067

Sran, S. S., & Sivia, J. S. (2023, July). Design of a Novel Wearable Hybrid Fractal Antenna for Wi-Fi, Bluetooth, and WiMAX Applications. *Wireless Personal Communications*, *132*(4), 737–755. DOI: 10.1007/s11277-023-10635-6

SRRC. (2008) General Micro-power (short range) radio transmission equipment, Regulation -SRRC type approval-radio certificate of P.R China. Available at: http://www.srrccn.org/srrc-approval-new2.htm

Stallings, W. (2014). Data and computer communications (Tenth edition) India: Pearson Education.

Stoica, I., & Zhang, H. (1999). Providing guaranteed service without per flow management. *Computer Communication Review*, *29*(4), 81–94. DOI: 10.1145/316194.316208

Sundan, A. P., Jha, R. K., & Gupta, A. (2020). Energy and Spectral Efficiency Optimization Using Probabilistic Based Spectrum Slicing (PBSS) in Different Zones of 5G Wireless Communication Network. *Telecommunication Systems*, *73*(1), 59–73. DOI: 10.1007/s11235-019-00598-0

Su, R., Zhang, D., Venkatesan, R., Gong, Z., Li, C., Ding, F., Jiang, F., & Zhu, Z. (2019). Resource Allocation for Network Slicing in 5G Telecommunication Networks: A Survey of Principles and Models. *IEEE Network*, *33*(6), 172–179. DOI: 10.1109/MNET.2019.1900024

Swetha, S., Vasudevan, K., Meenambal, B., & Sherene, J. (2024, April). Compact Wideband CPW Fed Metamaterial Antenna for V2X Communication. In *2024 International Conference on Recent Advances in Electrical, Electronics, Ubiquitous Communication, and Computational Intelligence (RAEEUCCI)* (pp. 1-6). IEEE.

Tam, S. W., Razzaghi, A., Wong, A., Narravula, S., Xu, W., Loo, T., Kambale, A., Lowrance, A., Carnu, O., Lin, Y., & Tsang, R. (2024). A 28 nm CMOS Dual-Band Concurrent WLAN and Narrow Band Transmitter With On-Chip Feedforward TX-to-TX Interference Cancellation Path for Low Antenna-to-Antenna Isolation in IoT Devices. *IEEE Journal of Solid-State Circuits*, *59*(5), 1301–1311. DOI: 10.1109/JSSC.2024.3352532

Tanenbaum, A. S., Feamster, N., & Wetherall, D. (2021). *Computer networks* (6th ed.). Pearson.

Taner, C. (2018). *Fundamentals of network administration* (3rd ed.). Abaküs Publications.

Tektronix (2018) Fundamentals of radar measurement and signal analysis -- part 1, Fundamentals of radar measurement and signal analysis. Available at: https://www.tek.com/en/blog/fundamentals-radar-measurement-and-signal-analysis-part-1

Texas Instruments (Ed.). (2023) Texas Instruments Radar academy, Ti developer zone. Available at: https://dev.ti.com/tirex/explore/node?node=A__AEIJm0rwIeU.2P1OBWwlaA__RADAR-ACADEMY__GwxShWe__LATEST&placeholder=true

Thangam, S., & Kirubhakaran, E. (2012). QoS-aware Service Selection architecture for pervasive environment. *European Journal of Scientific Research*, *80*(2), 159–168.

Tsiftsis, T. A., Valagiannopoulos, C., Liu, H., Boulogeorgos, A.-A. A., & Miridakis, N. I. (2020). Metasurface-coated devices: A new paradigm for energy-efficient and secure 6G communications. *IEEE Open Journal of the Communications Society*, *1*, 1446–1455.

Uluslu, A. (2022). Chameleon Swarm algorithm assisted optimization of U-Slot patch antenna for Quad-Band applications. *IEEE Access : Practical Innovations, Open Solutions*, *10*, 74152–74163. DOI: 10.1109/ACCESS.2022.3190378

Valiente, R., Raftari, A., Mahjoub, H. N., Razzaghpour, M., Mahmud, S. K., & Fallah, Y. P. (2023). Context-aware target classification with hybrid Gaussian process prediction for cooperative vehicle safety systems. *IET Intelligent Transport Systems*, *17*(7), 1344–1362. DOI: 10.1049/itr2.12327

Vector Informatik Gmb, H. (2022) XCP – the standard protocol for ECU development. Available at: https://cdn.vector.com/cms/content/application-areas/ecu-calibration/xcp/XCP_ReferenceBook_V3.0_EN.pdf

Verulkar, S., Rochkari, A., Trimukhe, M., Bodade, V., & Gupta, R. (2024, February). High Gain Compact Dual Band Antenna Using Frequency Selective Surface for 5G and WLAN Applications. *Progress in Electromagnetics Research C. Pier C*, *142*, 1–11. DOI: 10.2528/PIERC24010101

Vinoth Kumar, P., & Venkatesh, K. (2024). Hybrid seagull and whale optimization algorithm-based dynamic clustering protocol for improving network longevity in wireless sensor networks. *China Communications*.

Walrand, J. (2023). A concise tutorial on traffic shaping and scheduling in time-sensitive networks. *IEEE Communications Surveys and Tutorials*, *25*(3), 1941–1953. DOI: 10.1109/COMST.2023.3275038

Wang, Q., Long, B., Wang, W., Lv, K., Tang, Y., & Xie, M. (2024). "An Enhanced Hard-Decision-Based ML Receiver for UsIM-SS Intra-Body Systems," *2024 IEEE 10th International Symposium on Microwave, Antenna, Propagation and EMC Technologies for Wireless Communications (MAPE)*, Guangzhou, China,, pp. 1-3, DOI: 10.1109/MAPE62875.2024.10813890

Wang, C.-X., Haider, F., Gao, X., You, X.-H., Yang, Y., Yuan, D., Aggoune, H. M., Haas, H., Fletcher, S., & Hepsaydir, E. (2014). Cellular architecture and key technologies for 5G wireless communication networks. *IEEE Communications Magazine*, *52*(2), 122–130. DOI: 10.1109/MCOM.2014.6736752

Wang, E., Liu, X., & Chang, H. (2023, July). En. C. Wang, X. F. Liu and H. Chang, "Wideband Circular Polarized Fractal Antenna for RFID/WiMAX/WLAN Applications,". *Progress in Electromagnetics Research Letters*, *111*, 111–120. DOI: 10.2528/PIERL23032601

Wang, Z.-G., You, R., Yang, M., Zhou, J., & Wang, M. (2024, March). Design of a Monopole Antenna for WiFi-UWB Based on Characteristic Mode Theory. *Progress in Electromagnetics Research M. Pier M*, *125*, 107–116. DOI: 10.2528/PIERM24012611

Wassie, G., Ding, J., & Wondie, Y. (2024). Detecting and predicting models for QoS optimization in SDN. *Journal of Computer Networks and Communications, 2024*(1). DOI: 10.1155/2024/3073388

White, P. P. (1997). RSVP and integrated services in the Internet: A tutorial. *IEEE Communications Magazine*, *35*(5), 100–106. DOI: 10.1109/35.592102

Wong, K., Huang, Y., & Li, W. (2024). Compact 8-Port 2 × 2 Array Based on Dual-Polarized Patch Antennas with Modified Cavity Field Distribution for Enhanced Port Isolation for 5G IoT Device MIMO Antennas. *IEEE Access : Practical Innovations, Open Solutions*, *12*, 79311–79326. DOI: 10.1109/ACCESS.2024.3409355

Xu, R., & Wunsch, D.II. (2005). Survey of clustering algorithms. *IEEE Transactions on Neural Networks*, *16*(3), 645–678. DOI: 10.1109/TNN.2005.845141 PMID: 15940994

Yahya, M. S., Soeung, S., Rahim, S. K. A., Musa, U., Ba Hashwan, S. S., & Haque, M. A. (2024). Machine Learning-Optimized Compact Frequency Reconfigurable Antenna With RSSI Enhancement for Long-Range Applications. *IEEE Access : Practical Innovations, Open Solutions*, *12*, 10970–10987. DOI: 10.1109/ACCESS.2024.3355145

Yang, K., & Zuo, C. (2022). Nanosheet lithium niobate acoustic resonator for mmWave frequencies. *IEEE Microwave and Wireless Components Letters*, *32*(5), 413–416.

Yıldıran, O., & Başdemir, H. D., (2017). Wideband Hexagonal Type Antenna Design for 5G Networks

Yoshihisa, K., Satoshi, S., & Satoshi, N. (2021). Trends and Target Implementations for 5G evolution & 6G. *NTT Technical Review*, *19*(11), 18–25. DOI: 10.53829/ntr202111fa1

You, X., Wang, C.-X., Huang, J., Gao, X., Zhang, Z., Wang, M., & Jiang, Y. (2021). Towards 6G wireless communication networks: Vision, enabling technologies, and new paradigm shifts. *Science China. Information Sciences*, *64*(1), 110301. DOI: 10.1007/s11432-020-2955-6

Yu, Z., Yu, J., Ran, X., & Zhu, C. (2017, June). A Novel Ancient Coin-like Fractal Multiband Antenna for Wireless Applications. *International Journal of Antennas and Propagation*, *2017*(1), 1–10. DOI: 10.1155/2017/6459286

Zade, P. L., & Khade, S. S. (2024, July). Miniaturized Novel Multi Resonance Monopole Planar Antenna with Slots, Slits, Split Ring Resonator. *Progress in Electromagnetics Research C. Pier C*, *145*, 75–82. DOI: 10.2528/PIERC24051102

Zhang, H., Chang, L., & Zhang, A. (2024). Extremely Low-Profile wideband patch antennas using electric- and Magnetic-Coupling for 5G mobile IoT devices application. *IEEE Antennas and Wireless Propagation Letters*, *23*(9), 2698–2702. DOI: 10.1109/LAWP.2024.3404779

Zhang, H., Liu, N., Chu, X., Long, K., Aghvami, A.-H., & Leung, V. C. M. (2017). Network Slicing Based 5G and Future Mobile Networks: Mobility, Resource Management, and Challenges. *IEEE Communications Magazine*, *55*(8), 138–145. DOI: 10.1109/MCOM.2017.1600940

Zhang, J., Xu, J., Chen, Q., & Li, H. (2023, August). Machine-Learning-Assisted Antenna Optimization With Data Augmentation. *IEEE Antennas and Wireless Propagation Letters*, *22*(8), 1932–1936. DOI: 10.1109/LAWP.2023.3269811

Zhang, S. (2019). An Overview of Network Slicing for 5G. *IEEE Wireless Communications*, *26*(3), 111–117. DOI: 10.1109/MWC.2019.1800234

Zhang, Z., Xiao, Y., Ma, Z., Xiao, M., Ding, Z., Lei, X., Karagiannidis, G. K., & Fan, P. (2019). 6G wireless networks: Vision, requirements, architecture, and key technologies. *IEEE Vehicular Technology Magazine*, *14*(3), 28–41. DOI: 10.1109/MVT.2019.2921208

Zhao, H., Gu, P. F., Fan, Z. H., He, Z., Cao, J., Tao, X., Wang, J. X., & Ding, D. Z. (2023). Ultrawideband dual circularly polarized waveguide array antenna for K- and Ka-Bands satellite communications. *IEEE Antennas and Wireless Propagation Letters*, *22*(8), 1808–1812. DOI: 10.1109/LAWP.2023.3266003

Zhao, Y., & Karypis, G. (2002). Evaluation of hierarchical clustering algorithms for document datasets. In *Proceedings of the eleventh international conference on Information and knowledge management*, pages 515–524, 2002. DOI: 10.21236/ADA439551

Zhu, J., Waltho, A., Yang, X., & Guo, X. (2007, August). Multi-radio coexistence: Challenges and opportunities. In *2007 16th International Conference on Computer Communications and Networks* (pp. 358-364). IEEE.

Zivkovic, M., & Zivkovic, T., Venkatachalam, K., & Bacanin, N. (2021). Enhanced dragonfly algorithm adapted for wireless sensor network lifetime optimization. In Data Intelligence and Cognitive Informatics: Proceedings of ICDICI 2020, pages 803–817. Springer.

Zong, B., Fan, C., Wang, X., Li, X., Niu, Z., & Li, P. (2019). 6G technologies: Key drivers, core requirements, system architectures, and enabling technologies. *IEEE Vehicular Technology Magazine*, *14*(3), 18–27. DOI: 10.1109/MVT.2019.2921398

About the Contributors

Eugene Ogbodo holds a Ph.D. in Electronic Engineering from the University of Greenwich, specializing in microwave engineering. His expertise lies in designing and developing advanced microwave devices for communication systems, including satellite communication and base station technology. Currently, he is affiliated with the University of Hertfordshire, where he serves as a researcher and academician. Dr. Ogbodo has made significant contributions to circuit designs, system integrations, and the miniaturization of passive microwave devices. His innovative work, showcased in numerous peer-reviewed publications, has enhanced the understanding and performance of high-frequency communication systems. His research focuses on improving the efficiency and integration of microwave circuits, benefiting both academic and industrial applications. His contributions have advanced satellite communication systems and base station technologies, solidifying his reputation as a leader in microwave engineering.

***

Dishan Narsimhachari Acharis pursuing his B.Tech in Amrita Vishwa Vidyapeetham. His research Interest are Networking, AI/ML, Cloud computing

Gajendra Kumar Ahirwar received Master degree from the School of Information Technology, RGPV Bhopal (2017) and BE degree in Computer Science and engineering from University Institute of technology, RGPV University Bhopal (2014). He is currently a Ph.D. scholar of Information Technology in University Institute of technology, RGPV University Bhopal. His research interests include Machine learning,Mobile ad hoc network, wireless sensor network and, Internet of things.

M. Praveena obtained her B.Tech. and M.Tech. degree from JNTUH in 2003 and 2007. She is pursuing her Ph.D. from JNTU University, Hyderabad in the field of image processing and VLSI implementation.

Meenambal Bose received her B.E degree in Electronics and Communication Engineering (2010) and M.E degree in Optical Communication (2013). She is currently pursuing the Ph.D. degree in Information and Communication Engineering, specializing in the field of antennas. Her research focuses on the development and optimization of antennas for various wireless communication applications. Her work involves exploring innovative designs and methodologies to enhance the performance and efficiency of antennas in modern communication systems. Her current research interests include metamaterial inspired antennas, mobile terminal antennas and multiple-input multiple-output (MIMO) antennas.

Ankit A. Bhurane has received his B.E. in Electronics and Communication, M.Tech. in Electronics, and Ph.D. in 2008, 2011, and 2016 respectively from SGBAU Amravati University, SGGSIET Nanded, and Indian Institute of Technology Bombay (IITB). He has 9+ years of experience teaching undergraduate and postgraduate courses at various institutes of national importances like Indian Institute of Information Technology, National Institute of Technology etc. He has published several papers in various International conferences and journals. He has been an active reviewer for reputed IEEE and Elsevier conferences and journals. Furthermore, he has on his name two Indian Patents Granted, and three filed with IPR office, India. His research interests include embedded systems, biomedical signal processing, scalable video coding and green Communication.

Kisalaya Chakrabarti has 24 years of teaching and research experience in the field of Electronics and Communication Engineering. His research domain covers different areas of Optical Communications. He has obtained Doctoral from University of Tsukuba, Japan and two Postdoctoral certificates from Utsunomiya University and National Institute for Material Science also from Japan. He has administrative experiences as Chair Professor in the department of Electronics and Communication Engineering at Bengal Institute of Technology and Management (Santiniketan), where he has also worked as Dean (R&D) for few months. He rendered his duties as Principal at Secom Engineering College situated at Sankrail, Howrah and Pailan College of Management and Technology (PCMT), Kolkata for around two years. The title of his PhD thesis was “An Investigation of Photonic Crystals and Time-Reversed Scattering using Nonstandard FDTD”. Presently he is working as a Professor in the Department of Electronics and Communication Engineering at Haldia Institute of Technology, ICARE Complex, HIT Campus,

Haldia, Purba Medinipur, India. He is the Senior Member of IEEE and he is the Life Fellow of Optical Society of India since 2005. He was the recipient of prestigious MONBUKAGAKUSHO (MEXT) Scholarship for his Doctoral Program at University of Tsukuba from the Japanese Government.

Ashish Chandelkar received a Master of Technology degree in Electronics from Pondicherry Central University, Puducherry, India. He was awarded JRF in 2015 and qualified for GATE 2013, GATE 2015, and Assistant Professor in December 2014 and June 2015. In 2015, he became a lecturer at the Government Polytechnic College, MP, India. He is currently pursuing a PhD in RF and microwave from Visvesvaraya National Institute of Technology (NIT Nagpur), Nagpur, India. His current research interest is in filtering antennas for 5G communication. E-mail: a shishchandelkarvnit@gmail.com

Gauri N. Damle completed her Bachelor of Engineering in Electronics and Telecommunication from RTM Nagpur University, Maharashtra. She earned her Master of Technology in Electronics and Communication Engineering from RTM Nagpur University. She has published several research papers in reputed national and international journals and conferences. She has around a decade of experience in teaching, research, and development.

Kiranbhai Ramabhai Dodiya is an experienced Cyber Forensic Expert and Assistant Professor specializing in cyber security and digital forensics. He has taught at prestigious institutions and is currently pursuing a PhD in Biochemistry and Forensic Science. With a strong background in ethical hacking and malware analysis, Kiranbhai is passionate about educating students and contributing to the field of cyber forensics. His expertise includes malware forensic, computer forensics, and data recovery software. Kiranbhai's commitment to professional growth is evident through his participation in workshops, seminars, and research publications. He is dedicated to shaping the next generation of professionals in the field and combatting cybercrimes.

Rama Lakshmi Gali is pursuing a Ph.D. from GITAM University, Hyderabad, and completed a Master from JNTUH in 2012 and a B.Tech from Pondicherry University in 2007.

Prema Ganesan is working as Assistant Professor(Sr.G) in ECE department, Mepco Schlenk Engineering College. She has 11 years of teaching experience. Her research interest include Image Processing, Deep learning, Antenna etc.

Selçuk Göçer has an MSc in cyber security from Ahmet Yesevi University. His current research interests include computer networking and cyber security.

Ramaharini J is a passionate Electronics and Communication Engineering student with a keen interest in technology and innovation.Always eager to explore new ideas and innovations.

Sherene Jacob is a dedicated research scholar at Thiagarajar College of Engineering, specializing in the field of antennas. With a strong background in electrical engineering, Ms. Jacob's research focuses on the development and optimization of antennas for various wireless communication applications. Her work involves exploring innovative designs and methodologies to enhance the performance and efficiency of antennas in modern communication systems. Ms. Jacob's passion for advancing technology and her commitment to academic excellence drive her pursuit of knowledge and innovation in the field of antennas.

Ayushi Jaiswal hold a bachelor's degree in Electronics and Communication Engineering and is presently pursuing a master's degree in Communication Systems from the Department of Electronics and Communication Engineering at Visvesvaraya National Institute of Technology, Nagpur. Her areas of interest include wireless communication, networking, and the Internet of Things (IoT). She is enthusiastic about exploring new technologies and applying theoretical concepts to practical applications in these fields.

Sunil Javali earned a Bachelor's degree in Electronics and Communication Engineering from B.M.S. Institute of Technology, Bangalore, India, and a Master's degree in Electrical Engineering (MSEE) from Michigan Technological University, USA. His MSEE research was later published at the 2017 IEEE International Conference on Wireless for Space and Extreme Environments (WiSEE), where it received the Best Paper Award. Sunil began his professional career as an Electrical Product Engineer at Aptiv (formerly Delphi Electronics & Safety), supporting the hardware design of automotive infotainment products. Additionally, he was actively involved in patent ideation within the Advanced Hardware Engineering organization, contributing to four published and granted patents at the USPTO and one trade secret. In 2020, Sunil transitioned to the Advanced Safety organization to contribute to the design and development of next-generation automotive radars. He is currently working as a Radar Systems Engineer.

Pranjali Jumle is working as a Professor in the Department of Electronics & Telecommunication Engineering, in the G H Raisoni College of Engineering Nagpur,

Maharashtra, India. Her research interest is on Biomedical Signal Processing, Microwave Communication

Naga Vishnu Vardhan Juturi is working as a Professor in the Department of ECE, BVRIT HYDERABAD College of Egineering for Women. 20 years of teaching experience and published 5 patents and 1 patent was granted. Published 10 + papers in various indexed Journals and Conferences. Organized 3 National (Funded by BRNS and DRDO)and 2 International Conferences (Springer). Successfully completed DST - TIDE project.

K. Kavitha received her Bachelor's degree in Electronics and Communication Engineering from Madurai Kamaraj University, Madurai, India and her Master's degree in Microwave & Optical Engineering from Madurai Kamaraj University, Madurai, India in She completed her doctoral degree at Anna University, Chennai, specializing in Hyperspectral Image Classification. Currently, she holds the position of Professor and Head in the Department of Electronics and Communication Engineering at Velammal College of Engineering and Technology in Madurai, Tamil Nadu. Her career spans over 23 years, during which she has made significant contributions in the realms of teaching and research. Her research interests encompass diverse areas, including Antenna Design, Hyperspectral Image Classification and Optical System Design. She has published and presented over 100 research articles in renowned international journals and international and national conferences. Presently, she is supervising the research works of ten Ph.D. scholars in various areas of her expertise. In recognition of her outstanding contributions to education, Dr.K.Kavitha was honored with Dr.A.P.K Abdul Kalam Teachers Excellence Award.

Samet Köse has an MSc in cyber security from Ahmet Yesevi University, Türkiye. His current research interests include industrial control systems and cyber security.

Arvind Kumar born in Jodhpur, India. He earned his Ph.D. in Microwave and RF in the Department of Electronics and Communication Engineering at the National Institute of Technology Tiruchirappalli, India, in 2019. He also completed a Master's degree (M.Tech.) with a specialization in Electronics from Pondicherry Central University and obtained his Bachelor's degree (B.Tech) in Electronics and Communication Engineering from Govt. engineering College Ajmer, affiliated with Rajasthan Technical University, Kota Rajasthan in 2012. He has achieved notable recognition as one of the Top 2% scientists in the World (all fields).: according to as survey conducted by Stanford University. E-mail: arvindkumar@ece.vnit.ac.in

Jesy Janet Kumari is a research scholar at Amrita School of Computing,Amrita Vishwa Vidyapeetham,Bengaluru,India. Her research Interests are VANET,Networks,IoT and AI.

Ananya Parameswaran received her BTech in Electronics and Communication Engineering from Toc H Institute of Science and Technology, CUSAT, Kerala, India in 2011 and MTech in Electronics from Pondicherry Central University, Pondicherry in 2014. She did her project internship at Vikram sarabhai Space Center (VSSC) ISRO Trivandrum. She earned her PhD in microwave engineering from Department of Electronics and Communication Engineering, National Institute of Technology Tiruchirappalli. She is the recipient of prestigious MeITY fellowship during her PhD. She is currently working as Assistant Professor with Department of ECE, SRM Institute of Science and Technology, Chennai, India. Her areas of interest are microwave engineering, filters, electromagnetic theory, antenna, microwave passive components and substrate-integrated waveguide. E-mail: ananyap@srmist.edu.in.

Ahila Priyadharshini is working as Professor in ECE department, Mepco Schlenk Engineering College. She has 25 years of teaching experience. Her research interest include Wireless communication,Image Processing, Deep learning etc.

Amireddy Shashank Reddy is pursuing his B.Tech in Computer Science and Engineering at Amrita Vishwa Vidyapeetham. His areas of interest include Cryptography, Blockchain Systems, Next-Generation Networks (5G/6G), Automata Theory, and Machine Learning.

Regina S completed her B.E. in Electronics and Communication Engineering from Pandian Saraswathi Yadav Engineering College, Anna University, Chennai, India, in 2005, her M.E. degree in Embedded System Technologies from Anna University, Tirunelveli, India in 2011 and her Ph.D degree from Anna University, Chennai, India, in 2022. She is currently working as an Assistant Professor in the Department of Electronics and Communication Engineering at Velammal College of Engineering and Technology, Madurai, India. Her research interests are mainly in the fields of Metamaterial Antenna, Wearable Antenna. She is a member of IEI.

Thangam S. currently serves as an Associate professor at the Amrita School of Computing, Bengaluru., Amrita Vishwa Vidyapeetham, India. Her area of interest in research includes Service Oriented Architecture, Networks, IoT, Artificial Intelligence, Data Structures and Cloud Computing. She completed her Ph. D. in Computer Science and Engineering from Anna University, Chennai. India. She has 26 years of teaching experience. She has published her research works in 25 international journals and conferences. She is a member of ISTE..

S. Raj Kumar received his bachelor's degree in Electronics and Communication Engineering from Anna University, Chennai, and his master's degree in VLSI Design from the same institution. He completed his doctoral degree at Anna University, Chennai, specializing in VLSI device modeling. Currently, he serves as an Assistant Professor in the Department of Electronics Engineering (VLSI Design and Technology) at Velammal College of Engineering and Technology, Madurai, Tamil Nadu, India. His career spans over 11 years, during which he has made significant contributions to teaching and research. His research interests include diverse areas such as modeling and simulation of analog VLSI. He has published over 10 research papers in international journals and presented 20 papers at international conferences. He has also received the award for "Best Outcome in Academic Performance".

S. Vasuki completed her B.E. degree in Electronics and Communication Engineering at Government College of Technology, Coimbatore in 1987 and M.E degree in Microwave and Optical Engineering at A.C.College of Engineering and Technology, Karaikudi in 1993. She successfully completed her Ph.D. degree in Color Image Processing at Anna University, Chennai. She has 30 years of teaching and research experience with 12 years in PSNA College of Engineering, Dindigul and 10 years in Mepco Schlenk Engineering College, Sivakasi. Presently, she is working as Professor and Head in Electronics and Communication Engineering Department at Velammal College of Engineering and Technology, Madurai. During these years, she shouldered a number of teaching, administrative and societal based assignments. Her research interests are in Computer Vision, Wavelet based Image analysis and Pattern recognition schemes. She has published 60 papers in International journals and 300 papers in International and National conferences. She is syllabus sub-committee member in Anna University, Chennai and also an approved supervisor for Ph.D. in Information and Communication Engineering, Anna University, Chennai. She is a Principal Investigator of 4 on-going research projects funded by various funding agencies like DRDO, DST and AICTE for a tune of 83.28 Lakhs. Also, receiver of Dr. A.P.J. Abdul Kalam Award for Scientific Excellence 2016 organized by M/s. Marina Research Laboratory, Chennai. She is a Reviewer for IEEE Transactions on Circuits and Systems for Video Technology, Defense Science Journal, International Journal of Technology, Knowledge and Society, SCOPUS Indexed Journals and Journal of Engineering Research. She is a Life Member of Indian Society of Technical Education (ISTE) and Fellow member of Institution of Electronics and Telecommunication Engineers (IETE) and Member of IEEE.

Prabira Kumar Sethy is an esteemed Associate Professor in the Department of Electronics and Communication Engineering at Guru Ghasidas Vishwavidyalaya

(Central University, Govt. of India), Bilaspur, Chhattisgarh, India since December 2023. Prior to this, he had served at Sambalpur University (State University, Govt. of Odisha) from February 2013 to December 2023, and worked as an Engineer in Doordarshan, Ministry of Broadcasting, Govt. in India, between August 2009 and February 2013. Holding a Ph.D. from Sambalpur University, M. Tech. from IIT Dhanbad, and B.E. from BPUT Odisha. Dr. Sethy is from small village, named Kapundi, situated on the banks of the Baitarani River in the Keonjhar District of Odisha. His early education, including primary and intermediate studies, took place in Keonjhar, Odisha. Dr. Sethy is an Editor of three reputable journals and serves as an editorial board member for the International Journal of Electrical and Computer Engineering and Ingénierie des Systèmes d'Information (IIETA). Additionally, he serves as an editorial member for Automation, Control and Intelligent Systems (Science Publishing Group) and PriMera Scientific Engineering (ISSN: 2834-2550). He also holds the position of Associate Editor of Onkologia I Radiotherapy. With two patents and one copyright to his name, Dr. Sethy has been recognized for his exceptional contributions. In 2020, he received the "InSc Young Achiever Award" for his research paper on "Detection of coronavirus (COVID-19) based on Deep Features and Support Vector Machine," organized by the Institute of Scholars, Ministry of MSME, Government of India. As a Senior Member of IEEE, he actively engages as a reviewer for various journals and takes on the role of session chair in international conferences. He is also within World Top 2% scientist, by Stanford University, in the year 2023 and 2024.

Ashish Sheikh is a Senior DSP Engineer at Cisco Systems, specializing in WLAN signal processing and interference detection. With extensive expertise in wireless communication systems, he has contributed to the field through multiple publications and patents. A Senior Member of IEEE, Ashish actively engages in advancing wireless technologies and signal processing innovations.

Shyam R. Sihare completed his Ph.D. at Raksha Shakti University in Ahmedabad, India. He holds a Master's degree in Computer Science from Nagpur University, Nagpur, India, which he obtained in 2003. Additionally, he attained an M. Phil. in Computer Science from Madurai Kamraj University, Madurai, India. In 2011 and 2018, he successfully cleared the Professor Eligibility Test GSLET (Gujarat) and MS-SET (Maharashtra) in India, respectively. Furthermore, he completed his MCA from IGNOU in New Delhi, India, in 2011. Currently, Dr. Sihare serves as an Assistant Professor in Computer Science and Application at Dr. APJ Abdul Kalam Govt. College in Silvassa, Dadra & Nagar Haveli (UT), India. His research interests encompass a wide range of areas including Quantum Computing, Quantum Algorithms, Quantum Cryptography, and Classical Computer Algorithms.

O.V. Sowmya, an Assistant Professor in Computer Science and Engineering, holds an M.Tech in C.S.E.Teaches at a renowned engineering college in Kadapa,K.S.R.M College of Engineering and inspiring students with my expertise and dedication. Passionate about education and research, focuses on areas like programming and emerging technologies. I was a committed academician who continuously motivates the students to achieve excellence.

Saleha Tabassum, M.Tech working as an assistant professor in the dept of Electrical and Electronics Engineering in KSRM COLLEGE OF ENGINEERING. She is pursuing Ph.D in the dept of Electrical and Electronics Engineering .

Gurkan Tuna is a Professor at the Department of Computer Programming of Trakya University (2006–. . ..). He has authored several papers in international conference proceedings and refereed journals, and has been actively serving as a reviewer for international journals and conferences. His current research interests include smart grid, ad hoc and sensor networks, cybersecurity, and machine learning.

Santhosh Kumar Veeramalla received the B.Tech. degree in electronics and communication engineering from the Jawaharlal Nehru Technological University, Hyderabad, India, in 2005 and the M.E. degree in electronics and communications from the University Visvesvaraya College of Engineering (UVCE), Bangalore, India in 2008. He received a Ph.D. in electronics and communication engineering from the National Institute of Technology, Warangal, India. Presently he is working as an associate professor at BVRIT Hyderabad College of Engineering for Women, Hyderabad. His research interests are in the area of biomedical signal processing and its implementation.

Index

Symbols

A

B

C

D

E

F

G

H

I

M

N

P

R

S

T

U

V

W

X

www.ingramcontent.com/pod-product-compliance
Lightning Source LLC
LaVergne TN
LVHW080902300926
849162LV00003B/527

* 9 7 9 8 3 3 7 3 0 8 5 2 4 *